Lecture Notes in Artificial Intelligence   1889

Subseries of Lecture Notes in Computer Science
Edited by J. G. Carbonell and J. Siekmann

Lecture Notes in Computer Science

Edited by G. Goos, J. Hartmanis and J. van Leeuwen

T0135045

Springer
Berlin
Heidelberg
New York
Barcelona
Hong Kong
London
Milan
Paris
Singapore
Tokyo

Michael Anderson   Peter Cheng
Volker Haarslev (Eds.)

Theory and Application of Diagrams

First International Conference, Diagrams 2000
Edinburgh, Scotland, UK, September 1-3, 2000
Proceedings

Springer

Series Editors

Jaime G. Carbonell, Carnegie Mellon University, Pittsburgh, PA, USA
Jörg Siekmann, University of Saarland, Saarbrücken, Germany

Volume Editors

Michael Anderson
University of Hartford, Department of Computer Science
Dana Hall 230, 200 Bloomfield Avenue
West Hartford, Connecticut 06117, USA
E-mail: anderson@hartford.edu

Peter Cheng
University of Nottingham, School of Psychology
ESRC Centre for Research in Development, Instruction and Training
University Park, Nottingham, NG7 2RD, UK
E-mail: peter.cheng@nottingham.ac.uk

Volker Haarslev
University of Hamburg, Computer Science Department
Vogt-Koelln-Str. 30, 22527 Hamburg, Germany
E-mail: haarslev@informatik.uni-hamburg.de

Cataloging-in-Publication Data applied for

Die Deutsche Bibliothek - CIP-Einheitsaufnahme

Theory and application of diagrams : first international conference,
diagrams 2000, Edinburgh, Scotland, UK, Sepember 1 - 3, 2000 ;
proceedings / Michael Anderson ... (ed.). - Berlin ; Heidelberg ; New
York ; Barcelona ; Hong Kong ; London ; Milan ; Paris ; Singapore ;
Tokyo : Springer, 2000
  (Lecture notes in computer science ; Vol. 1889 : Lecture notes in
  artificial intelligence)
  ISBN 3-540-67915-4

CR Subject Classification (1998): I.2, D.1.7, G.2, H.5, J.4, J.5

ISBN 3-540-67915-4 Springer-Verlag Berlin Heidelberg New York

Springer-Verlag Berlin Heidelberg New York
a member of BertelsmannSpringer Science+Business Media GmbH
© Springer-Verlag Berlin Heidelberg 2000
Printed in Germany

Typesetting: Camera-ready by author, data conversion by DA-TeX Gerd Blumenstein
Printed on acid-free paper   SPIN: 10722507   06/3142   5 4 3 2 1 0

layouts) through to more specialized forms of diagrams for particular purposes (e.g., representations of time, Celtic Knots).

In addition to the 31 technical paper presentations, in eight sessions, Diagrams 2000 included: an invited talk by Alan MacEachren (Representations to mediate geospatial collaborative reasoning: A cognitive-semiotic perspective); a talk in memory of Jon Barwise given by Keith Stenning; a tutorial by Kim Marriott on formal approaches to visual languages; a tutorial on cognitive approaches to diagrams co-presented by David Gooding, Hermi Schijf, and Jiajie Zhang; a session at which ten posters were presented.

The program co-chairs would like to thank all the members of the program committee for all their efforts towards making Diagrams 2000 a success. We are particularly grateful to Jo Calder, Alan Blackwell, Bernd Meyer, and Nigel Birch.

September 2000   Peter Cheng, Michael Anderson, Volker Haarslev

Organization

Diagrams 2000 was:

- funded by the Engineering and Physical Sciences Research Council (EPSRC grant no. GR/N08803)
- in cooperation with the American Association for Artificial Intelligence (AAAI)
- hosted by the Human Communication Research Centre (HCRC).

Program Chairs

Michael Anderson	University of Hartford (USA)
Peter Cheng	University of Nottingham (UK)
Volker Haarslev	University of Hamburg (Germany)

Program Committee

Tom Addis	University of Portmouth (UK)
Gerard Allwein	Indiana University (USA)
Nigel Birch	EPSRC (UK)
B. Chandrasekaran	Ohio State University (USA)
Maria Francesca Costabile	Università di Bari (Italy)
Gennaro Costagliola	Università di Salerno (Italy)
Ellen Yi-Luen Do	University of Washington (USA)
Max Egenhofer	University of Maine (USA)
George Furnas	University of Michigan (USA)
Janice Glasgow	Queens University (Canada)
David Gooding	University of Bath (UK)
Mark D. Gross	University of Washington (USA)
Pat Hayes	University of West Florida (USA)
Mary Hegarty	UCSB (USA)
Kathleen Hornsby	University of Maine (USA)
Mateja Jamnik	University of Birmingham (UK)
Stefano Levialdi	Università di Roma (Italy)
Robert Lindsay	University of Michigan (USA)
Ric Lowe	Curtin University of Technology (Australia)
Robert McCartney	University of Connecticut (USA)
N. Hari Narayanan	Auburn University (USA)
David Barker-Plummer	Stanford University (USA)
Clive Richards	Coventry University (UK)
Eric Saund	Xerox PARC (USA)
Barbara Tversky	Stanford University (USA)

Additional Referees

G. M. Bierman
P. Bottoni
M. Erwig
F. Ferrucci

T. Jansen
M. Matera
G. Polese
A. Shimjima

M. Wessel
H. Zeevat

Administration

Local Organization:	Jo Calder (University of Edinburgh)
	Corin Gurr (University of Edinburgh)
Publicity:	Bernd Meyer (Monash University)
Web Site:	Alan Blackwell (University of Cambridge)
Series Liaison:	Kim Marriott (Monash University)
	Patrick Olivier (University of York)

Table of Contents

Invited Talks

Tutorial 1 – Formal Approaches to Diagrams

Tutorial 2 – Cognitive Approach to Diagrams

Logic and Diagrams

Theoretical Concerns about Diagrams

Cognition and Diagrams

Human Communication with Diagrams

Diagrammatic Reasoning/Proof Systems

Diagrams for Systems, Systems for Diagrams

Posters

Invited Talk:
Representations to Mediate Geospatial Collaborative Reasoning: A Cognitive-Semiotic Perspective

Alan M. MacEachren

GeoVISTA Center (www.geovista.psu.edu), Department of Geography,
Penn State University, USA,
alan@geog.psu.edu
www.geovista.psu.edu/MacEachren/alanhome.html

Abstract. This presentation will address the representation of geospatial information in the context of group work. The focus is on visual representations that mediate between human collaborators who are participating in a joint reasoning process, within a place and/or space-based problem context. The perspective developed for addressing the challenges involved builds upon the cognitive-semiotic approach outlined in How Maps Work, extending it to consider the issues that underlie creation of maps and related diagrams that work in a group work context. This context requires representations that depict not only geospatial information but also individual perspectives on that information, the process of negotiation among those perspectives, and the behaviors (work) of individuals participating in that negotiation.

M. Anderson, P. Cheng, and V. Haarslev (Eds.): Diagrams 2000, LNAI 1889, p. 1, 2000.

Invited Talk:
Jon Barwise: A Heterogeneous Appreciation

Keith Stenning

Human Communication Research Centre, Edinburgh University
2, Buccleuch Place, Edinburgh EH8 9LW, Scotland
keith@cogsci.ed.ac.uk
www.hcrc.ed.ac.uk/Site/STENNIKE.html

Abstract. Jon Barwise was unique amongst logicians in leading engagement between logic and other disciplines, notably linguistics, computer science, and the several disciplines concerned with diagrams. My main contact with Jon was through working on cognitive analyses of the learning processes of students being taught logic using Hyperproof, the heterogeneous environment he and John Etchemendy designed. This talk will trace some of Jon's enthusiasms for interdisciplinary interactions in this area. I hope the audience will contribute as much or more than the speaker.

M. Anderson, P. Cheng, and V. Haarslev (Eds.): Diagrams 2000, LNAI 1889, p. 2, 2000.
© Springer-Verlag Berlin Heidelberg 2000

Tutorial 1:
Formal Approaches to Visual Language Specification and Understanding

Kim Marriott

School of Computer Science and Software Engineering
Monash University, Clayton Vic. 3168, Australia
marriott@csse.monash.edu.au
www.cs.monash.edu.au/~marriott/

Abstract. Two of the most fundamental questions in visual language research are how to specify a visual language and how to recognize and understand diagrams in a particular visual language. In this tutorial we survey the many formalisms which have been suggested over the last three decades for visual language specification, discuss computational approaches to diagram understanding based on these formalisms and indicate possible applications. We shall also review recent directions in visual language theory, notably efforts to develop an analogue of the Chomsky hierarchy for visual languages, the specification of diagrammatic reasoning, and cognitive models of visual language understanding.

M. Anderson, P. Cheng, and V. Haarslev (Eds.): Diagrams 2000, LNAI 1889, p. 3, 2000.
© Springer-Verlag Berlin Heidelberg 2000

Tutorial 2a:
Cognitive History of Science: The Roles of Diagrammatic Representations in Discovery and Modeling Discovery

David Gooding

Science Studies Centre, Department of Psychology
University of Bath, Bath BA2 7AY, UK
hssdcg@bath.ac.uk
www.bath.ac.uk/~hssdcg/home.html

Abstract. This session looks at some uses of diagrams in scientific discovery, particularly their role as intermediate representations which mediate between phenomena, descriptions which can be communicated and descriptions which are general. A range of examples will illustrate a variety of uses, including: the abstractive, generative role; diagrams as encoded knowledge; reasoning with diagrammatic representations in discovery; and communication (exposition and argumentation). The tutorial will encourage consideration of two issues: (a) whether, from a cognitive standpoint, diagrams are essential to reasoning about natural phenomena and processes, and (b) the relationship of diagrammatic reasoning to other types of visualisation and visual thinking in the sciences, including cognitive and computational modeling of discovery.

M. Anderson, P. Cheng, and V. Haarslev (Eds.): Diagrams 2000, LNAI 1889, p. 4, 2000.
© Springer-Verlag Berlin Heidelberg 2000

Tutorial 2b:
Cognitive (Production System) Modelling of How an Expert Uses a Cartesian Graph

Hermi Schijf

Dept. of Educational Sciences, Utrecht University
Heidelberglaan 2, 3584 CS De Uithof - Utrecht, The Netherlands
h.schijf@fss.uu.nl

Abstract. This tutorial covers, in brief, the road from observing behavior to the implementation of the observed behavior in a mixed rule-based and parallel network computer model. The emphasis will be on production system, or rule-based modelling. Rule-based modeling uses independently firing if-then rules to capture behavior. Why do this type of modeling, what types of data do you need, what are some advantages and limitations of the method? Simple examples of rule-based modeling will be given; these will be extended to a brief explanation of the mixed model. This model gives a theoretical explanation of the behavior of an expert using visual reasoning based on a Cartesian graph combined with verbal reasoning to teach Economics principles to students.

M. Anderson, P. Cheng, and V. Haarslev (Eds.): Diagrams 2000, LNAI 1889, p. 5, 2000.

Tutorial 2c:
The Coordination of External Representations and Internal Mental Representations in Display-Based Cognitive Tasks

Jiajie Zhang

Department of Health Informatics, University of Texas at Houston
7000 Fannin Street, Suite 600, Houston, TX 77030 USA
Jiajie.Zhang@uth.tmc.edu
acad88.sahs.uth.tmc.edu/

Abstract. Many cognitive tasks, whether in everyday cognition, scientific practice, or professional life, are distributed cognitive tasks–tasks that require integrative, interactive, and dynamical processing of information retrieved from internal representations and that perceived from external representations through the interplay between perception and cognition. The representational effect is the ubiquitous phenomenon that different representations of a common structure can generate dramatically different representational efficiencies, task complexities, and behavioral outcomes. A framework of distributed representations is proposed to account for the representational effect in distributed cognitive tasks. This framework considers internal and external representations as two indispensable components of a single system and suggests that the relative distribution of information across internal and external representations is the major factor of the representational effect in distributed cognitive tasks. A representational determinism is also proposed–the form of a representation determines what information can be perceived, what processes can be activated, and what structures can be learned and discovered from the specific representation. Applications of the framework of distributed representations will be described for three domains: problem solving, relational information displays, and numeration systems.

M. Anderson, P. Cheng, and V. Haarslev (Eds.): Diagrams 2000, LNAI 1889, p. 6, 2000.
© Springer-Verlag Berlin Heidelberg 2000

Positive Semantics of Projections in Venn-Euler Diagrams

Joseph (Yossi) Gil[1] [*],[**], John Howse[2][*,*,*], and Elena Tulchinsky[1][**]

[1] Department of Computer Science
Technion–Israel Institute of Technology
Technion City, Haifa 32000, Israel
[2] School of Computing and Mathematical Sciences
University of Brighton, UK

Abstract. Venn diagrams and Euler circles have long been used as a means of expressing relationships among sets using visual metaphors such as "disjointness" and "containment" of topological contours. Although the notation is effective in delivering a clear visual modeling of set theoretical relationships, it does not scale well. In this work we study "projection contours", a new means for presenting sets intersections, which is designed to reduce the clutter in such diagrams. Informally, a projected contour is a contour which describes a set of elements limited to a certain context. The challenge in introducing this notation is in producing precise and consistent semantics for the general case, including a diagram comprising several, possibly interacting, projections, which might even be of the same base set. The semantics investigated here assigns a "positive" meaning to a projection, i.e., based on the list of contours with which it interacts, where contours disjoint to it do not change its semantics. This semantics is produced by a novel Gaussian-like elimination process for solving set equations. In dealing with multiple projections of the same base set, we introduce yet another extension to Venn-Euler diagrams in which the same set can be described by multiple contours.

1 Introduction

Diagrammatic notations involving circles and other closed curves, which we will call *contours*, have been used to representate classical syllogisms since the Middle Ages [ ]. In the middle of the 18th century, the Swiss mathematician Leonhard Euler introduced the notation we now call *Euler circles* (or *Euler diagrams*) [ ] to illustrate relationships between sets. This notation uses the topological properties of *enclosure*, *exclusion* and *partial overlap* to represent the set-theoretic

---

[*] Work done in part during a sabbatical stay at the IBM T. J. Watson Research Center

[**] Research was supported by generous funding from the Bar-Nir Bergreen Software Technology Center of Excellence – the Software Technology Laboratory (STL), at the department of computer science, the Technion

[*,*,*] Research was supported by the UK EPSRC grant number GR/M02606

M. Anderson, P. Cheng, and V. Haarslev (Eds.): Diagrams 2000, LNAI 1889, pp. 7– , 2000.
© Springer-Verlag Berlin Heidelberg 2000

notions of *containment, disjointness,* and *intersection,* respectively. Another such notation is Venn-diagrams, named after their inventor, the 19th century logician John Venn [ ]. A *Venn diagram* contains n contours representing n sets, where the contours must intersect in such a way that they divide the plane into 2^n "zones", which are connected regions of the plane. For every subset of the contours, there must be a *zone* of the diagram, such that the contours in this subset intersect at exactly this zone. That is to say, the zone is contained in these contours and in no other contour. A shaded zone indicates that the particular intersection of sets it denotes is empty.

Venn diagrams are expressive as a visual notation for depicting constraints on sets and their relationships with other sets, but difficult to draw.

(a) A simple and symmetrical Venn diagram of four contours

(b) The simple symmetrical Venn diagram of five contours

Fig. 1.1: Venn diagrams of four and five contours.

Fig.   (a) shows a symmetrical Venn diagram of four contours, while Fig.   (b) is the only simple symmetric Venn diagram of five contours.

Examining these two figures, it is clear why it is so rare to see Venn diagrams of four or more contours used in visual formalism. Even in these simple and symmetrical diagrams, most regions take some pondering before it is clear which combination of contours they represent.

On the other hand, Euler circles are intuitive and much easier to draw, but are not as expressive as Venn diagrams for two reasons. First, restricting the shape of contours to ovals does not allow all possible combinations of sets to be drawn. Second, the lack of provisions for shading restricts expressiveness even further. For example, there is no Euler diagram equivalent to the Venn diagram depicted in Fig. 1.2.

It is therefore the case that an informal hybrid of the two notations is often used for teaching purposes. We use the term *Venn-Euler diagrams* for the notation obtained by a relaxation of the demand that all contours in Venn-diagrams must intersect or conversely, by introducing shading and non-oval contours into Euler diagrams. Gil, Howse and Kent [ ] provided formalism for Venn-Euler diagrams as part of the more general *spider diagrams* notation.

This paper investigates an extension to the Venn-Euler diagrams notation designed for the purpose of showing a set in a certain context. Intersection can be used for just this purpose: an intersection of A and B shows the set A in the context of B and vice-versa. A *projection* is a contour, which is used to denote an intersection of a set with a "context". In Fig.   (a) for example, the set Women is *projected* into the set of employees. The set Employees is called the *context* for the projection of Women. The corresponding traditional Venn-Euler diagram of Fig.   (a) is shown in Fig.   (b).

(a) using projections   (b) without projections

Fig. 1.3: Denoting the set of all women employees.

The projected contour (the dashed inner oval) represents the set of women employees; it doesn't say that all women are employees (as implied by Fig.   ). It is possible in this particular setting to draw a "dual" diagram, in which (taking a somewhat feminist approach) the set of employees could be projected into the set of all women. However, in general the context may comprise several sets, and such a dual representation does not seem trivial.

Fig. 1.4: A Venn-Euler diagram stating that all women are employees.

By convention, dashed iconic representation is used to distinguish projections from other contours. We can see that the use of projections potentially reduces the number of regions, with the benefit that regions that are not the focus of attention are not shown, resulting in less cluttered diagrams.

Several intriguing questions arise in trying to extend the notion of projections beyond simple examples.

Context The notation must have a well-defined semantics when a projection partially overlaps with a contour, and not only when it is disjoint to it or contained in it. Moreover, there are cases in which a projection is covered by a set of contours, rather than a single one. The question here is: What is the desired intuitive meaning of the context of a projection and how should this be computed from the topology of a diagram?

Interacting Projections A diagram may contain more than one projection that may intersect with each other, share part of a context, or interact in other subtle ways. Can a projection define a context for another projection?

---

[1] Our use of the term projections should not be confused with the projection operator of relational algebra.

[2] We view diagrams as formulae of first order predicate logic. Therefore, the term "semantics" is used here in the mathematical logic sense, i.e., the *interpretation* of a certain formula in the model of its dictionary.

More generally, what is the semantics of a general topological relationship among several projections?

Multi-Projections Generalizing even further, the same base set might be projected several times into the same diagram and these projections, nicknamed *multi-projections*, might interact. We now can ask whether *multi-projections* should be allowed to interact, and what is the semantics of such an interaction?

The projection concept was first suggested and used as part of the constraint diagram language [ ]. However, the questions raised above complicating matters were not dealt with. Instead, there was a tacit understanding that only "simple" use of projections, which avoided these problems, was permitted. The work reported here represents an attempt to systematically deal with the semantics of projections. Another such attempt that took another direction, which might be thought of as *negative* or *non-monotone* semantics is described in [ ]. The differences between the two approaches will be highlighted below. While selecting an approach might be a matter of personal preference, it is interesting to note how they lead to very different mathematical supporting theories.

Outline The remainder of this paper is structured as follows. The next Sec.   sets out notation, terminology and conventions. Sec.   makes the case of the projection notation.

Sec.   constitutes an informal exploration of the first issue, trying to establish an intuition for the notion of context. Based on this, the concept of context is precisely defined in Sec. . This makes it possible to give a precise semantics to diagrams with a single projection, or, more generally, diagrams in which no projection takes part in the context of another.

In Sec.   we consider interacting projections. It is shown there that the semantics of these can be computed by solving a system of set equations, using what can be thought of as a Gaussian elimination process applied to set theory. We believe that this process might be of independent interest.

Sec.   deals with multi-projections issue. It reaches the conclusion that it is better to treat multi-projections of the same base set as a single entity. Finally, Sec.   discusses related work and draws the lines of future research.

2   Preliminaries

To simplify mathematical formulae involving complex relationships between sets, we will use the arithmetical summation symbol $(+)$ to denote set union, and arithmetical product (denoted by catenation) to denote set intersection. This notation is consistent with Boolean algebra, and will make it easier to apply the commutative, distributive and associative rules. We will also use, with all due caution, the arithmetical minus symbol $(-)$ to denote set minus operation.

Even though we will ascribe precise meaning to all terms, it is convenient to make an initial agreement about the following, somewhat informal, terminology which is common in hand-waving referral to Venn and Euler diagrams. A *contour* is a simple closed plane curve which denotes a set. A *boundary contour* is a contour that contains all other contours. A *district* is the set of points in the plane enclosed by a contour, while a *zone* is a region of the plane which is not further divided by the contours.

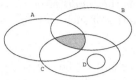

Fig. 2.1: A Venn-Euler diagram.

For example, the Venn-Euler diagram in Fig. has four non-boundary contours A, B, C, D and the boundary is omitted. Its interpretation includes $D \subseteq (C - B) - A$ and $ABC = \emptyset$.

Mathematical preciseness dictates that in referring to a diagram, a distinction should be made between a *contour* and the *set* which it represents. For example, in Fig. (which is a redraw of Fig. (a) using abstract names) the outer most contour is not the same as the set B which it depicts. (Ultimate preciseness might even distinguish between the *label* "B" and the set B.)

To streamline the discourse we forgo this preciseness and use B to refer also to the contour. In cases where the intended meaning is not obvious from context terms such as "set B" and "contour B" will be used.

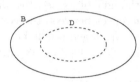

Fig. 2.2: A simple projection.

This convention works since the mapping from ordinary contours to sets is one-to-one. To deal with projections a small extension must be made. The inner oval in the diagram contour represents the *projection* of the set D. Let $\tilde{D}$ denote the set D after projection. We have $\tilde{D} = DB$.

In Fig. the base set D is projected twice. Since the label D is non-unique, *contour labels* f and g are used to distinguish between the two dashed contours labeled D. We insist that unprojected set names are upper case (mostly drawn from the beginning of

Fig. 2.3: Projection of the same base set into two different contexts.

the Latin alphabet), while contour labels are lower case. With this convention, set D confined to A (resp. B) is denoted by D^e (resp. D^f).

3   The Case for Projections

Consider Fig. (a) which was constructed using More's algorithm [ ] for six contours. It is hard to refrain from appreciating the complexity of the drawing or to imagine how it can contribute to clearer visualization of set relationships.

In sharp contrast stands Fig. (b) which shows the same 64 zones as Fig. (a), but in a much clearer fashion. (For technical reasons, which will become clear in Sec. , we needed to add a pseudo-boundary contour Q.) It is also quite easy to generalize the diagram to nine contours, by drawing a clover of three projections of three other sets in each of these 64 zones. Clearly, the process can be carried out indefinitely.

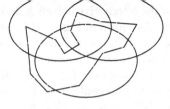
(a) without projections

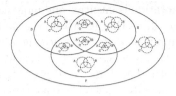
(b) using projections

Fig. 3.1: A Venn diagram with six contours.

Fig.   (b) demonstrates the natural need for interactions among projections and for multiple projections of the same base set.

Even Venn-Euler diagrams can get very cluttered when many contours are involved. The issue of clutter becomes even more crucial when such diagrams are used as a foundation for other, more advanced visual formalism. There are for example extensions of Venn-Euler diagrams in which elements of sets can be shown diagrammatically; these include Peirce diagrams [, ,] and spider diagrams [ ]. Yet another example is the *constraint diagrams* notation [ ] which uses arrows and other diagrammatic elements to model constraints not only on simple sets, but also on mathematical *relations*.

In order to reduce clutter, and to focus attention within the diagram appropriately, constraint diagrams have already made use of the projection notation. Consider for example the constraint diagram in Fig.   (a).

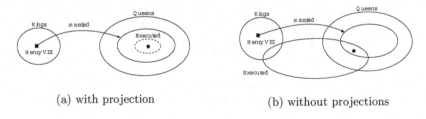

(a) with projection   (b) without projections

Fig. 3.2: A constraint diagrams.

This diagram states (among other things) that the sets Kings and Queens are disjoint, that the set Kings has an element named Henry VIII, that all women that Henry VIII married were queens and that there was at least one queen he married who was executed. The dashed contour is a projection of the set Executed; it is the set of all executed people projected into the set of people married to Henry VIII, that is, it gives all the queens who were married to Henry VIII and executed.

In the example, the inner most ellipse labeled "*Executed*" denotes the intersection of the set Executed with the set of women who were married to Henry VIII.

The notation is intuitive and more concise than the alternative, which is drawing a large ellipse that intersects the Queens contour. As shown in Fig. (b), this ellipse must also intersect with the Kings contour, or otherwise the diagram would imply that no kings were ever executed. To substantiate our claim that Fig. (a) is simpler than Fig. (b) we note that the four contours of Fig. (b) divide the plane into 8 disjoint areas. In contrast, the same four contours in Fig. (a) partition the plane into 5 such areas, which is the minimal possible number of non-overlapping zones.

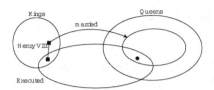

Fig. 3.3: Using a spider notation to preserve the semantics of Fig. (a) while eliminating projections from it.

Moreover, Fig. (a) does not specify whether Henry VIII was executed or not. Eliminating the projections from the figure requires delving into a history book and explicitly specifying this point as shown in Fig. (b). Alternatively, one could use what is known as a spider to refrain from stating whether or not Henry VIII was executed. As shown in Fig. , this alternative is even more cumbersome, and will probably draw the attention of the reader to an irrelevant point.

4 Intuitive Semantics of Projections

Taking a rather informal and intuitive approach this section explores the notion of the context of a projection, setting the guidelines for the precise definitions in the following section.

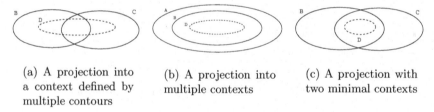

(a) A projection into a context defined by multiple contours

(b) A projection into multiple contexts

(c) A projection with two minimal contexts

Fig. 4.1: Variations in the context of a projection.

Perhaps the first thing to notice is that there are cases in which a projection is not into a single contour but rather into an area defined by multiple contours. This situation is depicted in Fig. (a), from which we surmise $\tilde{D} = D(B + C)$. This conclusion follows from the simple observation that the inner oval is not contained in the one labeled B, nor the one labeled C. Hence none of these two contours can define by itself a context for the projection of D.

Consider now Fig. (b). An awkward way to interpret the diagram that would still be consistent with our interpretation of Fig. is $\tilde{D} = DA \subseteq B$. This interpretation seems quite contrived because it does not select a minimal "context" for $\tilde{D}$, and stands in disagreement with our traditional mathematical and logical thinking. In trying to delineate the boundaries of any concept, or in making a definition, one strives to make the strongest possible constraint. The statement "platypusii live in Australia" is more informative and useful than "platypusii live on planet earth". Our semantics for the context notion follows this line of thought.

Conversely, suppose that we allow a non-minimal context to be chosen in defining a context for a projection. Then, how should a selection among many possible contexts be made in a consistent manner? If, for example, a "maximal" context is to be selected in all cases, then even Fig.   (a) loses its natural semantics, since the maximal context is the boundary contour. Projection where the boundary serves as a context is nonsensical, since the boundary is an ominous set which is supposed to contain the whole universe. Thus, a projection of the sort depicted in Fig.   is deemed illegal.

The alternative, and much more intuitive, semantics of Fig.   (b) is simply $\tilde{D} = DB$ with the additional requirement $B \subseteq A$. We have thus discovered that if there are multiple contexts for a projection only the minimal one should be considered. The understanding of a projection should

Fig. 4.2: An illegal projection.

not change if we add a contour that contains its context. However, as demonstrated in Fig.   (c), (which can be thought of as a dual to Fig.   (b)) there could be cases in which a contour has multiple minimal contexts.

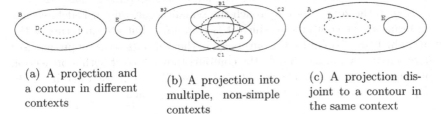

(a) A projection and a contour in different contexts

(b) A projection into multiple, non-simple contexts

(c) A projection disjoint to a contour in the same context

Fig. 4.3: More variations in the context of a projection.

Since D is contained in B, we would like to maintain that $\tilde{D} = DB$. This however contradicts the symmetrical demand that $\tilde{D} = DC$. From the natural language demand that "$\tilde{D}$ is the set of D's which are in the context of B, and in the context of C", we obtain that $\tilde{D} = D(BC)$, which is in agreement with the topology of the diagram, since the contour $\tilde{D}$ is constrained to the zone defined by the intersection of B and C.

Fig.   (a) gave an example in which several contours together provided a context for a projection. The minimality of the context is not only with respect to containment, but also with regard to the contours taking part in the context. For example, in Fig.   (a) only B is considered a context of D even though all of $\{B, D, E\}$, $\{B, E\}$, $\{B, D\}$, and $\{E, D\}$ cover B. Similarly, we are not interested in a context of a contour which comprises of the contour itself.

Fig. (b) may at first seem complicated. A moments reflection will reveal that it is nothing more than a generalization of Fig. (a) and Fig. (c). Indeed, the projection is contained in the union of B_1 and C_1 as well as the union of B_2 and C_2. Thus, $\tilde{D} = D((B_1 + C_1)(B_2 + C_2))$.

The semantics of projections was so far defined by what may be called a *contour-based approach*, which computes the projection with respect to all of its "minimal contexts", where a context is defined as a union of a collection of contours. An alternative definition follows from what might be called a *zone-based* approach. In this approach, the semantics of a projection is defined in terms of the zones in which it falls, and where each of these zones is appropriately defined by the set of contours which contain it, as well as the set of contours disjoint to it.

In some of the preceding examples, both approaches lead to the same semantics. For example, in Fig. (c), set D is projected into the zone BC, while in Fig. (a) it is projected into the union of three zones: BC, $B-C$, and $C-B$ and

$$\tilde{D} = D((BC) + (B - C) + (C - B))$$
$$= D(B + C).$$

A diagram which clearly distinguishes between the two approaches is depicted in Fig. (c). In the zone-based approach we have $\tilde{D} = D(A - E)$, while in the contour-based approach $\tilde{D} = DA$, with the additional statement that $\tilde{D}E = \emptyset$. We see that in the zone-based approach the semantics of contour is defined not only in terms of the contours in which it is contained, but also in terms of the contours which are disjoint to it. Conversely, in the contours-based approach uses "positive semantics", in which only the contours that intersect with a projection contribute to its semantics. The zone-based approach was investigated in [ ]. In this paper, we concentrate on the contours-based approach.

5   Covers and Contexts

Through a series of definitions, this section develops a formal notion of a context, $\kappa(c)$, of a contour c (Def.). The *territory* of c, denoted by $\tau(c)$ (Def.) is the entire set of contours which take part in the definition of a context.

Although all definitions carry a geometrical and topological intuition, for the sake of preciseness we rely on a pure, set theoretical mathematical model. In this model, a (Venn-Euler) diagram is nothing but a set of contours and zones.

Definition 1. *A diagram is a pair $\langle \mathcal{C}, \mathcal{Z} \rangle$ of a finite set $\mathcal{C}$ of objects, which we will call contours, and a set $\mathcal{Z}$ of non-empty subsets of $\mathcal{C}$, which we will call zones, such that $\forall c \in \mathcal{C}, \exists z \in \mathcal{Z}, c \in z$.*

The most simple diagram is an empty one: $\langle \emptyset, \emptyset \rangle$. A more interesting example is the pair

$$\langle \{A, B, C, D, E\}, \{z_1, z_2, z_3, z_4, z_5, z_6, z_7, z_8, z_9\} \rangle, \tag{1}$$

where

$$
\begin{array}{lll}
z_1 = \{A\} & z_2 = \{A, B\} & z_3 = \{A, C\} \\
z_4 = \{A, B, D\} & z_5 = \{A, C, D\} & z_6 = \{A, B, C, D\} \\
z_7 = \{A, B, C\} & z_8 = \{A, E\} & z_9 = \{E\}
\end{array}
\tag{2}
$$

is a diagram since $A \in z_1$, $B \in z_2$, $C \in z_3$, $D \in z_4$, and $E \in z_8$. This diagram will serve as our running example for the remainder of this section.

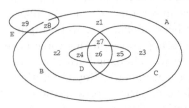

Fig. 5.1: A diagram with five contours and nine zones.

The geometrical interpretation of the Def. is that each zone is thought of as the set of contours in which it is contained. It is implicitly disjoint to the remaining contours. The requirement $\forall c \in \mathcal{C}, \exists z \in \mathcal{Z}, c \in z$ ensures that there are no degenerate contours. Def. does not include a boundary contour. Accordingly, we do not consider the zone which is outside all contours, and therefore zones must be non-empty sets of contours.

In many cases it is possible to obtain a *layout* of a diagram from this interpretation. One possible layout of our running example is given in Fig. . Note that in this particular layout zone z_7 is mapped to a noncontiguous area of the diagram. However, Fig. is useful for gaining insight into our definitions.

A *model* for a diagram is a set $\mathcal{U}$, called the *universe* and a *semantic function* $\psi : \mathcal{C} \to 2^{\mathcal{U}}$, mapping each contour to a subset of the universe. Let us extend the domain of ψ to include $\mathcal{Z}$ as follows. For all $z \in \mathcal{Z}$,

$$\psi(z) \equiv \prod_{c \in z} \psi(c) - \sum_{c \in \mathcal{C} - z} \psi(c). \tag{3}$$

In words, the semantics of z is the intersection of the semantics of the contours containing it, minus the semantics of all contours which do not contain it.

The *semantics* of the diagram is given in

$$\sum_{c \in \mathcal{C}} \psi(c) = \sum_{z \in \mathcal{Z}} \psi(z). \tag{4}$$

Equation ( ) is called in the literature [ ] *the plane tiling condition*, due to the geometrical intuition behind it. In words, this condition says that the list of zones is comprehensive i.e., every set member, must be a member of one of its zones.

Definition 2. *The* district *of a contour c is $d(c) = \{z \in Z | c \in z\}$. The district of a set of contours S is the union of the districts of its contours* $d(S) = \sum_{c \in S} d(c)$.

Preface

Diagrams 2000 is dedicated to the memory of Jon Barwise.

Diagrams 2000 was the first event in a new interdisciplinary conference series on the Theory and Application of Diagrams. It was held at the University of Edinburgh, Scotland, September 1-3, 2000.

Driven by the pervasiveness of diagrams in human communication and by the increasing availability of graphical environments in computerized work, the study of diagrammatic notations is emerging as a research field in its own right. This development has simultaneously taken place in several scientific disciplines, including, amongst others: cognitive science, artificial intelligence, and computer science. Consequently, a number of different workshop series on this topic have been successfully organized during the last few years: Thinking with Diagrams, Theory of Visual Languages, Reasoning with Diagrammatic Representations, and Formalizing Reasoning with Visual and Diagrammatic Representations.

Diagrams are simultaneously complex cognitive phenonema and sophisticated computational artifacts. So, to be successful and relevant the study of diagrams must as a whole be interdisciplinary in nature. Thus, the workshop series mentioned above decided to merge into Diagrams 2000, as the single interdisciplinary conference for this exciting new field. It is intended that Diagrams 2000 should become the premier international conference series in this area and provide a forum with sufficient breadth of scope to encompass researchers from all academic areas who are studying the nature of diagrammatic representations and their use by humans and in machines.

The call for papers and posters for Diagrams 2000 attracted submissions from a very wide variety of disciplines and departments, including: architecture, art and design, artificial intelligence, cognitive science, computer science, education, engineering, human computer interaction, information science, management, mathematics, medicine, philosophy, psychology, speech pathology, textile technology. Submissions were received from countries all over the world, including: Austria, Australia, Denmark, Canada, Finland, France, Germany, Japan, Israel, Italy, New Zealand, The Netherlands, Poland, The USA, Spain, Switzerland, and The UK.

The standard of the accepted papers was high, with an acceptance rate of about 30%. The papers covered a wide variety of topics and for the sake of imposing some organizational structure on the conference, the presented papers were classified into the following themes: logic and diagrams; theoretical concerns about diagrams; cognition and diagrams; human communication and diagrams; diagrammatic reasoning and proof systems; diagrams to support the development of software systems, and systems to support the development of diagrams. Cutting across these themes was a substantial variety of different types of diagrams. These ranged from classes of diagrams that are ubiquitous in this area of research (such as node-link formats, Euler/Venn diagrams, bar charts, design

A district is basically a collection of zones.

In our example, we have that $d(A) = \{z_1, z_2, z_3, z_4, z_5, z_6, z_7, z_8\}$. Also, the district of the set $\{B, C\}$ is the union of districts of its respective elements:

$$d(\{B, C\} = \{z_2, z_4, z_6, z_7\} + \{z_3, z_5, z_6, z_7\}$$
$$= \{z_2, z_3, z_4, z_5, z_6, z_7\}.$$

Although our objective is to develop a notion of context for a single contour, it is mathematically easier to do so for the more general case, which is a set of contours. The overloading of the term district in Def. makes it possible to treat both contours and sets of contours in a similar fashion, avoiding the cumbersome equating of a contour c with the singleton set $\{c\}$. For the remainder of this section, we let the variables X, Y and Z range over $C + 2^C$.

Using the district notion, we can define a partial order on C.

Definition 3. *We say that X is* covered by Y *if $d(X) \subseteq d(Y)$. We say that X is* strictly covered by Y *if the set containment in the above is strict.*

From a topological standpoint, a contour (or a set of contours) is (strictly) covered by another contour (or a set of contours), if the area defined by the first is (strictly) contained by the area defined by the second. In the above example, the contour A strictly covers contour D as well as the set $\{B, C, D\}$. The set $\{B, C, E\}$ strictly covers the set $\{D, E\}$, as well as the single contour B.

Some covering relationships are less "interesting" than others. Let $S, S' \subseteq 2^C$. Then, if S covers X, then the set $S + S'$ is (trivially) a cover of X. Stated differently, there might be some members of S which do not "contribute" to its coverage of X. In an extreme case, S could be a strict cover of X. But, when we remove from S those members of it whose district is disjoint to that of X, S ceases to be a strict cover of X (although it remains a cover of it). The following definition tries to capture a more meaningful notion of coverage.

Definition 4. *A set of contours S is a* reduced cover *of X if S strictly covers X, $XS = \emptyset$, and there is no $S' \subset S$ such that S' covers X.*

The natural extension of the above definition is that if a single contour is a strict cover of X then it is also a reduced cover of X. In our example, A (just as $\{A\}$) is a reduced cover of D and of $\{B, C\}$.

Lemma 1. *Let S, a set of contours, be a reduced cover of X. Then, for all $c \in S$, $d(c)d(X) \neq \emptyset$. (In words, no member of S can be disjoint to X.)*

In a sense, a reduced cover of X gives a context of X. The collection of these is called territory.

Definition 5. *The* territory *of X is $\tau(X) = \{S \subseteq C | S$ is a reduced cover of $X\}$.*

---

[3] The proof of this, and all other subsequent claims, is omitted for space reasons.

In our running example, we have that $\tau(D) = \{\{A\}, \{B, C\}\}$. We are now in a good position to define the context of a contour or of a collection of contours.

Definition 6. *The* context of X, $\tau(X) \neq \emptyset$ *is defined as*

$$\kappa(X) = \prod_{S \in \tau(X)} d(S) = \prod_{S \in \tau(X)} \sum_{c \in S} d(c).$$

If on the other hand $\tau(X) = \emptyset$, we say that X is context free.

The context of a contour (or a set of contours), when it exists, is a set of zones. In our example, the context of D is

$$\kappa(D) = d(A)d(\{B, C\}) = d(A)(d(B) + d(C)) = \{z_2, z_3, z_4, z_5, z_6, z_7\}.$$

Similarly, we have $\kappa(\{B, C\}) = d(A) = \{z_1, z_2, z_3, z_4, z_5, z_6, z_7, z_8\}$.

Now it should be clear why we used the set of all reduced covers, rather than simply the set of all strict covers in the definition of territory. If we were to compute the intersection of all districts of sets which strictly cover E, then this intersection would have been also over $\{E, B\}$ and $\{E, C\}$. Clearly, the result of such an operation would have been $d(E)$, which is hardly useful for defining a context for E. In our example, both A and E are context-free.

6 Interacting Projections

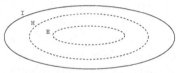

Fig. 6.1: Speakers of English among Hebrew speakers among Israeli Citizens.

The previous section solved the problem of computing the semantics of a diagram with only one projection in it. Rather informally, we say that the semantics of the projection is the intersection between the base set and this context. In this section we formalize this process, while dealing with the more general case, in which the context of a projection may contain other projections as well. In Fig. for example, we have that the set of English speakers, denoted E, is projected into the set formed by projecting the set of Hebrew speakers (denoted H) into the set of Israeli citizens (denoted I).

These conditions can be summarized in the following set of equations:

$$\tilde{H} = HI$$
$$\tilde{E} = E\tilde{H} = EHI.$$

A more challenging example is given in Fig. in which the context of each of the two projections is defined by the other projection. The figure gives rise to a system of equations

$$\tilde{H} = H(I + \tilde{E}) \tag{5}$$
$$\tilde{E} = E(U + \tilde{H}), \tag{6}$$

where U denotes the set of citizens of the UK. Substituting ( ) into ( ) gives a single equation with a single unknown $\tilde{H}$:

$$\begin{aligned}
\tilde{H} &= H(I + E(U + \tilde{H})) \\
&= HI + HEU + HE\tilde{H} \\
&= \alpha\tilde{H} + \beta,
\end{aligned} \tag{7}$$

where the coefficients

$$\begin{aligned}
\alpha &= HE \\
\beta &= HI + HEU = H(I + EU)
\end{aligned} \tag{8}$$

depend only on the given sets H, I, E, and U.

The following lemma is pertinent in solving ( ).

Lemma 2. *Let α and β be two given sets. Then, the equation*

$$x = \alpha x + \beta \tag{9}$$

holds if and only if

$$\beta \subseteq x \subseteq \alpha + \beta. \tag{10}$$

Applying Lemma   we see that ( ) has multiple solutions. We argue that only the minimal one, namely

$$\tilde{H} = \beta = H(I + EU) \tag{11}$$

makes sense in assigning semantics $\tilde{H}$. On the one hand, ( ) takes the format of projecting H into the context $I + EU$, i.e., the set of Hebrew speakers among either Israeli citizens or the English speaking citizens of the UK. This is in agreement with the topology of the contour denoted H in Fig.   . On the other hand, a non-minimal solution to ( ) must contain an element j, such that

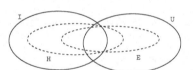

Fig. 6.2: Speakers of English and Hebrew among Israeli and UK Citizens.

$$j \in (\alpha - \beta) \tag{12}$$

It easily follows from ( ) and the definitions of α and β ( ) that $j \notin (I + U)$. In words, j is neither a citizen of Israel nor of the UK. This stands in contrast to the fact that the contours denoted I and E in Fig.   cover both projections.

The generalization of the path we just pursued is largely technical.

Definition 7. *A projections diagram is a diagram $\langle \mathcal{C}, \mathcal{Z} \rangle$, with some set $\mathcal{P} \subset \mathcal{C}$ of contours which are marked as projections. A projections diagram is legal only if all of its projections have a context (Def.   ).*

To give semantics to a projections diagram, we use a *labeling function* χ : $\mathcal{C} \to 2^{\mathcal{U}}$ which maps each of the contours in a projections diagram to some subset of the universe. For $c \in \mathcal{C} - \mathcal{P}$, the semantics function is defined simply as $\psi(c) = \chi(c)$. The problem is to associate a *projected semantics*, in the form of the function $\psi(c)$, with all $c \in \mathcal{P}$. (For all such c, $\chi(c)$ is the set being projected, and $\psi(c)$ is its projected meaning.)

After $\psi(c)$ is computed for all c, we can extend its domain to zones as before ( ), and then use the plane tiling condition ( ) to write the semantics of the diagram. Start-ing for example from Fig.   , then after solv-ing for $\psi(c)$ for all projected contours we can proceed with the *non-projection* diagram of Fig.   .

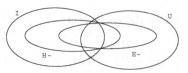

Fig. 6.3: A redraw of Fig.   after the semantics of the projected con-tours was computed.

It follows from the leading discussion that the projected semantics of a con-tour $p \in \mathcal{P}$ is given by

$$\psi(p) = \chi(p) \prod_{S \in \tau(p)} \sum_{c \in S} \psi(c). \tag{13}$$

Examining ( ) more closely, we see that it is in fact a set of polynomial (so to speak) equations over subsets of $\mathcal{U}$. The unknowns in this system are the values of $\psi(p)$ for $p \in \mathcal{P}$. For brevity, let us denote these as $x_1, \dots, x_n$. The constants in this system are the values of $\chi(c)$, where $c \in \mathcal{C}$, which we will denote as $\sigma_1, \dots \sigma_m$. The structure of the equations is determined by the "topology" of the diagram, i.e., the relationships between $\mathcal{Z}$ and $\mathcal{C}$, as expressed in the contexts of each of the projections contours. We can rewrite ( ) as

$$x_1 = P_1(\sigma_1, \dots, \sigma_m, x_2, \dots, x_n)$$
$$\vdots \tag{14}$$
$$x_n = P_n(\sigma_1, \dots, \sigma_m, x_1, \dots, x_{n-1})$$

where P_i, $i = 1, \dots, n$, is a multivariate positive set polynomial (i.e., a formal expression involving only set union and intersection) over $\sigma_1, \dots, \sigma_m$ and

$$x_1, \dots, x_{i-1}, x_{i+1}, \dots x_n.$$

The following lemma is pertinent in deriving a general process of solving ( ).

Lemma 3. *Every multivariate positive set polynomial P over variables $\delta_1, \dots, \delta_\ell, x$ can be rewritten as $P(\delta_1, \dots, \delta_\ell, x) = P_1(\delta_1, \dots, \delta_\ell)x + P_2(\delta_1, \dots, \delta_\ell)$.*

In the case $n = 2$, a solution for ( ) is obtained by substituting the equation defining x_2 in the equation defining x_1. By applying Lemma   we have that the result of the substitution is a linear equation in x_1. We then proceed to finding a minimal solution to this equation using Lemma   . To solve for x_2 we substitute into the equation defining it, the solution for x_1.

The procedure is only slightly more complicated for $n > 2$. In fact, it can be thought of as a Gaussian elimination process applied to set equations. Using Lemma   we rewrite the equation $x_n = P_n(\sigma_1, \ldots, \sigma_m, x_1, \ldots, x_{n-1})$ as a linear equation in x_n in the format of Lemma . We solve for x_n in terms of $\sigma_1, \ldots, \sigma_m$ and $x_1, \ldots, x_{n-1}$, using Lemma   and choosing as usual the minimal solution. We then substitute this solution into $P_1, \ldots, P_{n-1}$, whereby reducing the number of variables by one. The process is then iterated until there is only one equation with only one unknown x_1. We solved this and express x_1 in terms of the constants $\sigma_1, \ldots, \sigma_m$. In backtracking these steps, with appropriate substitutions of the values of the unknowns which were already we can write the minimal solution for $x_2, \ldots, x_n$ in order in terms of the constants only.

7   Multi-Projections

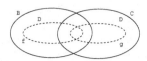 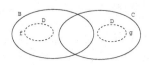

(a)   Intersecting   multi-projections into intersecting contexts

(b) Disjoint multi-projections into intersecting contexts

Fig. 7.1: Multi-projections into intersecting contexts.

So far, the discussion assumed that every base set is projected only once into a diagram. However, an additional expressive power of projections arises by introducing *multi-projections*, i.e., the ability to make several projections of the same base set into a diagram. Multi-projections enable the drawing of Venn diagrams with a large number of contours in a structured, visually-clear fashion, as demonstrated in the Venn diagrams of six contours Fig.   (b).

There are some simple cases in which the desired semantics of multi-projections is clear. In Fig.   (a) for example we have

$$D^f = DB$$
$$D^g = DC.$$

$$(15)$$

The same equations (   ) hold also for Fig.   (b). An additional constraint in this diagram is that, since contour f is disjoint to C in this diagram, $D^f C = \emptyset$, i.e.,

$$DBC = \emptyset. \qquad (16)$$

Equation (   ) follows also from the fact that g is disjoint to B, and also from f being disjoint to g.

Obtaining a semantics for simple examples such as Fig.   (a) and Fig.   (b) is easy to do with only a slight generalization to the semantics as described in the previous section. We can try to define a *multi-projection diagram* as a projection diagram, adorned with an equivalence relationship $\Lambda \subseteq \mathcal{P} \times \mathcal{P}$. With this definition we will say that contours p_1 and p_2 are *multi-projections*, i.e., projections of the same base set if and only if $\langle p_1, p_2 \rangle \in \Lambda$. The labeling function must then satisfy

$$\langle p_1, p_2 \rangle \in \Lambda \Rightarrow \chi(p_1) = \chi(p_2). \tag{17}$$

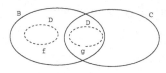

Fig. 7.2: Multi-projection into contained contexts.

Under this requirement, the system of equations (   ) is generated as usual and the semantics is computed by using the Gaussian elimination process as described above.

Unfortunately, this simple-minded approach leads to surprises in many cases. In Fig.   for example,

$$D^f = DB$$
$$D^g = D(BC). \tag{18}$$

However, since f and g are disjoint it follows that

$$\emptyset = D^f D^g$$
$$= DBC \tag{19}$$
$$= D^g,$$

even though the diagram may suggest that D^g is not necessarily empty.

A remedy to this, and other more intricate problems (which are not discussed here for space constraints), might be a demand that the contexts of multi-projections are disjoint, but this would render illegal Fig.   (b). Instead, we take an approach which allows not only projections to show up several times in a diagram, but also ordinary contours.

Recall that a diagram was defined (Def.   ) in abstract terms, which could be described topologically on the plane. In laying out the running example of Sec.   (the diagram defined by (   ) and (   )), in Fig.   , one zone (z_7) was mapped to a noncontiguous area. Let us now generalize in allowing layouts in which contours, and not only zones, can be mapped into noncontiguous areas, as was done in Fig.   .

The figure gives an alternative layout of the running example, in which the abstract contours E and D are each mapped into three noncontiguous areas of the plane. Contours E and D can therefore be thought of as "multi-contours" with several *fragments* in the diagram. Note that zones z_8 and z_9 become noncontiguous while z_7 is contiguous in this layout.

As revolutionary as multi-contours may look, at this stage they have absolutely no effect on our definitions and algorithms which relied on an abstract definition of a diagram, devoid of any geometrical or topological connotation. The main effect is on layout editors, such as *CDEditor* [ ], which must be smart enough to treat noncontiguous zones and contours as one. For example, if a zone is shaded, then this shading should be applied to all the areas it comprises.

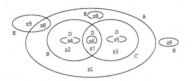

The following definition promotes multi-contours and fragments from the intuitive- to the precise conceptual level.

Fig. 7.3: A noncontiguous layout of
Fig.   .

Definition 8. *A multi-projection diagram is a diagram (Def.), adorned with a set $\mathcal{P} \subseteq \mathcal{C}$, of contours marked as projections, as well as a contour fragments function $\phi : \mathcal{C} \to 2^{2^{\mathcal{Z}} - \{\emptyset\}}$ such that*

$$\forall c \in \mathcal{C} \bullet \sum_{s \in \phi(c)} s = d(c) \tag{20}$$

In words, the function ϕ maps each contour into its fragments, where a fragment of a contour is defined as a non-empty set of zones. Fragments are distinguished by the set of zones in them. In addition, it is required that the contour is "the sum of its fragments", i.e., that the union of the zones defined in the fragments is the same as the zones in (the district of) the contour. If $\phi(c) = \{d(c)\}$ holds for a contour c, then we say, by a slight abuse of terminology, that c is *unfragmented.* If on the other hand $|\phi(c)| > 1$ we say that c is a multi-contour with each of the members of $\phi(c)$ considered its fragments.

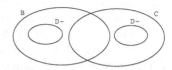

Fig. 7.4: Redrawing Fig.   (b) after computing the semantics of the multi-projection.

For example, the fragments function inferred from Fig.   is given by $\phi(D) = \{\{z_4\}, \{z_5\}, \{z_6\}\}$, and $\phi(E) = \{\{z_8\}, \{z_9\}, \{z_8, z_9\}\}$, while contours A, B, and C are unfragmented. It is important to keep in mind however, that although it is in many cases possible to produce a *layout* for a multi-projection diagram, such a diagram remains a pure set-theoretical creature.

With Def. , a multi-projection is a *single* projected contour which has multiple fragments. Taking this approach, the semantics of a multi-projection is strikingly simple, and is basically a rehash of the procedure described in Sec. , ignoring the fragments altogether!

1. Produce the system of equations ( ), while considering every multi-projection as a single contour. This is to say that the context of each multi-projection is determined with complete disregard to its fragments.
2. Compute the projected semantics of each multi-projection using the process describe above in Sec. .
3. Apply the plane tiling condition, ignoring the fragments of every multi-projection, those of any other multi-contour.

In applying the above to the diagram of Fig.   (b), we have that the D is a multi-projection contour with context $B + C$. Therefore the semantics is $\tilde{D} = D(B + C)$, which after replacement produces the multi-diagram without projection depicted in Fig.   .

Applying the plane tiling condition to Fig.   we obtain that $\tilde{D}(BC) = \emptyset$, which is the same as (   ), but without going through the unnecessary step of defining a semantics for each fragment as done in (   ).

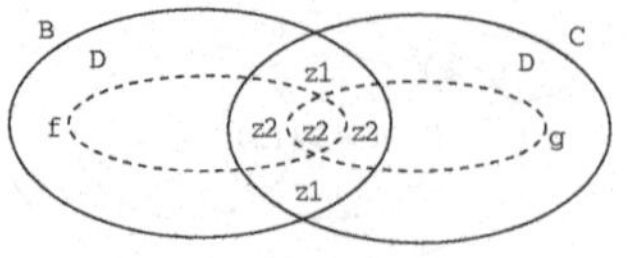

Fig. 7.5: Degenerate layout of zones in Fig.   (a).

It is important to notice that after the system of equations was solved, the extension of the function ψ to zones makes it also possible to assign semantics to each of the fragments of a multi-contours as follows. For all $f \in \phi(c)$, we extend ψ to it as $\psi(f) \equiv \sum_{z \in f} \psi(z)..$

It is not difficult to check that the semantics of a contour is the same as the union of the semantics of its fragments $\psi(c) = \sum_{f \in \phi(c)} \psi(f)$.

In Fig.   , for example, we find $\tilde{D} = DB$, without the incurred difficulty in (   ). In this case we have that the semantics of the fragment f is $D(B - C)$ while that of fragment g is $D(BC)$.

In concluding our discussion of multi-projections, it is interesting to reconsider Fig.   (a) in light of Def.   . Redrawing this figure while highlighting some of the odd shaped zones, we see (Fig.   ) that zone $z_1 = \{B, C\}$ occupies a non-contiguous area of the plane. Zone $z_2 = \{B, C, D\}$ has an even more degenerate layout, in which the area it occupies on the plane is somewhat artificially partitioned by contour fragments. This anomaly accentuates the fact that a "better" layout of the diagram would be as in Fig.   (a).

This situation could have been eliminated by requiring that for every multi-contour c, and for every two distinct $f_1, f_2 \in \phi(c)$, $f_1 f_2 = \emptyset$. We refrained from making this requirement since the aesthetic end it serves is not very essential.

8   Conclusions and Future Research

It remains open to find an efficient algorithm for producing and then solving this system of equations, or proving that no such algorithm exists. It is intriguing to determine whether the contour-based approach, or the zone-based approach [ ], or perhaps a third one, is found by users to be more intuitive.

Acknowledgments. We pay tribute to Yan Sorkin, the creator of *CDEditor*, the automatic tool which was used in generating virtually all diagrams presented here. The production of this manuscript would have impossible without his Sisyphean efforts in making updates to the editor as the research on semantics progressed.

References

1. L. Euler. *Lettres a Une Princesse d'Allemagne*, volume 2. 1761. Letters No. 102–108.

2. J. Gil, J. Howse, and S. Kent. Constraint diagrams: A step beyond UML. In *Proceedings of TOOLS USA '99*, 1999. ,

3. J. Gil, J. Howse, and S. Kent. Formalizing spider diagrams. In *Proceedings of IEEE Symposium on Visual Languages (VL99)*. IEEE Press, 1999. , ,

4. J. Gil, J. Howse, S. Kent, and J. Taylor. Projections in Venn-Euler diagrams. Manuscript, availalble from the first author., 2000. , ,

5. J. Gil and Y. Sorkin. Ensuring constraint diagram's consistency: the *cdeditor* user friendly approach. Manuscript, availalble from the second author; a copy of the editor is available as `http://www.geocities.com/ysorkin/cdeditor/`, Mar. 2000.

6. B. Grünbaum. Venn diagrams I. *Geombinatorics*, 1(2):5–12, 1992.

7. R. Lull. *Ars Magma*. Lyons, 1517.

8. T. More. On the construction of Venn diagrams. *Journal of Symbolic Logic*, 24, 1959.

9. C. Peirce. *Collected Papers*. Harvard University Press, 1933.

10. S.-J. Shin. *The Logical Status of Diagrams*. CUP, 1994.

11. J. Venn. On the diagrammatic and mechanical representation of propositions and reasonings. *Phil.Mag.*, 1880. 123.

On the Completeness and Expressiveness of Spider Diagram Systems

John Howse, Fernando Molina and John Taylor

School of Computing & Mathematical Sciences
University of Brighton, UK
{John.Howse,F.Molina,John.Taylor}@bton.ac.uk

Abstract. Spider diagram systems provide a visual language that extends the popular and intuitive Venn diagrams and Euler circles. Designed to complement object-oriented modelling notations in the specification of large software systems they can be used to reason diagrammatically about sets, their cardinalities and their relationships with other sets. A set of reasoning rules for a spider diagram system is shown to be sound and complete. We discuss the extension of this result to diagrammatically richer notations and also consider their expressiveness. Finally, we show that for a rich enough system we can diagrammatically express the negation of any diagram.

Keywords Diagrammatic reasoning, visual formalisms.

1 Introduction

Euler circles [2] is a graphical notation for representing relations between classes. This notation is based on the correspondence between the topological properties of enclosure, exclusion and intersection and the set-theoretic notions of subset, disjoint sets, and set intersection, respectively. Venn [14] modified this notation to illustrate all possible relations between classes by showing all possible intersections of contours and by introducing shading in a region to denote the empty set. However a disadvantage of this system is its inability to represent existential statements. Peirce [11] modified and extended the Venn system by introducing notation to represent empty and non-empty sets and disjunctive information. Recently, full formal semantics and inference rules have been developed for Venn-Peirce diagrams [13] and Euler diagrams [6]; see also [1, 5] for related work. Shin [13] proves soundness and completeness results for two systems of Venn-Peirce diagrams.

Spider diagrams [3, 7, 8, 9] emerged from work on constraint diagrams [4, 10] and extend the system of Venn-Peirce diagrams investigated by Shin. Constraint diagrams are a visual diagrammatic notation for expressing constraints such as invariants, preconditions and postconditions that can be used in conjunction with the Unified Modelling Language (UML) [12] for the development of object-oriented systems.

M. Anderson, P. Cheng, and V. Haarslev (Eds.): Diagrams 2000, LNAI 1889, pp. 26-41, 2000.
© Springer-Verlag Berlin Heidelberg 2000

In [8] we considered a system of spider diagrams called SD2 that extended the diagrammatic rules and enhanced the semantics of the second Venn-Peirce system that Shin investigated (i.e., Venn-II, see [13] Chapter 4) to include upper and lower bounds for the cardinality of represented sets. In the proof of completeness of SD2 we opted for a strategy in which the diagram that results from combining a set of diagrams and the diagram that is the consequence of that set are expanded in a way similar to the disjunctive normal form in symbolic logic. This strategy extends to other similar systems, including the one considered in this paper. In this paper, we extend SD2 by introducing new notation and extending the inference rules to cover this notation. This extended system is shown to be sound and complete.

A discussion of the system SD2 is conducted in section 2, where the main syntax and semantics of the notation is introduced and soundness and completeness results are given. In section 3 we introduce new notation into the SD2 system, extend the inference rules described in section 2 to include the new notation and show that the extended system is sound and complete. We also enrich the system by providing additional results for reasoning with more expressive diagrams. In section 4, we show that we can express diagammatically the negation of every extended SD2 diagram. Section 5 states the conclusions of this paper and details related, ongoing and future work. Throughout this paper, for space reasons, we omit most proofs.

2   Spider Diagrams: SD2

This section introduces the main syntax and semantics of SD2, a system of spider diagrams. For further details see [8].

2.1   Syntactic Elements of Unitary SD2 Diagrams

A *contour* is a simple closed plane curve. A *boundary rectangle* properly contains all other contours. A *district* (or *basic region*) is the bounded subset of the plane enclosed by a contour or by the boundary rectangle. A *region* is defined, recursively, as follows: any district is a region; if r_1 and r_2 are regions, then the union, intersection, or difference, of r_1 and r_2 are regions provided these are non-empty. A *zone* (or *minimal region*) is a region having no other region contained within it; a zone is uniquely defined by the contours containing it and the contours not containing it. Contours and regions denote sets.

A *spider* is a tree with nodes (called *feet*) placed in different zones; the connecting edges (called *legs*) are straight lines. A spider *touches* a zone if one of its feet appears in that region. A spider may touch a zone at most once. A spider is said to *inhabit* the region which is the union of the zones it touches. For any spider s, the *habitat* of s, denoted $\eta(s)$, is the region inhabited by s. The set of complete spiders within region r is denoted by $S(r)$. The set of spiders touching region r is denoted by $T(r)$. A spider denotes the existence of an element in the set denoted by the habitat of the spider. Two distinct spiders denote distinct elements.

Every region is a union of zones. A region is *shaded* if each of its component zones is shaded. A shaded region denotes the empty set if it is not touched by any spider. A *unitary SD2 diagram* is a single boundary rectangle together with a finite collection of contours (all possible intersections of contours must occur, i.e., the underlying diagram is a Venn diagram), spiders and shaded regions. Each contour must be labelled and no two contours in the same unitary diagram can have the same label. The labelling of spiders is optional. For any unitary diagram D, we use $C = C(D)$, $Z = Z(D)$, $Z^* = Z^*(D)$, $R = R(D)$, $R^* = R^*(D)$, $L = L(D)$ and $S = S(D)$ to denote the sets of contours, zones, shaded zones, regions, shaded regions, contour labels and spiders of D, respectively.

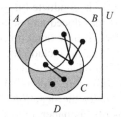

D

Figure 1

The SD2 diagram D in Figure 1 can be interpreted as:

$$A - (B \cup C) = \varnothing \ \wedge \ 1 \leq |C - (A \cup B)| \leq 2 \ \wedge \ 2 \leq |C - B| \ \wedge \ 1 \leq |B|.$$

2.2   Semantics of Unitary SD2 Diagrams

A *model* for a unitary SD2 diagram D is a pair $m = (\mathbf{U}, \Psi)$ where $\mathbf{U}$ is a set and $\Psi : C \rightarrow Set \ \mathbf{U}$, where $Set \ \mathbf{U}$ denotes the power set of $\mathbf{U}$, is a function mapping contours to subsets of $\mathbf{U}$. The boundary rectangle U is interpreted as $\mathbf{U}$, $\Psi(U) = \mathbf{U}$.

The intuitive interpretation of a zone is the intersection of the sets denoted by those contours containing it and the complements of the sets denoted by those contours not containing it. We extend the domain of Ψ to interpret regions as subsets of $\mathbf{U}$. First define $\Psi : Z \rightarrow Set \ \mathbf{U}$ by

$$\Psi(z) = \bigcap_{c \in C^+(z)} \Psi(c) \cap \bigcap_{c \in C^-(z)} \overline{\Psi(c)}$$

where $C^+(z)$ is the set of contours containing the zone z, $C^-(z)$ is the set of contours not containing z and $\overline{\Psi(c)} = \mathbf{U} - \Psi(c)$, the *complement* of $\Psi(c)$. Since any region is a union of zones, we may define $\Psi : R \rightarrow Set \ \mathbf{U}$ by

$$\Psi(r) = \bigcup_{z \in Z(r)} \Psi(z)$$

where, for any region r, $Z(r)$ is the set of zones contained in r.

The semantics predicate $P_D(m)$ of a unitary diagram D is the conjunction of the following two conditions:

Distinct Spiders Condition: The cardinality of the set denoted by region r of unitary diagram D is greater than or equal to the number of complete spiders in r:

$$\bigwedge_{r \in R} |\Psi(r)| \geq |S(r)|$$

Shading Condition: The cardinality of the set denoted by a shaded region r of unitary diagram D is less than or equal to the number of spiders touching r:

$$\bigwedge_{r \in R^*} |\Psi(r)| \leq |T(r)|$$

2.3   Compound Diagrams and Multi-diagrams

Given two unitary diagrams D_1 and D_2, we can *connect* D_1 and D_2 with a straight line to produce a diagram $D = D_1-D_2$. If a diagram has more than one rectangle, then it is a *compound* diagram. The 'connection operation' is commutative, $D_1-D_2 = D_2-D_1$. Hence, if a diagram has n unitary components, then these components can be placed in any order.

The semantics predicate of a compound diagram D is the disjunction of the semantics predicates of its component unitary diagrams; the boundary rectangles of the component unitary diagrams are interpreted as the same set **U**.

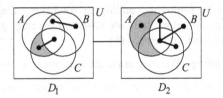

$$D_1 \qquad\qquad D_2$$

Figure 2

The compound diagram $D = D_1-D_2$ in Figure 2 asserts that:

$$(\exists x, y \bullet \ x \in A \cap C \ \wedge \ y \in B - C \ \wedge \ |A \cap C - B| \leq 1) \ \vee$$
$$(\exists x, y \bullet \ x \in B \ \wedge \ y \in A - (B \cup C) \ \wedge \ |A - B| = 1).$$

A spider *multi-diagram* is a finite collection Δ of spider diagrams. The semantics predicate of a multi-diagram is the conjunction of the semantics predicates of the individual diagrams; the boundary rectangles of all diagrams are interpreted as the same set **U**. Contours with the same labels in different individual diagrams of a multi-diagram Δ are interpreted as the same set.

In [8] we describe how to compare regions across diagrams. This formalizes the intuitively clear notion of 'corresponding regions' in different diagrams. For example, in figure 3, the region $z = z_1 \cup z_2$ in D corresponds to the zone z' in D' since both represent the set $B - A$.

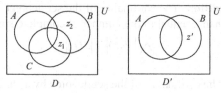

Figure 3

2.4   Compliance and Consistency

A model $m = (U, \Psi)$ *complies* with diagram D if it satisfies its semantic predicate $P_D(m)$. We write $m \models D$. That is, $m \models D \Leftrightarrow P_D(m)$. Similarly, a model m *complies* with multi-diagram Δ if it satisfies its semantic predicate $P_\Delta(m)$. That is, $m \models \Delta \Leftrightarrow P_\Delta(m)$. A diagram is *consistent* if and only if it has a compliant model. All SD2 diagrams are consistent. However, there exist inconsistent SD2 *multi*-diagrams [8].

2.5   Rules of Transformation for SD2

In [8], we introduced rules that allow us to obtain one unitary diagram from a given unitary diagram by removing, adding or modifying diagrammatic elements. These rules are summarised below; they are based on the rules given by Shin in [13], which developed earlier work of Peirce [11].

Rule 1: Erasure of shading. We may erase the shading in an entire zone.

Rule 2: Erasure of a spider. We may erase a complete spider on any non-shaded region.

Rule 3: Erasure of a contour. We may erase a contour. When a contour is erased:
- any shading remaining in only a part of a zone should also be erased.
- if a spider has feet in two regions which combine to form a single zone with the erasure of the contour, then these feet are replaced with a single foot connected to the rest of the spider.

Rule 4: Spreading the feet of a spider. If a diagram has a spider s, then we may draw a node in any non-shaded zone z that does not contain a foot of s and connect it to s.

Rule 5: Introduction of a contour. A new contour may be drawn interior to the bounding rectangle observing the partial-overlapping rule: each zone splits into two zones with the introduction of the new contour. Each foot of a spider is replaced with a connected pair of feet, one in each new zone. Shaded zones become corresponding shaded regions.

Rule 6: Splitting spiders. If a unitary diagram D has a spider s whose habitat is formed by n zones, then we may replace D with a connection of n unitary diagrams

D_1- ... $-D_n$ where each foot of the spider s touches a different corresponding zone in each diagram D_i.

Rule 7: Rule of excluded middle. If a unitary diagram D has a non-shaded zone z where $|S(z)| = n$, then we may replace D with D_1-D_2, where D_1 and D_2 are unitary and one of the corresponding zones of z is shaded with $|S(z)| = n$ and the other is not shaded with $|S(z)| = n+1$.

Rule 8: The rule of connecting a diagram. For a given diagram D, we may connect any diagram D' to D.

Rule 9: The rule of construction. Given a diagram D_1- ... $-D_n$, we may transform it into D if each $D_1,...,D_n$ may be transformed into D by a sequence of the first eight transformation rules.

2.6   Consistency of a Multi-diagram and Combining Diagrams

Definition: An α diagram is a diagram in which no spider's legs appear; that is, the habitat of any spider is a zone.

Any SD2 diagram D can be transformed into an α diagram by repeated application of rule 6, splitting spiders.

Two unitary α diagrams with the same contour set are consistent if and only if for all zones

(i) corresponding shaded zones contain the same number of spiders;

(ii) when a shaded zone in one diagram corresponds to a non-shaded zone in the other, the shaded zone contains at least as many spiders as the non-shaded zone.

Two diagrams D^1 and D^2 are consistent if they can be transformed into α diagrams with the same number of contours D^{1a} and D^{2b} (by rules 5 and 6) and there exist unitary components D_i^{1a} of D^{1a} and D_j^{2b} of D^{2b} such that D_i^{1a} and D_j^{2b} are consistent. See [8] for details.

Intuitively, the diagrammatic conditions (i) and (ii) would prevent the case in which two corresponding zones denote two sets whose cardinalities are inconsistent; this is the only case in which a pair of unitary α diagrams can be inconsistent.

Given two consistent diagrams, D^1 and D^2, we can combine them to produce a diagram $D = D^1 * D^2$, losing no semantic information in the process. Given D^1 and D^2, first transform them into α diagrams D^{1a} and D^{2b} with the same number of contours (using rules 5 and 6). Then the combined diagram $D = D^1 * D^2$ is the compound diagram formed by combining each component D_i^{1a} of D^{1a} with each component D_j^{2b} of D^{2b}; where two components are inconsistent, we do not obtain a corresponding component in D. In forming $D_i^{1a} * D_j^{2b}$, the number of spiders in a zone z is equal to the maximum of the numbers of spiders in the corresponding zones of D_i^{1a} and D_j^{2b}, and z is shaded if and only if at least one of the corresponding zones in D_i^{1a} or D_j^{2b} is shaded – see [8] for details.

The associativity of $*$ allows us to define the combination of the components of a multi-diagram $\Delta = \{D^1, D^2, ..., D^n\}$ unambiguously as $D* = D^1 * D^2 * ... * D^n$. If Δ is inconsistent, the result will be no diagram; $D*$ is only defined when Δ is consistent. A test for the consistency of Δ is to try to evaluate $D*$. Note that there exist inconsistent multi-diagrams each of whose proper subsets are consistent.

Rule 10: The rule of inconsistency. Given an inconsistent multi-diagram Δ, we may replace Δ with any multi-diagram.

Rule 11: The rule of combining diagrams. A consistent multi-diagram $\Delta = \{D^1, D^2, ..., D^n\}$ may be replaced by the combined diagram $D* = D^1 * D^2 * ... * D^n$.

2.7 Soundness and Completeness

D' is a consequence of D, denoted by $D ! D'$, if every compliant model for D is also a compliant model for D'. A rule is *valid* if, whenever a diagram D' is obtained from a diagram D by a single application of the rule then $D ! D'$. We write $\Delta " D'$ to denote that diagram D' is obtained from multi-diagram Δ by applying a sequence of transformations. We write $D " D'$ to mean $\{D\} " D'$, etc.

For space reasons, we omit the proofs of the validity of rules 1 to 11. These rules are similar to those of the Venn-II system given in [13] and the proofs are fairly straightforward. It can be noted that rules 5, introduction of a contour, 6, splitting spiders, 7, excluded middle, and 11, combining diagrams do not lose any semantic information; this is useful for the proof completeness.

Theorem 1   Soundness Theorem Let Δ be a multi-diagram and D' a diagram.

Then $\Delta " D' \# \Delta ! D'$.

The result follows by induction from the validity of the rules. To prove completeness we show that if diagram D' is a consequence of multi-diagram Δ, then Δ can be transformed into D' by a finite sequence of applications of the rules. That is, $\Delta ! D' \# \Delta " D'$. The proof of the completeness theorem, together with an explanation of the strategy of the proof, can be found in [8].

Theorem 2   Completeness Theorem Let Δ be a multi-diagram and let D' be a diagram. Then $\Delta ! D' \# \Delta " D'$.

3   Extending the Notation

In this section we introduce new notation into the SD2 system, extend the transformation rules to include the new notation and show that the extended system is sound and complete.

In fact the new syntactic elements we introduce do not increase the formal expressiveness of the system as a whole. However, they do increase the expressive power of unitary diagrams so that information can be represented more compactly and naturally using the extended notation.

3.1   Extending the Notation

The notion of a *strand* was introduced into spider diagrams (see [3]) to provide a means for denoting that spiders may represent the same element should they occur in the same zone. A *strand* is a wavy line connecting two nodes, from different spiders, placed in the same zone. The *web* of spiders s and t, written $\zeta(s, t)$, is the union of zones z having the property that there is a sequence of spiders $s = s_0, s_1, s_2, \ldots, s_n = t$ such that, for $i = 0, \ldots, n-1$, s_i and s_{i+1} are connected by a strand in z. Two spiders with a non-empty web are referred to as *friends*. Two spiders s and t may (but not necessarily must) denote the same element if that element is in the set denoted by the web of s and t. In Figure 4, it is possible that if the elements denoted by s and t happen to be in $A \cap B$ then they may be the same element.

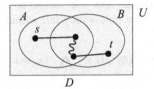

Figure 4

A *Schrödinger spider* is a spider each of whose feet is represented by a small open circle. A Schrödinger spider denotes a set whose size is zero or one: rather like Schrödinger's cat, one is not sure whether the element represented by a Schrödinger spider exists or not. Because of this, a Schrödinger spider in a non-shaded region does not assert anything; however, in shaded regions they are useful for specifying bounds for the cardinality of the set denoted by the region. They are also useful in representing the negation of a diagram (see next section). The set of Schrödinger spiders in diagram D is denoted by $S^* = S^*(D)$. We also let $T^*(r)$ denote the set of Schrödinger spiders touching region r and $S^*(r)$ denote the set of complete Schrödinger spiders in r. From Figure 5, we can deduce $|A - B| \leq 1$, $1 \leq |A \cap B| \leq 2$, $1 \leq |A| \leq 2$ and $|B - A| \leq 1$.

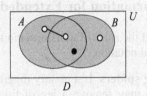

Figure 5

SD2 is based on Venn diagrams; that is, all possible intersection of districts must occur. In general, spider diagrams are based on Euler diagrams, in which information regarding set containment and disjointness is given visually (in terms of enclosure and exclusion). A spider diagram based on a Venn diagram is said to be in *Venn form*; otherwise, it is in *Euler form*. Figure 6 shows a spider diagram in Euler form.

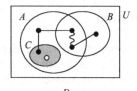

D

Figure 6

Extending SD2 by including strands and Schrödinger spiders and basing it on Euler diagrams we can express the system's semantics as the conjunction of the following conditions.

Spiders Condition: A non-Schrödinger spider denotes the existence of an element in the set denoted by its habitat and the elements denoted by two distinct non-Schrödinger spiders are distinct unless they fall within the set denoted by the zone in the spiders' web:

$$\exists x_1,...,\exists x_n \bullet \left[\bigwedge_{i=1...n} x_i \in \Psi(\eta(s_i)) \wedge \bigwedge_{\substack{i,j=1...n \\ i \neq j}} (x_i = x_j \# \bigvee_{z \in \zeta(s_i,s_j)} x_i, x_j \in \Psi(z)) \right]$$

where $S = \{s_1, ..., s_n\}$.

Plane Tiling Condition: All elements fall within sets denoted by zones:

$$\bigsqcup_{z \in Z} \Psi(z) = \mathbf{U}$$

Shading Condition: The set denoted by a shaded region contains no elements other than those denoted by spiders (including Schrödinger spiders):

$$\bigwedge_{r \in R^*} |\Psi(r)| \leq |T(r)| + |T^*(r)|$$

3.2   Rules of Transformation for Extended Notation

We can adapt and extend the rules of transformation given in section 2.6 to include rules involving the extended notation.

Adapted Rule 6: Splitting spiders. If a unitary diagram D has a spider s formed by two or more feet then we may remove a leg of this spider and replace D with a connection of two unitary diagrams D_1–D_2, each containing a different component of the split spider. If splitting a spider disconnects any component of the 'strand graph' in a zone, then the components so formed should be reconnected using one or more strands to restore the original component. The rule is reversible: if a compound diagrams contains diagram D_1 and D_2 as just described, then D_1 and D_2 can be replaced by diagram D.

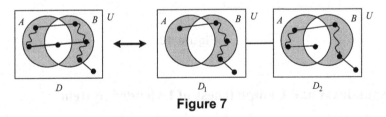

Figure 7

Rule 12: The rule of strand equivalence. A unitary α diagram D (that is, a unitary diagram in which each spider is single-footed) containing a strand in a zone can be replaced by a pair of connected unitary diagrams D_1 and D_2 which are copies of D. In D_1 the strand is deleted and in D_2 the two spiders connected by the strand are deleted and replaced by a single-footed spider in the zone originally containing the strand. Again, the rule is reversible.

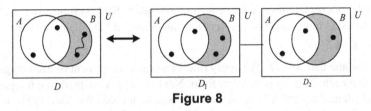

Figure 8

Rule 13: The rule of Schrödinger spider equivalence. A unitary diagram D containing a Schrödinger spider strand can be replaced by a pair of connected unitary diagrams D_1 and D_2 which are copies of D. In D_1 the Schrödinger spider is deleted and in D_2 the Schrödinger spider is deleted and replaced by non-Schrödinger spider. Again, the rule is reversible.

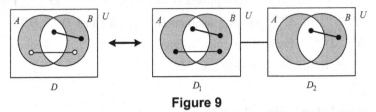

Figure 9

Rule 14: The rule of equivalence of Venn and Euler forms. For a given unitary diagram in Euler form there is an equivalent unitary diagram in Venn form.

Figure 10 shows equivalent diagrams in Euler (left) and Venn form.

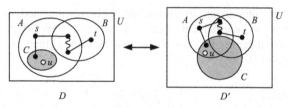

Figure 10

3.3   Soundness and Completeness of Extended System

The validity of the rules stated above is given in Theorem 3.

Theorem 3   If D' is derived from D by a sequence of applications of adapted rule 6 or rule 12 or rule 13 or rule 14, then $D \vdash D'$ and $D' \vdash D$.

The proof is omitted for space reasons. The soundness of the extended SD2 system is derived by induction from the validity of each of the rules.

Theorem 4   Completeness Theorem for extended SD2 system. Let Δ be an extended SD2 multi-diagram and let D' be an extended SD2 diagram. Then

$$\Delta \vdash D' \# \Delta \vDash D'.$$

Proof.   Assume that $\Delta \vdash D'$. We apply rule 14 to each component of each diagram in Δ, so that each diagram is in Venn form. Next, we apply adapted rule 6 repeatedly to each diagram until all the spiders are single footed. We then apply rule 12 repeatedly until we have removed all strands and then rule 13 repeatedly to remove all Schrödinger spiders. The resulting multi-diagram Δ^{SD2} is a set of SD2 diagrams. We apply the same strategy to D' to produce D'^{SD2} an SD2 diagram. By Theorem 3 $\Delta^{SD2} \vdash \Delta$, and $D' \vdash D'^{SD2}$. Hence by transitivity, $\Delta^{SD2} \vdash D'^{SD2}$. So, by Theorem 2 (completeness of SD2), $\Delta^{SD2} \vDash D'^{SD2}$. For each diagram D^{SD2} in Δ^{SD2} there is a diagram D in Δ such that $D \vDash D^{SD2}$, so we have, $\Delta \vDash D'^{SD2}$. Each of the rules adapted rule 6, 12, 13, 14 is reversible, so $D'^{SD2} \vDash D'$. Hence, by transitivity, $\Delta \vDash D'$.

3.4   Derived Reasoning Results

In this section we enrich the system by providing additional results for reasoning with diagrams containing strands or Schrödinger spiders. Several of the results are extensions of the given rules to include diagrams with strands. Each lemma is illustrated in a figure immediately following the lemma. Their proofs are omitted.

Lemma 1 (Rule 2): Erasure of a spider. We may erase a complete spider on any non-shaded region and any strand connected to it. If removing a spider disconnects any component of the 'strand graph' in a zone, then the components so formed should be reconnected using one or more strands to restore the original component.

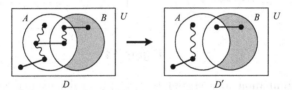

Figure 11

Lemma 2 (Rule 5): Introduction of a contour. A new contour may be drawn interior to the bounding rectangle observing the partial-overlapping rule: each zone splits into two zones with the introduction of the new contour. Each foot of a spider is replaced with a connected pair of feet, one in each new zone. Likewise, each strand bifurcates and becomes a pair of strands, one in each new zone. Shaded zones become corresponding shaded regions.

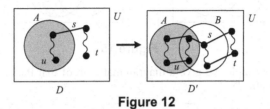

Figure 12

Lemma 3 (Rule 7): Rule of excluded middle. If a unitary diagram D has a non-shaded zone z, then we may replace D with D_1–D_2. D_1 is copy of D where zone z is shaded and D_2 is copy of D where zone z contains an extra single-footed non-Schrödinger spider.

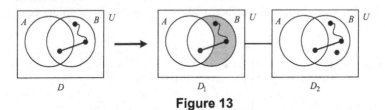

Figure 13

Lemma 4: Elimination of a strand I. Let D be a unitary diagram containing a single-footed non-Schrödinger spider s in an unshaded zone z connected by a strand to a non-Schrödinger spider t whose habitat is unshaded. Then D is equivalent to a diagram D' in which the spider t has been deleted.

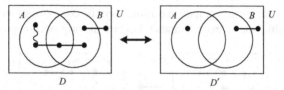

Figure 14

Lemma 5: Elimination of a strand II. Let D be a unitary diagram containing a single-footed non-Schrödinger spider s in an unshaded zone z connected by a strand to a non-Schrödinger spider t. Then D is equivalent to a diagram D' in which the part of t lying in an unshaded region is deleted and that part of t lying in a shaded region is replaced by a Schrödinger spider.

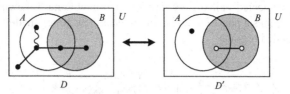

Figure 15

Example. In this example we give a diagrammatic proof that the combination of D_1 and D_2 is equivalent to D.

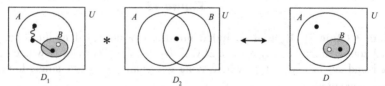

First, transform D_1 into its equivalent α SD2 diagram by lemma 5, elimination of a strand, rule 14, Venn-Euler equivalence, and rule 13, Schrödinger spider equivalence, as follows

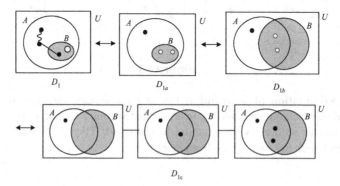

Combining D_{1c} and D_2 by rule 11 we obtain D'.

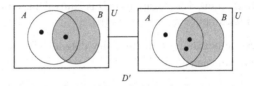

$$D'$$

Again, by rule 13, Schrödinger spider equivalence, we obtain a unitary diagram D''.

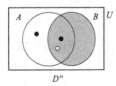

$$D''$$

Finally, since we can transform the Euler form D into the Venn form D'' above, we can complete proof and obtain D from D''.

It is worth noting that, given an SD2 diagram in Euler form, the transformation to its corresponding Venn form is algorithmic. However, in general transforming from Venn to Euler forms is not mechanical.

4   Negation

One of the important properties of the SD2 system is that it is syntactically rich enough to express the negation of any diagram D in a reasonably natural manner. We describe the construction of the negation of D, a diagram which may include Schrödinger spiders, in several stages as follows.

(i) D is an α unitary diagram with n zones which are shaded or contain spiders. The negation of D gives a (compound) diagram with m components ($m \geq n$). Any non-shaded zone z with p spiders gives a unitary component where its corresponding zone z_1 is shaded and contains $p - 1$ Schrödinger spiders. Any shaded zone z with q Schrödinger spiders and r ($r \geq 1$) non-Schrödinger spiders gives two unitary components where its corresponding zones z_1 and z_2 contain $q + r + 1$ non-Schrödinger spiders and $r - 1$ Schrödinger spiders respectively and z_2 is shaded. (When $r = 0$ we obtain a single unitary component whose corresponding zone z_1 contains $q + 1$ non-Schrödinger spiders). If D is an α unitary diagram where no zone is shaded or contains spiders, its negation is any inconsistent multi-diagram.

(ii) D is a compound diagram with n α unitary components. The negation of D gives a multi-diagram formed by n (compound) diagrams being each member of the collection the negation of each α unitary component as in case (i)

(iii) D is any (compound) diagram. We transform it into its α diagram D^α and negate D^α as in (ii).

(iv) D is any multi-diagram. The negation of D is equivalent to negate D^*, the result of combining the components of D, as in (ii).

Figure 16 illustrates the negation of an α unitary diagram. We use the diagrammatic notation $\overline{D}$ to denote the negation of D.

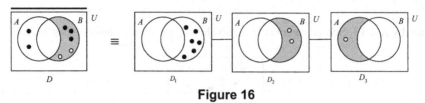

Figure 16

5   Conclusion and Related Work

We have extended the syntax and inference rules of the system of spider diagrams we call SD2 and have shown that this extended system is sound and complete. We have given a number of derived reasoning rules to aid reasoning in the extended notation and shown that we can syntactically give the inverse of any diagram in this system. This extended system contains most of the syntactic elements of spider diagrams given in [3].

Our longer term aim is to prove similar results for constraint diagrams, and to provide the necessary mathematical underpinning for the development of software tools to aid the diagrammatic reasoning process.

References

[1]   Allwein, G., Barwise, J. (1996) *Logical Reasoning with Diagrams*, OUP.

[2]   Euler, L. (1761) *Lettres a Une Princesse d'Allemagne*. Vol. 2, Letters No. 102-108.

[3]   Gil, Y., Howse, J., Kent, S. (1999) Formalizing Spider Diagrams, *Proceedings of IEEE Symposium on Visual Languages* (VL99), IEEE Computer Society Press.

[4]   Gil, Y., Howse, J., Kent, S. (1999) Constraint Diagrams: a step beyond UML, *Proceedings of TOOLS USA 1999*, IEEE Computer Society Press.

[5]   Glasgow, J, Narayanan, N, Chandrasekaran, B (1995) *Diagrammatic Reasoning*, MIT Press.

[6]   Hammer, E. M. (1995) *Logic and Visual Information*, CSLI Publications.

[7]   Howse, J., Molina, F., Taylor, J., (2000) A Sound and Complete Diagrammatic Reasoning System, accepted for ASC 2000, IASTED Conference on Artificial Intelligence and Soft Computing.

[8] Howse, J., Molina, F., Taylor, J., (2000) SD2: A Sound and Complete Diagrammatic Reasoning System, accepted for VL2000, IEEE Symposium on Visual Languages.

[9] Howse, J., Molina, F., Taylor, J., Kent, S. (1999) Reasoning with Spider Diagrams, *Proceedings of IEEE Symposium on Visual Languages* (VL99), IEEE Computer Society Press.

[10] Kent, S. (1997) Constraint Diagrams: Visualising Invariants in Object Oriented Models. *Proceedings of OOPSLA 97.*

[11] Peirce, C. (1933) *Collected Papers.* Vol. 4. Harvard University Press.

[12] Rumbaugh, J., Jacobson, I., Booch, G. (1999) *Unified Modeling Language Reference Manual.* Addison-Wesley.

[13] Shin, S.-J. (1994) *The Logical Status of Diagrams.* CUP.

[14] Venn, J. (1880) On the Diagrammatic and Mechanical Representation of Propositions and Reasonings, *Phil. Mag.* 123.

Non-standard Logics for Diagram Interpretation

Kim Marriott and Bernd Meyer

School of Computer Science & Software Engineering, Monash University, Australia
marriott@csse.monash.edu.au
bernd.meyer@acm.org

Abstract. A key component of computational diagrammatic reasoning
is the automated interpretation of diagram notations. One common and
successful approach to this is based on attributed multiset grammars.
The disadvantages of grammars are, however, that they do not allow
ready integration of semantic information and that the underlying theory is not strongly developed. Therefore, embeddings of grammars into
first-order logic have been investigated. Unfortunately, these are unsatisfactory: Either they are complex and unnatural or else, because of the
monotonicity of classical first-order logic, cannot handle diagrammatic
reasoning. We investigate the use of two non-standard logics, namely
linear logic and situation theory, for the formalization of diagram interpretation and reasoning. The chief advantage of linear logic is that it
is a resource-oriented logic, which renders the embedding of grammars
straightforward. Situation theory, on the other hand, has been designed
for capturing the semantics of natural language and offers powerful methods for modelling more complex aspects of language, such as incomplete
views of the world. The paper illustrates embeddings of grammar-based
interpretation into both formalisms and also discusses their integration.

1 Introduction

Diagram understanding is a basic prerequisite for diagrammatic reasoning. Thus,
a key component of computational diagrammatic reasoning is the automated
interpretation of diagram notations such as state charts, weather maps or structural chemical formula. The syntax-oriented approach to diagram understanding
can be split into two tasks: *parsing* and *interpretation* proper [   ]. Parsing is the process of recognizing the (usually hierarchical) syntactic structure of
a diagram from a flat representation of its components, while interpretation is
the process of building a representation of the meaning of the diagram from its
syntactic structure.

Computational approaches to syntax-based diagram interpretation are most
commonly built on multi-dimensional grammars. Grammar-based parsing is relatively well-understood, having been investigated for over 30 years [   ].
Grammars can also be used to formalize limited types of diagrammatic reasoning
by means of diagram transformations.

Consider reasoning about whether a particular string is in the language of the
automaton represented by a state transition diagram. As illustrated in Figure

M. Anderson, P. Cheng, and V. Haarslev (Eds.): Diagrams 2000, LNAI 1889, pp. 42–   , 2000.
© Springer-Verlag Berlin Heidelberg 2000

we can perform this reasoning by specifying the behaviour of the automaton in terms of syntactic transformations of the state transition diagram. Essentially such transformations detail an animated execution of the automaton. Note that in Figure the remaining input and the current state are denoted by a text label that is always moved to below the current state.

In general, such diagram transformations detail how one diagram can *syntactically* be transformed into another diagram that directly and explicitly exhibits additional information that is either not present or only implicit and hidden in the prior diagram. This can be understood as a form of diagrammatic reasoning by simulation. Unrestricted grammars have been used to express such transformation systems [   ].

Multi-dimensional grammars, however, have two significant disadvantages. First, grammars do not allow ready integration of other domain information. Second, their underlying theory is not well developed, meaning that it is difficult, for instance, to prove the equivalence of grammars.

In part for these reasons, embeddings of grammars into first-order logic have been investigated. Unfortunately, these embeddings are unsatisfactory. The natural embedding in which diagram elements are modelled by predicates is unsatisfactory for diagrammatic reasoning. This is because the combination of a single, flat universe and monotonicity in classical first-order logic means that any piece of information is either true throughout the universe or not at all. It is not possible to model a transformation in a natural manner if this transformation changes the diagram in a non-monotonic way, for instance by deletion of an element. Of course, this can be achieved in first-order logic by modelling diagrams as complex terms. However, this approach leads to a complex and convoluted modelling which does not leverage from the underlying logical inference rules.

Fig. 1. State Transition Diagram

In this paper, we investigate the formalization and computational implementation of diagram interpretation and reasoning with two non-standard logics that are more suitable for these tasks, namely linear logic [   ] and situation theory [   ]. The chief advantage of linear logic is that it is a resource-oriented logic, which renders the embedding of grammars straight-forward. While linear logic eliminates a major technical problem (monotonicity) and is able to model change over time, it offers no adequate way to model partial and incomplete knowledge of the world. In contrast, a sub-divided universe is an inherent notion in situation theory. Situation theory was explicitly developed to model natural language semantics and discourse. It is therefore a good candidate for modelling more complex semantic aspects of diagrammatic reasoning and may also offer a better basis to account for cognitive phenomena in diagram perception, such as the focus of attention.

---

[1] Whenever we mention first-order logic in the remainder of the paper we are referring to *classical* first-order logic.

The main contributions of the current paper are: We detail the problems with embedding multi-dimensional grammatical formalisms into first-order logic (Section). We exemplify how to naturally encode attributed multiset grammars in linear logic so that parsing and transformation can be modelled (Section). We demonstrate how standard situation-theoretic grammar can be extended to capture attributed multiset grammars (Section). Finally, we speculate how a combination of linear logic with situation theory can provide a powerful integrated formalism for diagram interpretation and reasoning (Section).

The work described in the paper is very much in progress. As a consequence, our argument proceeds mainly by means of examples. We intend to detail the formal theory behind these concepts in the future.

Various researchers have directly used first-order logic for diagram interpretation, see e.g. [,   ,   ]. In [   ] description logic [   ] is used as the formal basis, while [   ] is based on a specialized spatial logic. Explicit embeddings for multi-dimensional grammars into first-order logic are given in [,   ,   ,   ,   ]. The current paper is the first to critically evaluate the use of first-order logic for diagram interpretation. To the best of our knowledge, it is also the first to look at diagram interpretation and transformation in terms of linear logic. The use of situation theory for formalizing diagrammatic reasoning is not new. Indeed, some of the most interesting accounts of diagrammatic reasoning have been given in a situation-theoretic setting [,   ,   ]. However, the work in these papers focuses on particular notations [,   ] or particular cognitive aspects [   ] and they do not attempt to provide a general framework for diagram interpretation or diagram transformation.

2 Diagram Interpretation with Attributed Multiset Grammars

As we have indicated, computational approaches to diagram interpretation are usually based on multi-dimensional grammars. Here we review a particular type of attributed multiset grammars, termed constraint multiset grammars (CMGs) [   ], which have been used by a number of researchers for reasonably complex tasks such as the interpretation of state diagrams and mathematical equations. In [   ] a precise formal treatment is given, and we review only the basic notions here.

CMGs can be viewed as an extension of classical string grammars. In contrast to strings, which are built by linear concatenation of tokens, diagrams must be defined in terms of more complex spatial arrangements of graphical tokens. In CMGs, this is achieved by equipping the tokens with geometric attributes and by making the spatial relations explicit in the productions. Productions have the form

---

[2] We note that these example languages can all be unambiguously interpreted on the basis of their syntax only: Thus interpretation is simpler than in the general case since contextual information need not be considered.

$$P ::= P_1, \ldots, P_n \; where \; (C) \; \{E\}$$

indicating that the non-terminal symbol P can be recognized from the symbols $P_1, \ldots, P_n$ whenever the attributes of $P_1, \ldots, P_n$ satisfy the constraints C. The attributes of P are computed using the assignment expression E. The constraints enable information about spatial layout and relationships to be naturally encoded in the grammar. The terms *terminal* and *non-terminal* are used analogously to the case in string languages. The only difference lies in the fact that terminal types in CMGs refer to graphic primitives, such as *line* and *circle*, instead of textual tokens and each of these *symbol types* has a set of one or more attributes, typically used to describe its geometric properties. A *symbol* is an instance of a symbol type. In each grammar, there is a distinguished non-terminal symbol type called the *start* type. Interpretation of the diagram is achieved by equipping the productions with additional semantic attributes in which the interpretation is built up during the parsing process.

As a running example for diagram interpretation, consider the language of state transition diagrams. A typical example was given in Figure . The terminal symbol type declarations for this grammar and their associated attributes are: *arrow(start:point,mid:point,end:point), text(mid:point,label:string), circle(mid:point,radius:real)*. The non-terminal symbol types for these diagrams are *std(ss:states,ts:transitions), arc(start:point,mid:point,end:point,label:string), state(mid:point,radius:real,label:string,kind:string)* and *transition(start:string, tran:string,end:string)* where *std* is the start type.

As an example of a production in a CMG consider that for recognizing a final state. This is made up of two circles *C1* and *C2* and text *T* satisfying three geometric relationships: the mid-points of the circle *C1* and the circle *C2* are the same; the mid-points of the circle *C1* and the text *T* are the same; and *C2* is the outermost circle. Note how the semantic attributes *kind* and *label* are used to construct an interpretation. The production is:

S:state ::= C1:circle,C2:circle,T:text where (
  C1.mid = C2.mid and C1.mid = T.mid and C1.radius ≤ C2.radius) {
  S.mid = C1.mid and S.radius = C2.radius and
  S.label = T.label and S.kind = final }

Using similar productions we can define normal states and start states as well as an *arc* which is an arrow with text above its midpoint. CMGs also include context-sensitive productions. Context symbols, i.e. symbols that are not consumed when a production is applied, are existentially quantified in a production. The following context-sensitive production recognizes a transition:

T:transition ::= A:arc
  exist S1:state,S2:state where (
    OnCircle(A.start,S1.mid,S1.radius) and OnCircle(A.end,S2.mid,S2.radius))
  { T.start = S1.label and T.tran = A.label and T.end = S2.label }

Grammatical formalisms have also been used for specifying limited forms of diagrammatic reasoning. The key is to regard diagrammatic reasoning as a syntactic transformation between diagram configurations. Unrestricted (type-0) variants of graph and multiset grammars can be used to specify these diagram transformations. As an example, consider the following production which specifies how a single character recognition step in a finite state automaton is executed:

> T:transition, $S1$:state, $S2$:state, $L1$:text ::=
> T:transition, $S1$:state, $S2$:state, $L0$:text where (
> $first(L0.label)=T.tran$ and $L0.mid=S1.mid - (0,S1.radius)$ and
> $T.start=S1.label$ and $T.end=S2.label$) {
> $L1.mid=S2.mid - (0,S2.radius)$ and $L1.label=rest(L0.label)$ }

3 Embedding Grammars into First-Order Logic

The great advantage of multi-dimensional grammatical formalisms is that considerable research has focused on finding efficient parsing algorithms. However, they also have two severe limitations. The first is that they are a relatively weak computational formalism which is not well-suited to encoding non-diagrammatic domain information. The second limitation is that multi-dimensional grammars are not well-understood. Parsing with them is understood, but many of their formal properties are not. In particular, there is no deductive calculus for grammars. Therefore we have no methods to prove important properties, such as equivalence of two grammars.

For these reasons, there has been interest in embedding attribute multiset grammars into first-order logic. The motivation is that first-order logic allows ready encoding of non-diagrammatic domain information and that it will allow one to prove properties of the embedded grammars. There have been two main approaches: In the first approach the diagram is encoded as a single term which models a hierarchical collection of the graphical objects in the diagram. In contrast, the second approach encodes the graphical objects directly as predicates. The first approach has been demonstrated in different variants in [,   ,   ,   ], while the second approach is used in [,   ,   ].

The first approach models parsing explicitly in the logic, much as parsing of string grammars is encoded in the DCG transformation used in Prolog [   ]. The idea is to model a diagram as a single hierarchical term. The key predicate is $reduce(D, D')$ which holds if diagram D can be reduced to D' using one of the productions in the grammar. To handle spatial conditions, we assume a first-order theory C which models the relevant aspects of geometry. C is assumed to be an additional set of axioms in our inferences.

For the state transition diagram grammar we can define $reduce_{fs}(D, D')$ which holds if D' can be obtained from D by recognizing a final state:

$$reduce_{fs}(D, D') \leftrightarrow \widetilde{\exists}\{D, D'\}.$$
$$circle(C^1_{mid}, C^1_{radius}) \in D \wedge circle(C^2_{mid}, C^2_{radius}) \in D \wedge$$
$$text(T_{mid}, T_{label}) \in D \wedge$$
$$C^1_{mid} = C^2_{mid} \wedge C^1_{mid} = T_{mid} \wedge C^1_{radius} \leq C^2_{radius} \wedge$$
$$S_{mid} = C^1_{mid} \wedge S_{radius} = C^2_{radius} \wedge S_{label} = T_{label} \wedge S_{kind} = final \wedge$$
$$D' = D \setminus \{circle(C^1_{mid}, C^1_{radius}), circle(C^2_{mid}, C^2_{radius}), text(T_{mid}, T_{label})\}$$
$$\uplus \{state(S_{mid}, S_{radius}, S_{label}, S_{kind})\}$$

where we use the Prolog convention of using identifiers starting with an upper-case letter for variables and those starting with a lowercase letter for predicates and functions, the functions $\uplus$, $\setminus$ and $\in$ work on multisets, and $\widetilde{\exists}\{D, D'\}$ indicates existential closure over all variables but D and D'.

It is also straightforward to model simple diagram transformation in this way. Our example transformation is mapped to:

$$transform(D, D') \leftrightarrow \widetilde{\exists}\{D, D'\}.$$
$$state(S^1_{mid}, S^1_{radius}, S^1_{label}, S^1_{kind}) \in D \wedge$$
$$state(S^2_{mid}, S^2_{radius}, S^2_{label}, S^2_{kind}) \in D \wedge$$
$$trans(T_{start}, T_{tran}, T_{end}) \in D \wedge text(L^0_{mid}, L^0_{label}) \in D \wedge$$
$$first(L^0_{label}) = T_{tran} \wedge L^0_{mid} = S^1_{mid} - (0, S^1_{radius}) \wedge$$
$$T_{start} = S^1_{label} \wedge T_{end} = S^2_{label} \wedge$$
$$L^1_{mid} = S^2_{mid} - (0, S^2_{radius}) \wedge L^1_{label} = rest(T_{label}) \wedge$$
$$D' = D \setminus \{text(L^0_{mid}, L^0_{label})\} \uplus \{text(L^1_{mid}, L^1_{label})\}$$

Certainly this encoding of parsing and diagram transformation is faithful—it exactly captures rewriting with a CMG production. However, the drawback is that the encoding does not greatly leverage from first-order logic. Ideally we would like reduction in the grammar to be directly modelled by implication in the logic.

This idea is the basis for the second approach in which elements in a diagram are identified with predicates rather than terms. For the state transition diagram grammar we can define the predicate *state* which holds if D' can be obtained from D by recognizing a final state:

$$state(S_{mid}, S_{radius}, S_{label}, S_{kind}) \leftrightarrow \widetilde{\exists}\{S_{mid}, S_{radius}, S_{label}, S_{kind}\}.$$
$$circle(C^1_{mid}, C^1_{radius}) \wedge circle(C^2_{mid}, C^2_{radius}) \wedge text(T_{mid}, T_{label}) \wedge$$
$$C^1_{mid} = C^2_{mid} \wedge C^1_{mid} = T_{mid} \wedge C^1_{radius} \leq C^2_{radius} \wedge$$
$$S_{mid} = C^1_{mid} \wedge S_{radius} = C^2_{radius} \wedge S_{label} = T_{label} \wedge S_{kind} = final$$

The advantage of this embedding is that implication directly corresponds to reduction. The disadvantage is that there is a fundamental mismatch between implication in first-order logic and grammatical reduction. The problem is that first-order logic is monotonic—something that has been proven true can never become untrue. Therefore, there is no way of determining if a particular element in the diagram has already been reduced or if all elements in the original diagram have been reduced to the start symbol. Thus parsing cannot truthfully be captured in this form.

Even more seriously, monotonicity means that there is no way to model diagram transformation, since this requires the diagram to change, with valid propositions becoming invalid. Consider the example of the state automaton. An execution step requires moving the current input marker from the current state to the next state. This clearly requires removing it from its current position, but, because of monotonicity of inference, this is impossible to model in classical first-order logic.

4   Embedding Grammars into Linear Logic

As we have seen, direct modelling of CMGs is not possible in first-order logic. The problem is that first-order logic is designed to model eternal laws and *stable* truths [   ]: If X is true and $X \rightarrow Y$ then Y is true and X continues to hold. However, in many physical situations, implication is better understood as an action which consumes the resource X to make Y. For instance, we can use $10 to buy a (mediocre) bottle of wine, but after the purchase the $10 is gone. Linear logic was introduced by Girard [   ] in 1987 to formalize such implication. It is a logic of *resources* and *actions* and has been widely used in computer science to formalize such diverse things as Petri nets and functional language implementation.

The notation $X \multimap Y$ (linear implication) is used to indicate that resource X can be consumed to give Y. As an example, if A is "$10" and B is "bottle of wine" then $A \multimap B$ captures the rule that we can purchase a bottle of wine for $10, but once we do so the money is gone. Clearly, $\multimap$ provides a natural means of formalizing reduction in grammars. The linear connective $X \otimes Y$ is used to indicate that we have both the resource X and the resource Y. If C is a "six-pack of beer" then $A \otimes A \multimap B \otimes C$ states that for 2×10 you can buy both a bottle of wine and a six-pack of beer. Notice that A is not the same as $A \otimes A$—money does not just materialize out of thin air.

The first step of our modelling is to encode the diagram itself. The connective $\otimes$ precisely captures the required combination of diagram elements since it is commutative and associative. In essence it combines resources into a multiset of resources, whereas classical first-order logic is essentially set-based.

Consider the encoding of a diagram consisting only of a final state. An example with arbitrarily chosen concrete coordinates is represented by the following linear logic formula: $\widetilde{\exists}(C_1 \otimes C_1^{att} \otimes C_2 \otimes C_2^{att} \otimes T \otimes T^{att})$

- T is $text(T_{mid}, T_{label})$;
- C_i is $circle(C_{mid}^i, C_{radius}^i)$, for $i = 1, 2$;
- C_1^{att} is $!(C_{mid}^1 = (1,1)) \otimes !(C_{radius}^1 = 2)$;

---

[3] For simplicity we use the first-order logic notation $\forall$ and $\exists$ for quantifiers, rather than Girard's notation ($\wedge$ and $\vee$, respectively).

- C_2^{att} is $!(C_{mid}^2 = (1,1)) \otimes !(C_{radius}^2 = 1)$; and
- T^{att} is $!(T_{mid} = (1,1)) \otimes !(T_{label} = \text{``a''})$.

Note the use of the exponential operator (!) in the attribute formulas. This allows unlimited copying of a resource, so it will enable us to inspect the attributes of an object as often as we like.

It is also relatively straightforward to encode a CMG rule in linear logic. If S represents a final state, C a circle and T some text, then $T \otimes C \otimes C \multimap S$ captures that we can recognize a final state if there are two circles and some text. As in the case of first-order logic, the requisite geometric relationships can be captured by appropriately constraining the geometric attributes.

This is possible because linear logic strictly subsumes first-order logic. Any first-order logic formula, modulo appropriate rewriting of the connectives and application of the exponential operator, is also a linear logic formula. Again, we assume an axiomatic definition of the underlying theory of geometry. We shall use C^{ll} to refer to the linear logic encoding of C.

Consider the CMG production P for recognizing a final state. The linear logic encoding of this production, P^{ll} is

$$\widetilde{\forall}(T \otimes C_1 \otimes C_2 \otimes R \multimap S \otimes A)$$

- T is $text(T_{mid}, T_{label})$;
- C_i is $circle(C_{mid}^i, C_{radius}^i)$, for $i = 1, 2$;
- S is $state(S_{mid}, S_{radius}, S_{label}, S_{kind})$,
- R is $(C_{mid}^1 = C_{mid}^2) \otimes (C_{mid}^1 = T_{mid}) \otimes (C_{radius}^1 \le C_{radius}^2)$, and
- A is $!(S_{mid} = C_{mid}^1) \otimes !(S_{radius} = C_{radius}^2) \otimes$
  $!(S_{label} = T_{label}) \otimes !(S_{kind} = final)$.

Notice that we consume the circles and the text as well as the geometric relationships between the circles and text, and generate unlimited copies of the attribute assignment to S.

On the basis of this encoding we can now embed complete CMGs into linear logic. Assume that the grammar G has productions $P_1, \ldots, P_n$ and let $P_1^{ll}, \ldots, P_n^{ll}$ be their encoding in linear logic. Then the linear logic encoding G^{ll} of G is

$$!P_1^{ll} \otimes \cdots \otimes !P_n^{ll}.$$

The exponential ! is required because each P_i^{ll} is a resource. Without the !, it could only be used exactly once in a derivation while $!P_i^{ll}$ means that the grammar rule can be copied and used as many times as necessary.

Clearly, linear logic can also be used to encode diagram transformation by unrestricted CMG rules: The animated execution of a state transition diagram can, for instance, be defined by the linear rule

$$\widetilde{\forall}(T \otimes S_1 \otimes S_2 \otimes L_0 \otimes R \multimap T \otimes S_1 \otimes S_2 \otimes L_1 \otimes A)$$

- T is $transition(T_{start}, T_{tran}, T_{end})$;
- S_i is $state(S_{mid}^i, S_{radius}^i, S_{label}^i, S_{kind}^i)$;

- L_i is $text(L^i_{mid}, L^i_{label})$, for $i = 0, 1$;
- R is $(first(L^0_{label}) = T_{tran}) \otimes (L^0_{mid} = S^1_{mid} - (0, S^1_{radius})) \otimes (T_{start} = S^1_{label}) \otimes (T_{end} = s^2_{label})$; and
- A is $!(L^1_{mid} = S^2_{mid} - (0, S^2_{radius})) \otimes !(L^1_{label} = rest(T_{label})$.

We are now set to interpret parsing as deduction in linear logic. The statement $F_1 \vdash F_2$ denotes that the linear formula F_2 is a consequence of F_1. Assume that we have a CMG G with start symbol s and an initial diagram D, encoded in linear logic as D^{ll}. Then D is in the language of G if

$$C^{ll}, G^{ll} \vdash D^{ll} \multimap \tilde{\exists} s.$$

Intuitively, this states that, in the context of the theory of geometry and the grammar, the initial diagram can be rewritten to s by applying the formulas that result from the encoding of the grammar G.

This idea clearly extends to understanding diagram transformation as linear deduction. Assume that we have CMG transformation rules T and an initial diagram D_1, with linear logic encodings T^{ll} and D^{ll} respectively. Then if D_1 can be transformed to D_2 using rules in T and D^{ll}_2 is the linear encoding of D_2, then

$$C^{ll}, T^{ll} \vdash D^{ll}_1 \multimap D^{ll}_2.$$

Thus we have shown that we can naturally embed diagram parsing and transformation into linear logic. But what have we gained by this, apart from an alternative formalization?

One clear advantage over grammars is that, because linear logic is an extension of first-order logic, we can encode non-diagrammatic domain knowledge in it and integrate this with diagram understanding and reasoning. Thus we have met the basic prerequisite for being able to resolve ambiguities and for integrating parsing with semantic theories.

The other big advantage is that linear logic has a well-developed proof and model theory that we can leverage from. This can be used to prove abstract properties of grammars, for example, that two grammars are equivalent or that a particular state transition diagram recognizes precisely a certain class of strings. In how far such proofs can be constructed automatically is, of course, a matter for future research due to the undecideability of linear logic. However, this situation is not different from the use of first-order logic for specification purposes. The possibility of constructing proofs also raises the interpretation of diagrammatic reasoning as diagram transformation to a new status. Whereas a pure grammar encoding allowed us only to reason about a diagram by simulation (execution of the transformation), the embedding into linear logic allows us to reason about the behavior of diagrams in the abstract. For example, we could construct a proof that every state transition diagram in which there is a cycle through a start state and an end state will accept words of unbounded length.

5 Embedding Grammars into Situation Theory

In this section we briefly introduce situation theory and sketch one way of handling diagram interpretation in situation theory. As we have indicated in the introduction, the main reason for our interest in situation theory is that, unlike first-order logic and linear logic which inherently model a flat world without subdivisions, situation theory is designed to model a partitioned universe composed of different "situations" each reflecting the viewpoint of an agent at some point in time. This is important because (1) when we view diagrammatic reasoning as the process of transforming diagram configurations, it is natural to regard each diagram configuration as a separate situation, and (2) it may allow us to account for cognitive phenomena, such as the focus of attention, diagrammatic discourse etc.

Situation theory was introduced by Barwise and Perry in 1983 [   ]. Their motivation was to develop a richer, more honest semantic theory for natural language understanding. Since then it has been a fertile research field for logicians, philosophers, linguists and cognitive scientists. Situation theory is a complex field and we can only sketch the basic ideas here. For a complete introduction the interested reader is referred to [   ].

A key insight underlying situation theory is that the truth of a proposition is relative to an observed scene at some point in time, i.e. a *situation*. For instance, truth of the proposition that Paul drinks wine is relative to some particular event in time and space. In situation theory we write $s \models \ll drinks, p, wine; 1 \gg$. This is read as: "the *infon*, i.e. the unit of information, $\ll drinks, p, wine; 1 \gg$ holds in situation s", where the object p is a unique identification for the person named "Paul". The first part of the infon details the relation and its arguments, in this case $drinks(p, wine)$, while the last argument is its *polarity*, which is 1 if the relation is true, 0 if it is false.

Situations are designed to handle partial, incomplete information about the world. Such incompleteness can arise, for example, from an agent's limited view of the world or when a situation is used to model a limited aspect of the world, such as a particular event or a particular slice of space-time. An important aspect of situation theory is the ability to relate such partial situations and to combine them into more complete situations. Inferences in situation theory therefore exist on two different levels: So-called *local* inference works only within a particular situation, whereas *global* inference combines knowledge from different situations, i.e. it works between situations.

A common approach to formalizing local inference is to use classical logic within situations. On the basis of such a logic, important concepts like the *coherence* of situations can be formalized. Coherence means that infons that are logically implied by the infons holding in a situation must also hold in the situation.

As indicated above, one of the motivations for situation theory was to model natural language. The definition and analysis of language structures by means of grammars is fundamental to such a modelling. Situation-theoretic grammar was introduced by Cooper in [   ] and later put into a computational framework by Black [   ]. We will now sketch Cooper's approach and show how it is straightforward to extend it to handle multi-dimensional grammars.

The purpose of a natural language grammar is to define the structure and meaning of spoken or written linguistic expressions, respectively corresponding to parsing and interpretation proper. Cooper's model starts from an *utterance*, i.e. the act of speaking or writing an expression. To define the meaning of an utterance he introduces a relation $gr(u, i)$ which relates an utterance u to its meaning given by an infon i. For example,

$$gr(\underline{\text{Paul smiles}}, \ll smiles, p; 1 \gg)$$

where the underlined text represents the utterance of the underlined expression.

More precisely, an utterance of the expression e, written as $\underline{e}$, can be modelled as a situation u such that $u \models \ll use\text{-}of, u, e; 1 \gg$ where $\ll use\text{-}of, u, e; 1 \gg$ indicates that the expression e has been written or spoken in situation u.

There is some subtlety in this treatment of meaning: The relation gr relates two different kinds of situations: A situation in which the expression is uttered and another situation which is described by this expression. Thus the relation gr must be defined so that if the expression u describes situation s correctly and $gr(\underline{u}, \sigma)$ then we have that $s \models \sigma$.

One of the key ideas in Cooper's approach is that compound utterances are modelled using hierarchical situations. In a sense the situation hierarchy mirrors the expression's syntax or parse tree. The grammatical structure is captured in a special bracket notation for the phrase structure: $[u_1, \ldots, u_n]_X$ denotes a structure of category X consisting of constituents of the categories $u_1 \ldots u_n$ in the corresponding order. For example, if our simple English grammar dictates that a sentence is made of a noun phrase followed by verb and another noun phrase, then the utterance of a sentence s is given by the phrase structure

$$\underline{[NP_1, V, NP_2]}_{sent}$$

The notation of utterances is extended to compound expression in the following way: $[u_1, \ldots, u_n]_X$ means that an utterance $\underline{X}$ consists of the utterances $\underline{u_1}, \ldots, \underline{u_n}$ such that for each i: $X \models \ll precede, \underline{u_i}, \underline{u_{i+1}}; 1 \gg$.

Thus, $u = \underline{[NP_1, V, NP_2]}_{sent}$ is a hierarchical situation such that

$$u \models \ll category, u, \text{sent}; 1 \gg$$
$$u \models \ll constituents, u, \underline{NP_1}, \underline{V}, \underline{NP_2}; 1 \gg$$
$$u \models \ll precede, \underline{NP_1}, \underline{V}; 1 \gg$$
$$u \models \ll precede, \underline{V}, \underline{NP_2}; 1 \gg$$

---

[4] Cooper presented two versions of situation-theoretic grammar. We will focus on the first version, since the extensions with parametric objects that are introduced in the second version do not appear to have immediate relevance for diagrammatic languages.

[5] Note that u is not well-founded. The capability to model such reflective situations is an important feature of situation theory.

The phrase structure notation can be interpreted in two different ways: (1) It can be read as the analogue of a grammar production, defining the general structure of the grammatical category *sent* or, (2) if the constituents NP_1, NP_2 and V are instantiated with concrete utterances, it can be read as a concrete sentence together with its parse tree.

The relation gr, which relates such compound utterances to their meaning, is defined in terms of relations for each possible type of constituent, such as sentence, phrase, verb, noun etc.: $gr(u, i)$ iff $sent(u, i) \lor np(u, i) \lor v(u, i) \lor \ldots$

As an example, let us inspect the relation *sent* which is part of gr and defines the meaning of a simple English sentence:

$$sent([NP_1, V, NP_2]_{sent}, \ll r, x, y; 1 \gg) \text{ if } np(\underline{NP_1}, x) \land v(\underline{V}, r) \land np(\underline{NP_2}, y).$$

The definition of the different components of gr must, of course, be anchored with a lexicon that defines the usable vocabulary. A lexicon contains one entry for each word, defining its grammatical category and meaning. For example,

$$n(u, p) \text{ if } u \models \begin{bmatrix} use\text{-}of: & \text{``Paul''} \\ category: \text{n} \end{bmatrix}$$

Since u obviously is a use of the word "Paul" this is equivalent to $n(\underline{Paul}, p)$. The semantics of this utterance is therefore defined to be the object p. Note that the bracket notation used above indicates that the infons in the brackets are functional, i.e. they define a unique value for a feature of the situation.

To complete the definition of gr we also need rules that create the basic phrase structure for the words in the lexicon: If c is a grammatical category defined in the lexicon this is given by $c([u]_c, i)$ if $c(\underline{u}, i)$.

We will now modify the situation-theoretic grammar approach to handle multi-dimensional languages. The basic requirement is to allow multi-dimensional (geometric) connectives instead of only a linear connective (concatenation). As before, we base the modelling on a set of primitive geometric shapes. We collect all geometric properties of an object o into a single record loc_o, so for a circle c, for example, we have $loc_c =< mid_c, radius_c >$. A "graphical utterance" can now be defined as above, provided we extend the infons with the spatial information. For example, the use of a circle c is thus defined as a situation u such that

$$u \models \ll use\text{-}of, u, circle, < mid_c, radius_c >; 1 \gg$$

This is abbreviated as $\underline{circle}_{loc_c}$. The definition is directly inline with the standard situation theory framework, as spatio-temporal locations for infons are already an integral part of basic situation theory. For convenience, we allow reference to the components of loc by name, for example $loc_c.mid$ is mid_c.

We can build up hierarchical situations corresponding to complex expressions in almost the same way as before. The main difference is that we need to model the spatial relations explicitly instead of using the *precede* relation only. This can be done by including the spatial relations in the bracket notation for the phrase structure. We also use identifiers for the individual components instead of indexing them like Cooper does. Analogously to Cooper's model, the utterance of a final state in our running example can be defined as a situation u_f where

$$u_f = \begin{cases} [T: \textit{text, C1: circle, C2: circle}]_{S:state} \text{ such that} \\ \quad S.mid = C1.mid \text{ and } S.radius = C2.radius \text{ and} \\ \quad C1.mid = C2.mid \text{ and } C1.mid = T.mid \text{ and } C1.radius \leq C2.radius. \end{cases}$$

$V : C$ is an abbreviation for $\underline{C}_{loc_V}$ and $V.a = loc_V.a$. This defines u_f as a situation such that

$$\begin{aligned} u_f &\models \ll category, u_f, state; 1 \gg \\ u_f &\models \ll constituents, u_f, \underline{text}_{loc_T}, \underline{circle}_{loc_{C1}}, \underline{circle}_{loc_{C2}}; 1 \gg \\ u_f &\models \ll =, loc_{C1}.mid, loc_{C2}.mid; 1 \gg \\ u_f &\models \ll =, loc_T.mid, loc_{C1}.mid; 1 \gg \\ u_f &\models \ll \leq, loc_{C1}.radius, loc_{C2}.radius; 1 \gg \\ u_f &\models [S.mid = C1.mid, S.radius = C2.radius] \end{aligned}$$

As in the case of linear logic, we require an underlying geometric constraint theory C which defines the arithmetic relations, such as equality of points. In situation theory parlance, C has the status of "necessary" constraints, i.e. it forms a background knowledge about the world as such and C is valid in all situations. Note that C does not model inference between situations, but only within each individual situation. Therefore it is an extension to the local logic that is used for inference within a situation. Assuming that we use some local theory which subsumes first-order logic (in particular first-order logic itself or linear logic), C can be captured truthfully.

What remains to be done is to specify the meaning of an utterance by using an analogue of the gr relation. Again we introduce a relation for each type of constituent. In the case of final states and assuming u_f is defined as above we have:

$$state(u_f, \ll kind, final; 1 \gg \wedge \ll label, TL; 1 \gg)$$
$$\text{if } text(\underline{T : text}, \ll label, TL; 1 \gg) \wedge circle(\underline{C1 : circle}) \wedge circle(\underline{C2 : circle})$$

Let us now look at diagram parsing in the context of diagrammatic reasoning. Previously we have outlined how a diagrammatic reasoning process can be understood as a chain of diagram transformations τ_i that successively transform a diagram $\mathcal{D}_0$ into a new diagram $\mathcal{D}_n$, which directly and explicitly exhibits the conclusions of the reasoning process. Each of the successive diagram configurations can be captured in a separate situation $\mathcal{S}_i$ so that our diagrammatic inference chain amounts to a global inference on the level of the situation structure. Using the technique described in this section, diagrammatic parsing can likewise be regarded as a process of transformation in which the phrase structure is gradually built up. This corresponds to modelling the stages of a bottom-up parse as individual situations.

6   Towards a Hybrid Approach

We have shown how Cooper's situation theoretic grammar can be extended to capture diagrammatic notations. Embedding the specification of diagrams into situation theory gives us a powerful logical formalism for discussing semantic and some cognitive phenomenon in diagram interpretation and reasoning.

In the way we have defined them above, grammars specify a global theory, since each reduction step in parsing gives rise to a new situation. However, this is not the only way, nor necessarily the best way, to model parsing in situation theory. The main problem with this approach is that a formal framework for logical inference in situation theory is not yet as well developed as in linear logic. In particular, a general syntactic calculus for inference on the global level (i.e., for inferences involving more than one situation) has yet to be developed. This is not surprising given the expressiveness of situation theory and its consequent complexity, but it means that one cannot readily prove properties of grammars if they are encoded in terms of a global theory.

On the other hand, local inference within a single situation is far better understood. We can exploit this in a different approach to performing diagram interpretation in situation theory. The idea is to employ global inference only at the topmost level to handle the transformations involved in true diagrammatic reasoning. Simpler tasks, such as parsing and basic syntactic interpretation, can be contracted into a single situation and handled by local inference. We have illustrated how these tasks can be formalized in linear logic. It appears possible to achieve the integration of both approaches by using linear logic as the calculus of local inference within situation theory.

Such a hybrid approach potentially offers the advantages of both embeddings. The use of linear logic for local inference in situation theory could bridge the gap between an executable, computationally-oriented approach to diagram interpretation and a general semantic account of diagrammatic reasoning. However, integrating linear logic into situation theory as a local calculus is by no means a small endeavour. Channel Theory [   ], an offspring of situation theory which attempts to formalize the collaboration of different types of information manipulating systems in a single composite system, may turn out to provide a general key for this integration as well as for the integration of diagrammatic reasoning with other kinds of reasoning.

We feel that it is well worth exploring the hybrid approach since it has the potential to combine an executable logic, proof theory and semantics into a single unifying framework for diagram interpretation and diagrammatic reasoning.

References

Bla92.   A.W. Black. *A Situation Theoretic Approach to Computational Semantics.* PhD thesis, Dept. of Artificial Intelligence, University of Edinburgh, 1992.

BP83. J. Barwise and J. Perry. *Situations and Attitudes*. MIT Press, 1983.

BS97. J. Barwise and J. Seligman. *Information Flow: The Logic of Distributed Systems*. Cambridge University Press, New York, 1997.

Coo90. R. Cooper. Three lectures on situation theoretic grammar. In *Natural Language Processing: EAIA 90*, pages 102–140, Guarda, October 1990.

GC96. J. M. Gooday and A. G. Cohn. Using spatial logic to describe visual programming languages. *Artificial Intelligence Review*, 10:171–186, 1996.

Gir87. J.-Y. Girard. Linear logic. *Theoretical Computer Science*, 50:1–102, 1987.

Gir91. J.-Y. Girard. Linear logic: A survey. Technical report, Int. Summer School on Logic and Algebra of Specification, 1991.

Haa98. V. Haarslev. A fully formalized theory for describing visual notations. In K. Marriott and B. Meyer, editors, *Visual Language Theory*, pages 261–292. Springer, New York, 1998.

HM91. R. Helm and K. Marriott. A declarative specification and semantics for visual languages. *Journal of Visual Languages and Computing*, 2:311–331, 1991.

HMO91. R. Helm, K. Marriott, and M. Odersky. Building visual language parsers. In *ACM Conf. Human Factors in Computing*, pages 118–125, 1991.

Mar94. K. Marriott. Constraint multiset grammars. In *IEEE Symposium on Visual Languages*, pages 118–125. IEEE Computer Society Press, 1994.

Mey97. B. Meyer. Formalization of visual mathematical notations. In M. Anderson, editor, *AAAI Symposium on Diagrammatic Reasoning (DR-II)*, pages 58–68, Boston/MA, November 1997. AAAI Press.

Mey00. B. Meyer. A constraint-based framework for diagrammatic reasoning. *Applied Artificial Intelligence: An International Journal. Special Issue on Constraint Handling Rules*, 4(14):327–344, 2000.

MM97. B. Meyer and K. Marriott. Specifying diagram animation with rewrite systems. In *International Workshop on Theory of Visual Languages (TVL'97)*, pages 85–96, Capri, Italy, September 1997.

MMA00. B. Meyer, K. Marriott, and G. Allwein. Intelligent diagrammatic interfaces: State of the art. In P. Olivier, M. Anderson, and B. Meyer, editors, *Diagrammatic Representation and Reasoning*. Springer, London, 2000. To appear.

MMW98. K. Marriott, B. Meyer, and K. Wittenburg. A survey of visual language specification and recognition. In K. Marriott and B. Meyer, editors, *Visual Language Theory*, pages 5–85. Springer, New York, 1998.

PW80. F. C. N. Pereira and David H. D. Warren. Definite clause grammars for language analysis - a survey of the formalism and a comparison with augmented transition networks. *Artificial Intelligence*, 13:231–278, 1980.

RM89. R. Reiter and A. K. Mackworth. A logical framework for depiction and image interpretation. *Artificial Intelligence*, 41:125–155, 1989.

Shi96a. A. Shimojima. Operational constraints in diagrammatic reasoning. In G. Allwein and J. Barwise, editors, *Logical Reasoning with Diagrams*, pages 27–48. Oxford University Press, New York, 1996.

Shi96b. S.-J. Shin. A situation theoretic account of valid reasoning with Venn diagrams. In G. Allwein and J. Barwise, editors, *Logical Reasoning with Diagrams*, pages 81–108. Oxford University Press, New York, 1996.

SM97. J. Seligman and L. S. Moss. Situation theory. In J. van Benthem and A. ter Meulen, editors, *Handbook of Logic & Language*. Elsevier, Amsterdam, 1997.

Tan91.   T. Tanaka. Definite clause set grammars: A formalism for problem solving. *Journal of Logic Programming*, 10:1–17, 1991.   ,

Wol98.   M. Wollowski. *DIAPLAN: A decidable diagrammatic proof system for planning in the blocks world.* PhD thesis, Indiana University, Bloomington, 1998.

WS92.   W. A. Woods and J. G. Schmolze. The KL-ONE family. In F. Lehmann, editor, *Semantic Networks in Artificial Intelligence*, pages 133–177. Pergamon Press, Oxford, 1992.

Reviving the Iconicity of Beta Graphs

Sun-Joo Shin

University of Notre Dame, Notre Dame, IN 46556, USA

Abstract. By devising a new reading method for Peirce's Existential
Graphs (EG), this paper moves away from the traditional method of evaluating diagrammatic systems against the criteria appropriate to symbolic
systems. As is well-known, symbolic systems have long been preferred to
diagrammatic systems and the distinction between the two types of systems has not been well defined. This state of affairs has resulted in a
vicious circle: because the unique strengths of visual systems have not
been discovered, diagrammatic systems have been criticized for lacking
the properties of a symbolic system, which, in turn, reinforces the existing prejudice against non-symbolic systems. Peirce's EG is a classic
example of this vicious circle.
Logicians commonly complain that EG is too complicated to put to
actual use. This paper locates a main source of this criticism in the
traditional reading methods of EG, none of which fully exploits the visual
features of the system. By taking full advantage of the iconicity of EG, I
present a much more transparent and useful reading of the Beta graphs.
I pursue this project by (i) uncovering important visual features of EG,
and (ii) implementing Peirce's original intuitions for the system.

1   Introduction

One shared view among diagrammatic reasoning researchers is that symbolic
systems have been strongly preferred to diagrammatic systems throughout the
history of modern logic and mathematics. This belief clearly presupposes that
symbols and diagrams are different from each other. However, no clear distinction
has been made among different kinds of representation systems. I would like to
focus on a circular relation formed between the prejudice against diagrammatic
systems and the non-clarified distinction of diagrams from symbols.

Without a solid theoretical background for the distinction between symbolic
and diagrammatic systems, we easily overlook different kinds of strengths and
weaknesses each type of system exhibits. Then, with the general predominance
of symbolic systems, we tend to try to understand and evaluate diagrammatic
systems against the criteria of a symbolic system. As a result, without its own
strengths being discovered, a diagrammatic system has been criticized mainly
because it lacks the properties of a symbolic system. In turn, this unfortunate
result, stemming partly from a prejudice against diagrams and partly from an
unclear distinction between symbols and diagrams, only reinforces the existing
prejudice against non-symbolic systems. I claim that the reception of Peirce's

M. Anderson, P. Cheng, and V. Haarslev (Eds.): Diagrams 2000, LNAI 1889, pp. 58– , 2000.
© Springer-Verlag Berlin Heidelberg 2000

Existential Graphic System (EG) is a classic example of this vicious circle, and I propose EG as a case study to show how to break this undesirable circular relation.

Extensive research has been performed on this impressive graphic system. However, many logicians strongly prefer symbolic systems over this graphic system. One of the main reasons for this preference is that Peirce's Beta graphs have been considered to be difficult to read off. In this paper, by uncovering visual features of EG (in the second section) and by reviving Peirce's original intuitions for the system (in the third section), I present a new reading algorithm in the fourth section. In the fifth section, I illustrate through examples the main differences between this new reading and existing ones. In the conclusion, I discuss the scope of this project and further work in this direction.

2   Conjunctive and Disjunctive Juxtaposition

The Beta system has four kinds of basic vocabulary: cut, juxtaposition, lines and predicate symbols. After a brief introduction of the meaning of cut and juxtaposition, I identify overlooked visual properties that represent more syntactic distinctions in EG than are made on the traditional approach. I do not introduce new syntactic devices into EG, but only observe significant visual differences already present in EG.

A cut represents the negation of whatever is written inside the cut, and juxtaposition the conjunction among whatever is juxtaposed. In the following example,

Graph (1) is read off as "S is true," (2) "S is true and W is true," (3) "It is not the case that both S and W are true," (4) "Neither S nor W is true," and (5) "It is not the case that neither S nor W is true," which is the same as "Either S or W is true."

It has long been believed that Peirce's EG has only two kinds of syntactic operation corresponding to logical connectives in a symbolic system, cut and juxtaposition. Therefore, Don Roberts emphasizes the importance of the order of negation and conjunction for the translation of the following Peircean graph:

---

[1] Zeman, Roberts, Sowa, Burch and Hammer.

Notice that we do *not* read it: 'Q is true and P is false', even though Q is evenly enclosed and P is oddly enclosed. . . . we read the graph from the outside (or at least enclosed part) and we proceed inwardly . . . .

Roberts suggests the following passage from Peirce as evidence that this method, "a method to which Peirce gave the name 'endoporeutic'," was what Peirce originally had in mind:

The interpretation of existential graphs is *endoporeutic*, that is proceeds inwardly; so that a nest sucks the meaning from without inwards unto its centre, as a sponge absorbs water.

I claim that this reading method has prevented us from benefiting from the visual power of the system. My main criticism is that this method does not reflect visually clear facts in the system. As Roberts points out in the above quotation, in the graph it is true that Q is evenly enclosed and P is oddly enclosed. However, the endoporeutic reading does not reflect directly this visually clear fact at all, and what is worse, it leaves the impression that this visual fact is misleading. To put this criticism in a more general way: this method forces us to read a graph only in one way. For example, we are supposed to read the above graph in the following way: This graph is a cut of the juxtaposition of P and a cut of Q. However, there are other possible readings. For example, we might say that this graph has two cuts with Q enclosed by both cuts and P enclosed by the outer cut only. Or, we might say that this graph has two letters, P and Q, and the juxtaposition of P and a cut of Q is enclosed by a cut (hence, P is enclosed by one cut and Q enclosed by two cuts). These readings are not reflected directly in the endoporeutic reading.

I suggest that the following two features be interpreted *directly* in a new reading: (i) the visual distinction between what is asserted in an evenly enclosed area and what is asserted in an oddly enclosed area, and (ii) the visual fact that some juxtapositions occur in an evenly enclosed area and some in an oddly enclosed area. These suggestions will be reflected in the new reading method presented in the next section in the following way: The juxtaposition occurs in an evenly enclosed area is read off as conjunction, and the juxtaposition in an oddly enclosed area as disjunction. Also, a negation is added by reading off a visual fact that a symbol is written in an oddly enclosed area.

---

[2] X is *evenly* enclosed iff X is enclosed by an even number of cuts, and X is *oddly* enclosed iff X is enclosed by an odd number of cuts. (My footnote)

[3] Roberts, 1973, p. 39.

[4] Roberts, 1973, p. 39.

[5] Peirce, Ms 650, quoted by Roberts, 1973, p. 39, n.13.

Let me illustrate how to read off these features through the example Roberts cited above. P is written in an oddly enclosed area, and Q in an evenly enclosed area. Hence, we obtain $\neg P$ and Q. The juxtaposition between P and a cut of Q occurs in an oddly enclosed area. Therefore, we easily get a sentence '$\neg P \vee Q$,' which does not require any further manipulation.

3 Peirce's Intuition for Iconicity

Another syntactic device in the Beta system, a line, called 'a line of identity,' (henceforth, 'LI') denotes the existence of an object, and it can have many branches, called 'ligature.' For example, the first graph says (i) "Something good is ugly," the second (ii) "Something good is useful and ugly," and the third (iii) "It is not the case that some good thing is not ugly," which is the same as "Everything good is ugly."

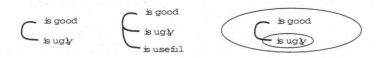

In this section, we will discuss Peirce's three important intuitions related to LI's which make this Beta system graphic, and accordingly, distinct from other predicate symbolic languages. I claim that all these three valuable intuitions should be reflected in our new reading algorithm.

3.1 Lines of Identity

It is well known that relations are harder to represent in a graphic system than properties. Hence, Peirce's first motivation for the Beta system is to represent relations graphically. His example is the proposition that A is greater than something that is greater than B. After suggesting unsuccessful graphs, finally he says

---

[6] Clause 4 in the new reading in section 4 formalizes this process.

[7] Another problem with the traditional reading is that in many cases we obtain a complicated-looking sentence, i. e. one with nested negations and conjunctions, which requires further syntactic manipulations.

[8] Two of these examples are borrowed from Roberts 1973, p. 51.

[9] For example, well-known Euler or Venn diagrams are limited to the representation of properties. (Refer to Shin 1994.) In the discussion of Venn diagrams, which precedes the writing on his EG, Peirce points out this aspect as one of the problems of the Venn system: "It [The Venn system] does not extend to the logic of relatives." (Peirce, 4.356.)

[10] "In many reasonings it becomes necessary to write a copulative proposition in which two members relate to the same individual so as to distinguish these members." (Peirce, 4.442.)

[I]t is necessary that the signs of them should be connected in fact. No way of doing this can be more perfectly iconic than that exemplified in Fig. 78 [the following graph].

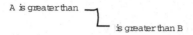

This syntactic device, i.e. the line connecting two predicates, represents one and the same object than which A is greater and at the same time which is greater than B. This is why Peirce calls this line a *line of identity* (henceforth, 'LI'). This intuition is captured in the following convention:

A heavy line, called a **line of identity**, *shall be a graph asserting the numerical identity of the individuals denoted by its two extremities.*

The introduction of lines adds expressive power to this system: it allows us to represent individuals and relations, while his Alpha system cannot. At the same time, this device makes a clear distinction between Beta graphs as a graphic system and other predicate languages as symbolic systems. While tokens of the same type of letter represent the numerical identity of individuals each token denotes, one network of lines represents the numerical identity among individuals denoted by each branch. That is, Peirce's system *graphically* represents the numerical identity by one connected network. Visual clearness is obtained by this aspect of the system. For example, in the first-order sentence, $\exists x \exists y [Tall(y) \wedge Love(x, y) \wedge Walk(y) \wedge Teacher(x) \wedge Smile(y) \wedge Woman(x)]$, we need to keep track of tokens of x and y more carefully than when the identity of each individual is represented by graphical lines in the following Beta graph:

3.2 Existential versus Universal Quantifiers

The most striking aspect of the Beta system compared with other predicate languages is that it does not introduce a syntactic device which corresponds to

[11] Peirce, 4.442.
[12] Peirce, 4.442.
[13] Peirce, 4.442.

quantifiers of symbolic languages. Then, how can the Beta system be about the logic of quantifiers? Rather than adopting one more syntactic device, Peirce relies on visual features which already exist in a graph:

> ... *any line of identity whose outermost part is evenly enclosed refers to* **something,** *and any one whose outermost part is oddly enclosed refers to* **anything** *there may be.*

Examples are in order:

Since the outermost part of the line is in an evenly enclosed in the graph on the left side, it says that something good is ugly. On the other hand, the graph on the right side says that everything good is ugly.

That is, existential and universal quantifiers are represented *visually* by whether the outermost part of an LI is evenly or oddly enclosed, without any additional diagrammatic or symbolic object introduced. I take this aspect of the system to be one of the most important ones, which makes the system graphic.

3.3 Scope of Quantifiers

Graphic systems are in general not linear expressions. In symbolic systems which are linear, the scope of syntactic operations is clarified by the linear order of the symbols, or sometimes along with the use of parentheses. Now, the question in the Beta system is how to represent the order among different quantifiers. Fore example, how does this system make a distinction between the proposition '$\forall x \exists y L(x, y)$' and the proposition '$\exists y \forall x L(x, y)$'?

Again, Peirce solves this problem in the Beta system by appealing to the *visual* feature about how many times lines' outermost parts are enclosed: The less enclosed the outermost part of a line is, the bigger scope the line gets. As an example, let us compare the following graphs:

---

[14] As Roberts starts with his chapter on the Beta system, this system is "a treatment of the functional or predicate calculus, the logic of quantification." (Roberts 1973, p. 47.)

[15] Peirce, 4.458. The *outermost* part of an LI is the part which is the least enclosed.

[16] Borrowed from Roberts 1973, p. 51.

[17] In the graph on the right side, some part of the LI is enclosed by one cut and some part by two cuts. Therefore, the part enclosed by one cut is the outermost part of the LI.

[18] Roberts 1973, p. 52.

In the first graph, the line whose outermost part is written in an oddly enclosed area is less enclosed than the line whose outermost part is in an evenly enclosed area. Therefore, a universal quantifier has a bigger scope than an existential quantifier. In the second graph, the opposite is true. While '$\forall x \exists y (Catholic(x) \rightarrow (Adores(x,y) \wedge Woman(y)))$' is what the first graph expresses, the second says '$\exists y \forall x (Woman(y) \wedge (Catholic(x) \rightarrow Adores(x,y)))$.'

It should be noticed that Peirce takes an advantage of a homomorphic relation between the position of lines and their scopes: If the outermost part of line l_1 contains every area the outermost part of line l_2 contains, then the reading of l_1 has a bigger scope than the reading of l_2. Without linear order or without additional graphic notation, Peirce disambiguates the scope relation among quantifiers. The way he accomplishes this goal, that is, reading off a simple feature visualized in a graph, clearly makes this system more graphic.

4 New Reading

In this section, I present a new algorithm for reading Beta graphs, which reflects two kinds of juxtapositions in the second section and Peirce's intuitions discussed in the third section.

For further discussion, let us examine more carefully how LI's are used in this system. In the following two graphs, a simple identity line says that something is P and Q:

Suppose we need to express that something is P, Q and R. This is where an LI needs to branch, as follows:

In this case, every branch of one and the same LI functions as one identify line, that is, every branch denotes the same object. We call this connected web of branches an *LI network*. In the above graphs, each graph has one LI network, just as a first-order language requires *one* type of variable under the same scope of one existential quantifier. Hence, there seems to be a strong connection between an LI in Beta graphs and a variable in first-order languages, and we will implement this connection in our reading method, with the following exception.

Suppose we translate the proposition that there is something which is P and something which is Q and these two are not identical with each other. "$\exists x \exists y (Px \wedge Qy \wedge x \neq y)$" is a translation in a first-order language. Two variables are used under the scope of two existential quantifiers. However, if we use two LI's attached to predicates P and Q respectively as follows, we do not obtain what we want:

What this graph says is that something is P and something is Q, i.e. ($\exists x Px \wedge \exists y Qy$). That is, it fails to express a non-identical relation between two objects. We need to *negate* an identical relation itself, by using a cut. Any of the following graphs does the job:

Therefore, when an LI crosses an odd number of cuts entirely, the generalization that each LI network refers to the same object is not true. On the contrary, we need to bring in (at least) two different objects in this case. If a portion of an LI crosses an odd number of cuts, we say that that line is *clipped*. Notice that an LI network may be clipped into more than two parts. Both graphs in the following have only one LI network, but the LI network in the graph on the left is clipped into two parts, while the LI network in the graph on the right is clipped into three parts by one cut:

We will implement these discussions of LI networks in the second step of the reading method presented below. Next, we will write atomic wffs at the

---

[19] Another main difference is that an LI denotes an object while a variable does not.

end of an LI. In the fourth step, we form a complex wff by reading off two kinds of juxtaposition. Finally, when we make a sentence out of this complex wff, whether the outermost part of an LI network is evenly or oddly enclosed determines whether an existential or a universal quantifier will be adopted.

Reading off Beta Graphs

1. Erase a double cut.
2. Assign variables to an LI network:
   (a) If no portion of an LI network crosses an odd number of cuts entirely, then assign a new variable to the outermost part (i.e. its least enclosed part) of the LI networks.
   (b) If a portion of an LI network crosses an odd number of cuts entirely (i.e. an odd number of cuts clips an LI into more than one part), then
      i. assign a different type of a variable to the *outermost* part of each clipped part of the network, and
      ii. at each joint of branches inside the *innermost* cut of these odd number of cuts, write $v_i = v_j$, where v_i and v_j are assigned to each branch into which the line is clipped by the cuts.
3. Write atomic wffs   for LI's: (Let us call it a quasi-Beta graph.)
   (a) For an end of an LI with a predicate, say P, replace P with $Pv_1 \ldots v_n$, respecting the order of hooks,   where v_1 (or $[v_1]$),..., and v_n (or $[v_n]$) are assigned to the lines hooked to P.
   (b) For each loose end of an LI, i.e. an end without a predicate, write $\top$.
   (c) For an LI which does not get any atomic wff or $\top$, i. e. a cycle whose part is not clipped, write $\top$ at the outermost part and at the innermost part of the LI.
4. Obtain a complex wff:
   Let G be a simple quasi-Beta graph.   The following f is a basic function:

   $$f(G) = \alpha \quad \text{if } G \text{ is atomic wff } \alpha.$$
   $$f(G) = \neg\alpha \quad \text{if } G \text{ is a single cut of atomic wff } \alpha.$$
   $$f(G) = \top \quad \text{if } G \text{ is } \top.$$

[20] Note that in order to read off available visualinformation of a graph as much as possible, we will not erase any part of a graph in the reading process, which makes a contrast between Zeman's and this new reading methods. (Zeman, pp. 79-82.)

[21] A scroll [two cuts, one inside the other] is a *double cut* if and only if nothing is written in the outer area of the scroll [i. e. the area between the inner and the outer cuts] except the portions of LI's which cross the outer area entirely.

[22] The term 'wff' is an abbreviation for a well-formed formula.

[23] In a clockwise direction.

[24] In order to keep track of one unclipped LI network, sometimes we might want to write down variables in brackets. Hence, in the step (5), the location of the variables in brackets should be ignored.

[25] A simple quasi-Beta graph is an atomic wff, a single cut of atomic wff, $\top$, an empty space or an empty cut.

$f(G) = \neg \top$ if G is a single cut of $\top$.
$f(G) = \top$ if G is an empty space.
$f(G) = \neg \top$ if G is an empty cut.

Now, we extend this function f to $\overline{f}$ to translate a quasi-Beta graph into a complex wff:

(a) $\overline{f}(G) = f(G)$ if G is a simple quasi-Beta graph.
(b) $\overline{f}([[G]]) = \overline{f}(G)$.
(c) $\overline{f}(G_1 \ldots G_n) = \overline{f}(G_1) \wedge \ldots \wedge \overline{f}(G_n)$.
(d) $\overline{f}([G_1 \ldots G_n]) = \overline{f}([G_1]) \vee \ldots \vee \overline{f}([G_n])$.

5. Obtain sentences. For each variable v_i in the wff obtained by the previous step,

(a) if v_i is written in an evenly enclosed area in step 2, then add $\exists v_i$ immediately in front of the smallest subformula containing all occurrences of v_i,   and

(b) if v_i is written in an oddly enclosed area in step 2, then add $\forall v_i$ immediately in front of the smallest subformula containing all occurrences of v_i.   and

We will illustrate how each clause works through examples.
Clause 1 tells us to make the following changes:

Clause 2(a) respects Peirce's intuition on LI. That is, a continuity of an identity line, however it may branch, visualizes a single object. 2(a) is easily applied to get the following stages:

But, there is an exception where this intuition breaks down. This is what clause 2(b) is about. This clause suggests that a different variable be assigned to each part:

---

[26] In this linear notation, cuts are represented by brackets.
[27] If there is no subformula containing v_i, this step is skipped.
[28] If there is no subformula containing v_i, this step is skipped.

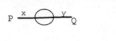

Both in 2(a) and 2(b), it is important to assign a variable to the least enclosed part. The location of each variable will play a crucial role when we assign a quantifier. Suppose the following graph is given: The first cut clips this network

into two parts. Two variables, x and y, should be assigned to the least enclosed part of each clipped part. Hence, in the following, the first one is correct, while the second is not:

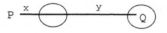 

Notice that clause 2(b) has one more requirement, that is, to write down $v_i = v_j$ inside the cut. This identity statement represents the visual fact that these two clipped portions are connected with each other. The following are the result of this clause being applied:

Let's move to clause 3. For each end of an LI network, if there is a predicate P, then replace this predicate with atomic wff $Pv_1 \ldots v_n$, and if there is no predicate, then write ⊤. The following shows how this rule is applied:

The next step, clause 4, shows a departure from Peirce's endoporeutic reading which all the existing reading methods follow. This endoporeutic reading prevents us from implementing Peirce's visual distinction between universal and existential quantifiers. Since all the existing methods for reading off EG adopt the endoporeutic method without any exception, we should not be surprised to

---

[29] Notice that the second cut does not clip this line.

find out that no current method could reflect Peirce's original intention for this insightful visual distinction.

Step 5 tells us how to obtain a final sentence out of the complex wff formed in the previous step, by reading off a visual distinction between *some* and *all*. Let's recall that in the step 2, variables are written at the outermost parts of LI's. Depending on whether a variable is written in an evenly enclosed or oddly enclosed area, quantifier $\forall$ or $\exists$ is adopted respectively.

Suppose we have the following quasi-Beta graph by steps 2 and 3:

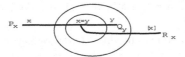

First, function $\overline{f}$ in clause 4 is applied:

$$\begin{aligned}
\overline{f}(Px[[x=y]Qy]Rx) &= \overline{f}(Px) \wedge \overline{f}([[x=y]Qy]) \wedge \overline{f}(Rx) & \text{by (c)}\\
&= \overline{f}(Px) \wedge (\overline{f}([[x=y]]) \vee \overline{f}([Qy])) \wedge \overline{f}(Rx) & \text{by (d)}\\
&= \overline{f}(Px) \wedge (\overline{f}(x=y) \vee \overline{f}([Qy])) \wedge \overline{f}(Rx) & \text{by (b)}\\
&= Px \wedge (x=y \vee \neg Qy) \wedge Rx & \text{by (a)}
\end{aligned}$$

Variable x is written in an evenly enclosed area and y in an oddly enclosed area. Therefore, step 5 tells us to introduce quantifiers to obtain $\exists x[Px \wedge \forall y(x = y \vee \neg Qy) \wedge Rx]$.

5 Comparison among Different Readings

Both Zeman's and Roberts' contribution to this subject cannot be overemphasized. While both of them provided brilliant methods for reading off EG, they accomplished their goal by taking very different routes, which result from their different emphases. Zeman aimed to come up with a formal and comprehensive translation algorithm for the system. By contrast, Roberts was interested in presenting Beta graphs at a more informal and intuitive level, which he achieved by a close reading of Peirce's manuscripts. Since space is limited, I will illustrate differences among these different methods through several examples.

For the following Beta graph,

Zeman's reading algorithm suggests the following process:

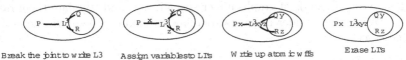

| Break the joint to write L3 | Assign variables to LI's | Write up atom in w ffs | Erase LI's |

---

[30] Zeman, Chapter 2 and Roberts 1973, Chapter 4.
[31] This graph says "Anything that is P is both Q and R."

Now, we obtain the following formula from the rightmost graph:
$$\neg(Px \wedge L^3xyz \wedge \neg(Qy \wedge Rz))$$
By adding existential quantifiers in front of the smallest subformula, we obtain the following sentence:
$$\neg\exists x(Px \wedge \exists y\exists z(L^3xyz \wedge \neg(Qy \wedge Rz)))$$
By replacing a temporary predicate, L^3, the following sentence is the final translation of the above Beta graph by Zeman's reading:
$$\neg\exists x(Px \wedge \exists y\exists z(x = y \wedge y = z \wedge \neg(Qy \wedge Rz)))$$
This quite complicated looking formula is logically equivalent to "$\forall x(Px \rightarrow (Qx \wedge Rx))$."

Now, our new algorithm comes up with the following process:

By 2 (a)   By 3 (a)

$$\neg Px \vee (Qx \wedge Rx) \qquad\qquad\qquad\qquad \text{by 4}$$
$$\forall x[\neg Px \vee (Qx \wedge Rx)] \qquad\qquad\qquad \text{by 5}$$

This reading process is simpler in the following sense. First, Zeman's reading adopts three types of variable while the new algorithm only one. Hence, Zeman's algorithm, unlike our new one, does not read off the visual fact that this Beta graph has one (*not* three) unclipped network of LI. Second, Zeman's reading method never has a chance to read off an important iconic feature about a distinction between universal and existential statements, but only gets existential quantifiers. Hence, universal statements are always expressed as a nested relation between negation and existential quantifiers. For these reasons, not only the final translation sentence is rather complicated looking, but also LI's become dispensable, which is why Zeman's method erases them in the process. This is quite ironic when we recall that the visualization of identity is one of the strengths of EG. While Zeman provides us with a very impressive and comprehensive reading method of this system, he fails to read off important iconic features of this graphic system.

Roberts' approach is quite different from Zeman's. Roberts respects Peirce's intuition about what LI's are supposed to mean in the system: Neither a breaking point nor an additional predicate is introduced. The numerical identity is emphasized as the interpretation of an LI, which works very well in some cases. However, how does Roberts' reading handle the cases in which an LI passes through an odd number of cuts? The price Roberts' project had to pay is to categorize certain kinds of graphs under special cases and stipulate them rather than covering them under a general reading method. These are some of his special cases:

---

[32] Recall that Zeman's reading requires certain LI's to be broken into parts and temporary predicates, L^1, L^2 and L^3, to be introduced.

[33] Roberts 1992, pp. 645-646.

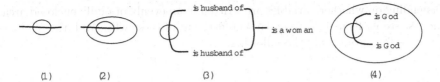

(1)   (2)   (3)   (4)

Under the heading *Special Cases*, Roberts gives us the following interpretations of the above graphs:

> Literally, this [graph (1)] means "There is an object and there is an object and it is false that these objects" – denoted by the two unenclosed parts of the line – "are identical." By means of this graph, we can say such things as "Some woman has two husbands" ([graph (3)]). Figure 28 [graph (2)] means "There is something, and it is false that there is something else non-identical to it" – that is, "Something is identical with everything." ... Figure 31 [graph (4)] expresses the Unitarian theology as summarized by Whitehead: "There is one God at most."

Zeman's comprehensive reading method does not treat these as special cases. After a broken points and temporary predicates being introduced, we obtain rather complicated-looking sentences as the translations of these graphs:

(1): $\exists x[x = x \land \exists y(x \neq y\ y = y)]$
(2): $\exists x[x = x \land \neg\exists y(x \neq y \land y = y)]$
(3): $\exists x \exists y[x \neq y \land \exists x \exists u \exists v(x = u \land u = v \land HusbandOf(x, z) \land HusbandOf(y, v) \land Woman(u))]$
(4): $\neg\exists x \exists y(God(x) \land God(y) \land x\ y)$

Unlike Roberts' method, our new reading algorithm covers these cases without treating them as special. At the same time, this new method yields the result more easily with simpler sentences than Zeman's. According to our algorithm, after applying clauses 1 and 2, each graph is transformed into the following:

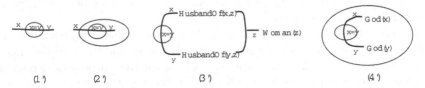

(1')   (2')   (3')   (4')

According to clause 3, the following complex wffs are obtained:

From (1)': $\top \land x \neq y \land \top$.
From (2)': $\top \land (x = y \lor \neg\top)$.
From (3)': $x \neq y \land HusbandOf(x, z) \land Woman(z) \land Husband(y, z)$.
From (4)': $x = y \lor \neg God(x) \lor \neg God(y)$.

---

[34] Roberts 1992, p. 645.

Depending on whether variables are written in an evenly or oddly enclosed area, the last step tells us how to make sentences out of these wffs. And, the readings for these four graphs are the following:

For (1): $\top \wedge \exists x \exists y x \neq y \wedge \top$.
  (I.e. At least two things exist.)
For (2): $\top \wedge (\exists x \forall y x = y \vee \neg \top)$.
  (I.e. Something is identical with everything.)
For (3): $\exists x \exists y (x \neq y \wedge \exists z [HusbandOf(x, z) \wedge Woman(z) \wedge Husband(y, z)])$.
  (I.e. Some woman has at least two husbands.)
For (4): $\forall x \forall y [x = y \vee \neg God(x) \vee \neg God(y)]$.
  (I.e. At most one God exists.)

6   Conclusion

We have explored a new reading algorithm of Peirce's Beta graphs. This project suggests the importance of several aspects of graphic representation systems. When we read off more fine-grained visual features present in graphs we obtain an easier and more direct reading method. The identity among individuals is visualized by one and the same LI in EG, while different tokens of the same type of variable are introduced in first-order languages. Also, a distinction between universal and existential statements is made not by different syntactic devices but by different ways to read off a graph. A similar distinction is made between conjunction and disjunction. Different ways of carving up a graph can give us different, but logically equivalent, translations. By contrast, no symbolic system allows us to carve up a given sentence in more than one way.

Two of the main reasons why these visual features have been ignored are, I believe, our biased attitude for symbolic systems and unclear distinctions among different forms of representation systems. When the logical equivalence between a symbolic system and a graphic system is established, many have thought that our investigation of the graphic system is accomplished. Hence, we are left with the wrong impression that graphic systems can become, at best, as logically powerful as symbolic systems.

On the other hand, we make a distinction among different symbolic logical systems. For example, an axiomatic system is chosen for the investigation of logical theories and a natural deductive system for the purpose of deduction. Similarly, if we can articulate the differences between symbolic and graphic systems, we may sometimes choose one over the other depending on our purpose. All this suggests that our investigation of graphic systems should go further than the proof of the logical equivalence between symbolic and graphic systems. This work illustrates one of the many steps we could take for further investigation of graphic representation.

By no means do I argue that EG is better than first-order languages. On the contrary, I do not believe in an absolute comparison among different logical systems. All I present in this limited space is a case study in approaching one graphic system based on its own strengths, not based on the criteria used for symbolic systems. I hope this approach will inspire others not only to find further visual aspects of Peirce's EG and other graphic systems, but also to find more concrete differences between graphic and symbolic systems. In this way, we will be able to identify different strengths of different kinds of representation systems so that we may take full advantage of these differences in research on the multi-modal reasoning.

Acknowledgment

I appreciate very helpful comments from two anonymous reviewers. This work is financially supported by the Institute for Scholars in the Liberal Arts, University of Notre Dame.

References

1. Burch, Robert: *Peircean Reduction Thesis: The Foundation of Topological Logic.* Lubbock, Tex.: Texas Tech University Press (1991)
2. Hammer, Eric: *Logic and Visual Information.* Stanford, CA: Center for the Study of Language and Information (1995)
3. Peirce, Charles S.: *Collected Papers of Charles Sanders Peirce,* ed. by Charles Hartshorne & Paul Weiss, Vol. 4. Cambridge, Mass.: Harvard University Press (1933)
4. Roberts, Don: *The Existential Graphs of Charles S. Peirce.* The Hague: Mouton (1973)
5. Roberts, Don: "The Existential Graphs," *Computers Math. Applications.* **23** (1992) 639-663
6. Shin, Sun-Joo: *The Logical Status of Diagrams.* New York: Cambridge University Press (1994)
7. Shin, Sun-Joo: "Reconstituting Beta Graphs into an Efficacious System," *Journal of Logic, Language and Information.* **8** (1999) 273-295
8. Shin, Sun-Joo: "Multiple readings of Peirce's Alpha graphs," *Thinking with Diagrams 98.* Heidelberg: Springer-Verlag (to appear)
9. Shin, Sun-Joo: *Iconicity in Peirce's Logical System.* Cambridge, Mass.: MIT Press (Bradford) (to appear)
10. Sowa, John: *Conceptual Structures: Information Processing in Mind and Machine.* Reading, Mass.: Addison-Wesley (1984)
11. Sowa, John: *Knowledge Representation: Logical, Philosophical, Computational Foundations.* Belmont, Calif.: Brooks/Cole (2000)
12. White, Richard: "Peirce's Alpha Graphs: The Completeness of Propositional Logic and the Fast Simplification of Truth-Function," *Transactions of the Charles S. Peirce Society.* **20** (1984) 351-361
13. Zeman, Jay: *The Graphical Logic of C. S. Peirce.* Ph. D. Diss. University of Chicago (1964)

Constraint Matching for Diagram Design: Qualitative Visual Languages

Ana von Klopp Lemon[1] and Oliver von Klopp Lemon[2]

[1] Sun Microsystems
901 San Antonio Road, Palo Alto, CA 94303
avk@eng.sun.com
[2] Center for the Study of Language and Information
Stanford University, CA 94305
lemon@csli.stanford.edu

Abstract. This paper examines diagrams which exploit qualitative spatial relations (QSRs) for representation. Our point of departure is the theory that such diagram systems are most effective when their formal properties match those of the domains that they represent (e.g. [, ,]). We argue that this is true in certain cases (e.g. when a user is constructing diagrammatic representations of a certain kind) but that formal properties cannot be studied in isolation from an account of the cognitive capacities of diagram users to detect and categorize diagram objects and relations.

We discuss a cognitively salient repertoire of elements in qualitative visual languages, which is different from the set of primitives in mathematical topology, and explore how this repertoire affects the expressivity of the languages in terms of their vocabulary and the possible spatial relations between diagram elements.

We then give a detailed analysis of the formal properties of relations between the diagram elements. It is shown that the analysis can be exploited systematically for the purposes of designing a diagram system and analysing expressivity. We demonstrate this methodology with reference to several domains, e.g. diagrams for file systems and set theory (see e.g. [ ]).

1 Introduction

This paper examines diagrams that rely on qualitative spatial relations (QSRs) – such as regions overlapping or lines intersecting – to stand for relations in a domain (i.e. the structure which is represented by the diagrams). We distinguish this class of diagrams from those which employ *quantitative* spatial relations; for example where a difference in size between different graphical objects is meaningful. Some practical examples of this class of diagram system are: UML [ ], category theory, statecharts [ ], flowcharts, file diagrams, topological maps, Euler Circles [ ], and Venn diagrams.

M. Anderson, P. Cheng, and V. Haarslev (Eds.): Diagrams 2000, LNAI 1889, pp. 74–   , 2000.
© Springer-Verlag Berlin Heidelberg 2000

Our objectives with this paper are to;

- further the programme of analysis of the expressive power of diagrams which employ QSRs;
- draw attention to the central role of cognitive factors in such a theory;
- propose a "constraint matching" methodology for designing effective diagram systems which employ QSRs (see section).

We will illustrate how our analysis can be used for design and evaluation with reference to some well-known domains and representation systems (directory tree representations, linear diagrams, web site maps, set-theory).

The formal study of diagrams is a relatively recent phenomenon. Current studies suggest a focus on structural properties of spatial relations, arguing that diagrams are more useful if the mathematical properties of the visual representations match those of the domain [, ,]. A particular instance of this argument is where formal properties of the visual language collude to restrict possible diagrammatic representations in such a way that impossible states of the domain cannot be represented (e.g. Euler Circles). This is considered to be of benefit in situations where users construct their own diagrams for problem solving (it is known as "self-consistency" [ ]).

This paper is a development and exploration of the formal perspective on diagrams introduced by [, , , , , , , ,]. The reader is referred to these papers for definitions and explanations of concepts such as isomorphic representation, soundness, correctness, self-consistency, and free-rides. As well as this semantic emphasis, our approach is distinguished from preceding formal approaches (e.g. [,]) in that we focus on a cognitively relevant vocabulary of graphical primitives (without translation into a spatial logic) and that we pay special attention to spatial constraints on diagrams such as planarity.

We recognize the formal perspective as a productive approach to a better understanding of diagrams; one which we can use to aid diagram design [,]. However, we also argue that concentrating solely on formal properties of spatial relations overlooks a non-trivial cognitive step. In order for the diagram user to be able to take advantage of a correspondence between formal properties of the domain and those of the visual language in general or of "free-rides" in particular, they first need to perform category detection and identification.

Being able to detect a spatial relation in a diagram and match it to a relation in the domain involves category detection and identification by the user: they need to identify categories of spatial objects in the visual representation and match them to categories in the domain. We argue that such processes, as well as formal properties of the relations, place restrictions on the expressivity of diagram systems.

2 Background

2.1 Scope and Terminology

The subject of this paper is diagrams which employ QUALITATIVE spatial relations (QSRs). For example, overlap is a binary spatial relation. If we consider overlap as a qualitative relation, we are concerned with whether any pair of regions overlap, but we are not concerned with comparing the *amount* of overlap between regions. Our objective is to begin a systematic categorization of possible QSRs for the purpose of aiding the process of designing a visual language for a particular domain. We will refer to a visual language that relies on QSRs as a qualitative visual language or QVL. A QVL consists of a REPERTOIRE OF GRAPHEMES, and a set of qualitative spatial RELATIONS that can hold between them. A grapheme is a spatial object (following [ ]); the repertoire of graphemes is a set of types of spatial objects (more on this below). In order to distinguish lines (very narrow spatial tokens) from other shapes, we will also introduce the term BLOB for shapes that are not lines or points.

2.2 Qualitative Spatial Relations vs. Topology

Formal topology is one useful tool for examining QSRs—results from topology have very direct consequences for many types of QVL, as will become apparent in section . However, it is important to realise that from the perspective of constructing diagrams, the starting point for establishing a repertoire of graphemes and the relations between them is not abstract mathematical theory, but the cognitive capacities of diagram users. Fundamental to the design of visual languages is human categorisation of spatial objects and our ability to distinguish relations between these objects (the available REPERTOIRE OF GRAPHEMES AND RELATIONS). Less fundamental, but nevertheless important for usability reasons is our ability to detect patterns and differences in a display that contains many tokens (SCALABILITY).

3 Graphemes

3.1 Repertoire

As noted in section , formal topology is a useful tool for analysing spatial relations. QVLs in general, however, typically rely on a different set of spatial primitives than those used in topological analysis. Several visual relations do not have a corresponding topological concept, or can only be described in ways that are difficult to understand or unintuitive to the lay person.

Topology formally recognises two types of objects: regions and points. For humans, the distinction between lines and blobs appears to be more intuitive or basic. That is, it is common for QVLs to exploit the distinction between different types of blob on the one hand, and different types of line on the other, to correspond to a fundamental category difference in the represented domain. For instance, file system and web site diagrams typically use blobs to stand for tokens in the domain (files and directories), and lines to denote (possibly different types of) relations between them. However, there is no easy way to distinguish between lines and blobs in topology (both are connected sets of points).

We argue that the distinction between line and blob is the most fundamental one with respect to graphemes in visual languages. These are the primitives, which stand for the primary categories recognized. In contrast, individual points (in the topological sense) are not available as primary objects from this cognitive design perspective. While they can be represented implicitly, as a point inside a region or as the point where two lines cross, they are difficult to represent explicitly. Points can be used as abstractions in the process of interpreting a diagram, but they are not primitives of the visual language.

Although we have no experimental evidence for this claim, it seems likely that there is a natural tendency to (attempt to) correlate perceived categories of graphical tokens with perceived categories in the represented domain. For example, if a particular language uses lines to denote relations between objects, it would be confusing to introduce another type of line to stand for an object. It seems likely that to maximize usability, the representation system ought to correlate graphical types that tend to be perceived as the same category (e.g. different types of line) with domain types that are also perceived to belong to the same category.

3.2 Scalability

One of the issues that needs to be considered when designing a representation system is how much domain information can be represented in a single view. Here, matching categories in a visual language to categories in the represented domain can become problematic. Unless the objects in the domain have an obvious visual representation (e.g. different types of animals), arbitrary shapes must be used, and in that case, achieving category correlation can be difficult.

A diagram designer can create visually recognizable categories of graphemes by using shape, color, pattern and color saturation, or by adding annotation such as mini-icons, arrowheads, or kinks on lines. But compared with textual languages this strategy is only available for a limited set of categories, since there are cognitive limits on the number of visual categories that can be effectively recognised.

If the distinguishing features are not iconic, then the diagram user must maintain the meanings of the annotations (e.g. "blue" means functional navigation link, "red" means broken link, etc.) in working memory (or else a key must be supplied and frequently referenced). This all adds to the effort of processing the diagram. Indeed, this shows that graphical representations may involve a type of processing that does arise with textual representations, where it is often possible to use mnemonic names.

The issue of preserving category structure is clearly recognized by application designers, as there are some common strategies for helping the user perform categorization tasks. One such strategy is to allow filtering, whereby the user can choose to hide graphemes representing domain objects that belong to certain

---

[1] A category in the represented domain could be a class of objects, a class of relations, or some other feature, like time.

categories. Another one is to add some textual annotation to spatial symbols, creating hybrid systems with complicated syntax.

4 Relation Analysis

This section examines structural properties of qualitative spatial relations between graphemes. Its objective is to support the process of designing a QVL, which consists of the following steps:

1. Identify categories and relations in the domain (i.e. in the represented world [ ]).
2. Evaluate the domain relations with respect to their structural properties.
3. Match each domain relation to a collection of qualitative spatial relations over graphemes whose properties are as similar (to those of the domain relation) as possible. (e.g. if the domain relation has properties x, y, z – find the set of qualitative spatial relations over graphemes that also have properties x, y, z.)
4. Choose a spatial relation for each domain relation is such a way that domain object categories are preserved (i.e. the categorization of the spatial objects matches the categorization of the objects in the represented world, as discussed in section).
5. Evaluate the resulting language with respect to other expressivity restrictions, such as whether planarity restrictions occur, or whether Helly's theorem (explained below, and see [ ]) applies.
6. Evaluate the resulting language with respect to scaling and other usability issues.

The properties of the relations will be reported in tables that provide an easy reference mechanism for step   above. Examples of how to use them will be provided in section .

Because of space restrictions, we have to limit our scope to binary relations. Even for these, our coverage is incomplete, though we have included the commonly used ones. As for ternary and higher order relations, it is worth noting that while it is easy to find examples of binary relations, relations that are strictly higher order are harder to come by, making them hard to represent. This becomes a practical problem e.g. in web site mapping [ ].

The spatial relations will be evaluated with respect to the following structural properties (see table):

---

[2] This may also require the user to perform an unusual type of processing task. [ ] explores the possibility that an advantage of at least some types of graphical representations is that users do not have to perform the equivalent of syntactic processing. This would make them easier to use than e.g. sentential representations. Clearly, any such advantage is lost if the visual language is augmented with annotation.

[3] Among the ones that have been left out are NEARNESS, TO THE LEFT OF, ON TOP OF, BESIDE.

[4] The Church-Rosser property stands for forced convergence.

Table 1. Definitions of structural properties

Property	Definition	Abbreviation
Reflexivity	$\forall a(aRa)$	Ref
Symmetry	$\forall a \forall b(aRb \leftrightarrow bRa)$	Sym
Transitivity	$\forall a \forall b \forall c(aRb \& bRc \rightarrow aRc)$	Trans
Irreflexivity	$\forall a \neg aRa$	Irref
Asymmetry	$\forall a \forall (aRb \rightarrow \neg bRa)$	Asym
Anti-symmetry	$\forall a \forall b((aRb \& bRa) \rightarrow (a = b))$	Anti-Sym
Atransitivity	$\forall a \forall b \forall c(aRb \& bRc \rightarrow \neg aRc)$	A-trans
Acyclicity	$\forall a \neg aR^n a$	Acyc
Non-convergency	$\forall a \forall b \forall c(aRb \& aRc \rightarrow \neg \exists d(bRd \& cRd))$	Non-conv
Church-Rosser	$\forall a \forall b \forall c(aRb \& aRc \rightarrow \exists d(bRd \& cRd))$	Ch-Ross
Arboresence	Acyc + Trans + Non-conv	Arb
Atransitive Tree	Acyc + Non-conv	AT Tree

1. Simple properties that apply to the relation itself, such as reflexivity and symmetry;
2. Properties that hold over several instances of the relation, such as transitivity, cyclicity, arboresence, and convergence.

In addition to constraints arising from structural properties, there is also a second level of constraints that arise from the use of the two-dimensional plane as a representational medium. These affect scaling (i.e. number of objects and relations which can be correctly represented) and are of two types:

1. Formal properties that arise from the number of relations the objects participate in and their patterns (Helly's theorem, planarity); and
2. Practical restrictions on how well the spatial relation allows depth and width of branching.

HELLY'S THEOREM concerns convex regions, i.e. convex blobs and straight lines for our primitives [ ]. The two-dimensional case states that given a set of four regions, if every triple out of the four has a non-empty intersection, then all four of the regions intersect. Say that a diagram has four blobs, and that out of these, each triple (there are four) has a non-empty overlap. Then a quadruple overlap between all four cannot be avoided. The same is true for the relation of touching along a line or at a point (since this means that the closure of the regions have a non-empty intersection).

PLANARITY restrictions arise where there are certain patterns of interconnection between simply-connected regions (i.e. one piece blobs or lines). Any graph containing a subgraph isomorphic to either Γ_5 (the graph with 5 nodes all connected to each other) or $\Gamma_{3,3}$ (where two groups of 3 nodes all connect) cannot be drawn in two dimensions without at least one forced intersection [ ] (more intersections are forced if only straight lines are used).

Helly's theorem and planarity constraints are important for the following reason: if a spatial language is used to represent a domain, certain patterns in

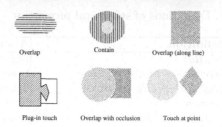

Fig. 1. Binary qualitative spatial relations between blobs

the domain *cause* spatial relations to be generated in the representation. This happens regardless of whether there is a corresponding relation in the domain (which is unlikely when representing anything other than a two-dimensional spatial domains).

We will start by examining binary relations holding between blobs. The relations discussed here are familiar ones such as OVERLAP, CONTAINMENT, and being SEPARATE. We distinguish between overlap where there is OCCLUSION or where both blobs are visible (have a degree of transparency), since these cases are visually distinct. We also include TOUCHING, which involves only the borders of the blobs. We distinguish three types: along a side (line), at a point, and plug-in touch (like pieces of a jigsaw puzzle; where it is significant that one plugs into the other).

It seems plausible that from the perspective of a user's perception the visual relations of overlap, containment, and touching might not be construed as reflexive (e.g. blobs do not overlap themselves). We will however include the reflexive interpretations (at the end of the table, in italics) for the sake of completeness.

The table below summarizes the structural properties of these relations – an X in box (n,m) denotes that the relation in row n has the property m, a C means that the relation has that property only if the blobs involved are convex. (e.g. 3rd row: overlap between blobs when one occludes the other is asymmetric, and is irreflexive if the blobs are convex, but does not have any of the other properties.

Blob-blob Relation	Ref	Sym	Trans	Irref	Asym	Anti-sym	A-trans
Overlap		x		c			
Overlap (occlusion)				c	c		
Contain			x	x	x		
Separated		x		x			
Touch (along line)		x		c			
Touch (point)		x		c			
Touch (plugin)							
Contain	x		x			x	
Overlap	x	x					
Touch (along line)	x	x					

---

[5] We treat containment as transitive, but other authors, e.g. [ ] have suggested that this is not so, arguing that whether one interprets it as transitive depends on what it

The next table shows the rest of the properties discussed above. Note that containment has different properties depending on whether any blobs contained are allowed to overlap.

Blob-blob Relation	Acyc	Non-Conv	Ch-Ross	Arb	AT Tree	Helly	Planar prob
Overlap			x			c	x
Overlap (occlusion)						c	x
Contain (overlap OK)	x						
Contain (no overlap)	x	x		x			
Separate			x				
Touch (along line)			x			c	x
Touch (point)			x			c	x
Touch (plugin)							x
Contain							x
Overlap			x			c	x
Touch (along line)			x			c	x

Next, we will evaluate spatial properties of binary relations between lines, listed in the tables below. As above, we distinguish two cases of relations such as overlap, one which is not reflexive, and one allowing for a reflexive interpretation (in italics at the end of the table). An s means that the relation has that property only if the lines involved are straight—straight lines cannot self-cross for instance. We say that curved lines are "parallel" if they follow the same contour.

Line-line Relation	Ref	Sym	Trans	Irref	Asym	Anti-sym	A-trans
Crossing (at a point only)		x		s			
Overlap (along line or point)		x		s			
Contain			x	x	x		
Separate		x		x			
Parallel		x	x	x			
Orthogonal (straight only)		x		x			x
Touch at a point		x		s			
Touch endpoints		x		s			
Contain	x		x			x	
Overlap (along line or point)	x	x					
Parallel	x	x	x				

---

is used to represent. This is a separate issue that we will address below (see section   ).
From a spatial perspective, what is available visually is transitive.

[6] Unfortunately, space restrictions only allow us to consider undirected lines here—introducing direction throws up some interesting combinations.

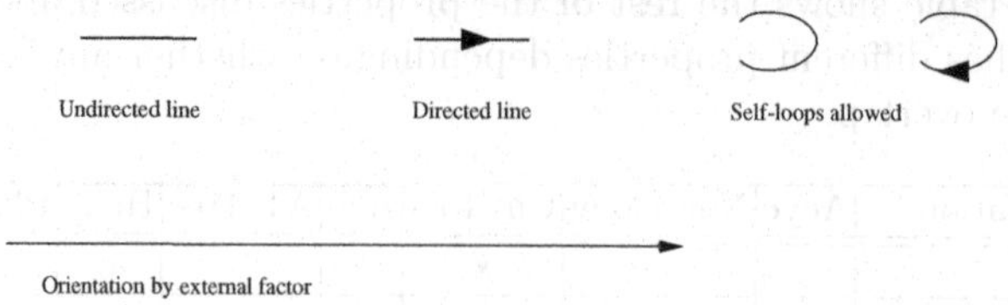

Fig. 2. Relations between two blobs joined by a line

Line-line Relation	Acyc	Non-Conv	Ch-Ross	Arb	AT Tree	Helly	Planar prob
Crossing (at a point only)			x			s	x
Overlap (along line or point)			x			s	x
Contain (overlap OK)	x						
Contain (no overlap)	x	x		x			
Separate			x				
Parallel			x				
Orthogonal			x				
Orthogonal + touch			x			s	x
Touch at point			x			s	x
Touch at endpoint			x			s	x
Contain							
Overlap (along line or point)	s		x			s	x
Parallel			x				

The final combination we will evaluate here is the use of lines to connect two blobs, where the blobs stand for objects in the domain and the line stands for a relation between them. This type of notation is common in practical applications of QVLs, e.g. file systems representations, or in class diagrams (e.g. in UML [ ]). Figure   shows the main categories of this type of relation. The distinguishing features are whether the line connecting the blob has direction, and whether the line can loop back to the blob it started from. Direction can be encoded either through annotation of the line, or through (implicit) orientation using an underlying directionality, e.g. left-to-right or top-to-bottom orientation of the plane. There are multiple variants of how lines can be annotated to show direction (e.g. the line could start at the boundary of one blob and finish inside another blob). These all have the same structural properties as lines with arrows.

Two blobs joined by	Ref	Sym	Trans	Irref	Asym	Anti-sym	Atrans
undirected line, self-loops allowed		x					
undirected line, no self-loops		x		x			
directed line, self-loops allowed							
directed line, no self-loops				x			
implicit orientation, 1 axis				x	x		
implicit orientation, 2 axes				x	x		

Two blobs joined by	Acyc	Non-Conv	Ch-Ross	Arb	AT Tree	Helly	Planar prob
undirected line, self-loops allowed			x				x
undirected line, no self-loops			x				x
directed line, self-loops allowed							x
directed line, no self loops							x
implicit orientation, 1 axis	x						x
implicit orientation, 2 axes	x	x			x		

5 Example Analyses of Domains and Diagram Systems

5.1 File System Diagrams

File system diagrams are frequently used in computer applications to show the user a visual representation of a file hierarchy and/or as a means of interacting with the files in the hierarchy (figure shows an example from Forte for Java .) The file system domain consists of two basic types of ob-

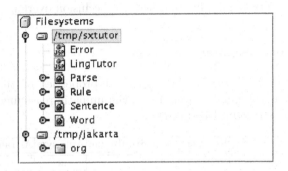

Fig. 3. File system representation

ject: directories (folders) and files. There is one relation, containment. A directory can contain zero, one, or more objects which can be either directories or files. The containment relation is binary, and has the following properties:

	Ref	Sym	Trans	Irref	Asym	Anti-sym	Atrans
Directory containment				x	x		x

	Acyc	Non-Conv	Ch-Ross	Arb	AT Tree	Helly	Planar prob
Directory containment	x	x			x	n/a	n/a

---

7 Forte for Java is a trademark of Sun Microsystems.

8 For the present purposes, we will ignore links and shortcuts.

9 The directory containment relation is atransitive from the perspective of a user whose goal is to locate or access a file, which is what these diagrams are used for.

Reading the tables in section   we find that blobs connected by lines that are organized through implicit orientation along two axes match this perfectly. This is also the relation used in figure , as well as in many other applications (some omit the connecting lines).

It will be interesting to consider some of the close matches as well. The best ones are BLOB CONTAINMENT (OVERLAP NOT ALLOWED), PLUG-IN TOUCH between blobs, OCCLUSION OVERLAP between blobs, BLOBS CONNECTED BY LINES WITH IMPLICIT ORIENTATION ALONG 1 AXIS and BLOBS JOINED BY DIRECTED LINES WITH NO SELF-LOOPS.

As for BLOB CONTAINMENT it fails to match perfectly because it can be interpreted as transitive, which it may be desirable to avoid. As for the others, one could argue that the particular constraints that these do not match, e.g. being acyclic, are less important than, say, symmetry in a situation where users do not themselves construct the diagram. If a user constructs a diagram, then it might be important to prevent them from constructing a cycle if cycles are not possible in the domain. If the diagram is constructed by somebody other than the user who has proper domain knowledge, then this can be controlled anyway.

Finally, given that there is no restriction on how many items a directory can contain, it is important that the chosen spatial relation scales well in terms of depth and width. For this reason, the relations involving blobs connected by lines are better suited than plug-in touch and occlusion overlap.

5.2   Set Theory Diagrams

The following table describes the structural properties of the fragment of set-theory that is commonly covered by diagrammatic reasoning systems such as Euler Circles and Venn Diagrams.

Relation	Refl	Sym	Trans	Irref	Asym	Anti-sym	A-trans
Set Inclusion: $a \subseteq b$	x		x			x	
Set Intersection: $a \cap b \neq \emptyset$	x	x					
Sets Disjoint: $a \cap b = \emptyset$		x		x			

	Acyc	Non-Conv	Ch-Ross	Arb	AT Tree	Helly	Planar prob
Set Inclusion: $a \subseteq b$			e				
Set Intersection: $a \cap b \neq \emptyset$			x				
Sets Disjoint: $a \cap b = \emptyset$			x				

As discussed in section , it is desirable to attempt to preserve category matching in the visual language and in the represented domain. Hence we should select either blobs or lines as representing sets, yielding two separate sets of possibilites. Starting with the upper table (Ref, Sym, etc.), the properties match as follows:

---

[10] This holds trivially for the emptyset. If the diagram system only represents non-empty sets – as Euler Circles and Englebretsen line diagrams [ ] do – then the property is lost.

BLOBS STAND FOR SETS
Set Inclusion: containment (reflexive)
Set Intersection: overlap (reflexive), touch along line (reflexive)
Sets Disjoint: separate, joined by undirected line, blob overlap (convex only), blobs touch along line or point (convex only).

LINES STAND FOR SETS
Set Inclusion: containment (reflexive)
Set Intersection: overlap (reflexive)
Sets Disjoint: separate, lines crossing (straight only), line overlap (straight only), lines touching at point or endpoint (straight only).

Moving on to the second set of relations (acyclicity etc.), the set-theoretic relations are not subject to the restrictions imposed by Helly's theorem, and the $\Gamma_{3,3}$ and Γ_5 patterns can occur without forcing any new domain relations. The former (Helly considerations) rules out convex regions, i.e. convex blobs or straight lines, and the latter (planarity restrictions) rules out connected regions (one piece blobs or lines). The other constraints match. This leaves us with the following:

BLOBS STAND FOR SETS, MAY BE CONCAVE OR DISCONNECTED
Set Inclusion: blob containment (reflexive)
Set Intersection: blob overlap (reflexive), blob touch along line (reflexive)
Sets Disjoint: Blobs separate, blobs joined by undirected line.

LINES STAND FOR SETS, MAY BE CURVED OR DISCONNECTED
Set Inclusion: line containment (reflexive)
Set Intersection: line overlap (reflexive)
Sets Disjoint: lines separate

These relations mostly seem to be fairly apt visual metaphors, with two possible exceptions. Firstly, having a line joining two blobs meaning that the two are separate might be unintuitive (and unnecessary in the circumstances). Also, if all the disconnected bits have to be marked this way, it reintroduces the planarity problem. Secondly, relying on touching to mean set-intersection might be unsuitable for practical reasons. It depends on how users of different levels of experience perceive touching in relation to the domain—do they see the blobs as sharing a border or just as being near each other?

These systems are perfect matches, but they do have a property that has been suggested as potentially problematic for cognitive reasons, and that is that they rely on disconnected blobs or lines. It has been argued (though not verified experimentally) that diagrams are easier to use if there is a token-token correspondence between the diagram and the domain (see [ ] on "specificity"). However, there is no requirement within set theory that sets be connected.

It is interesting to note that the two sets of relations described above differ somewhat from the ones employed by existing diagram systems for set theory. Euler circles use containment, overlap, and separateness, but require blobs to be convex and connected, thus failing to avoid the consequences of Helly's theorem and planarity restrictions. This means that Euler circles are correct only for fewer than four sets [ ]. Englebretsen's system [ ] which allows only straight lines uses line-containment for set-inclusion, line-crossing for set-intersection, and line-separation for set-disjointness. As can be seen above, these are all good matches except that they, like Euler Circles, also fall prey to Helly's theorem and of planarity problems [ ].

This short exploration illustrates the design potential of our framework, and raises yet more questions; which of the systems generated here is most useable? How could we represent set-membership? We leave such issues for another paper.

6 The Benefits of Constraint Matching

The objective of this study was to provide a methodology that facilitates the design of QVLs in which the diagram language matches (as closely as possible) the structural properties of the represented domain. It is important to reflect on what this means in practice. There is only a minimal claim that we can make with certainty: if the properties of the QVL precisely match those of the represented domain, then the diagram user cannot represent inconsistent states of the domain (self-consistency) and all possible states of the domain can be represented (completeness). If the match is not perfect (as is more often the case) and if the designer understands what the limits on expressivity are, then it is less likely that some domain states cannot be represented, or that the system yields representations that are misleading and can lead to incorrect inferences.

Various studies (e.g. [,]) have suggested that constraint matching is important for diagram systems to help problem solving. We can see how this is true if the diagram user creates the diagram—a well designed language could prevent them from expressing invalid states of the domain (self-consistency). If the diagram user interprets existing diagrams however, then the argument that constraint matching helps understand the domain is less persuasive. It would require that the user is able to identify the structural properties of the spatial relations and translate them into structural properties of the relations in the represented domain (or vice versa). That is not an assumption one would like to make in the absence of compelling experimental evidence. [ ] review various experiments and conclude that although there is some evidence that subjects perform better in problem solving tasks when constraint matching occurs, this does not appear to translate into a better understanding of the domain.

Fig. 4. Using blob containment to represent a relation that is not transitive

Here it will be useful to consider an example from [ ], where it is argued that domain knowledge can take precedence over properties of spatial relations. The relation of blob containment is usually considered to be transitive, but Wang argues that if it is used to represent a relation that the diagram user knows not to be transitive, then the transitivity property of the containment relation is easily discarded. This is illustrated in figure . When presented with the picture on the left and told that it stands for x being the father of y, it is natural to interpret the picture on the right hand side as stating that x is the grandfather of z, without also inferring that x is the father of z. In contrast, if the blob containment relation was used to represent set inclusion, the transitivity interpretation becomes natural.

7 Conclusion

We have, a) defined a class of qualitative visual languages; b) discussed the importance of cognitive factors and scalability in a formal analysis of diagrams; c) identified a collection of diagram objects (graphemes) and cognitively plausible qualitative relations between them; d) given an analysis of the structural properties of these relations; e) shown that spatial constraints on the relations restrict expressivity in many cases; f) used the analysis in diagram design for two domains: file-system structures and (a fragment of) set-theory; g) discussed the limits on expressivity of this class of visual languages (see [ ]).

Due to space limitations we have only been able to report only a portion of our results—we have examined other spatial relations and existing diagram systems (e.g. diagrams used in theoretical physics, computer science, and applications). We have also begun work applying this theory in designing graphical representations for other domains, such as dynamically generated web sites and modal logics.

References

1. Jon Barwise and Atsushi Shimojima. Surrogate Reasoning. *Cognitive Studies: Bulletin of Japanese Cognitive Science Society*, 4(2):7 – 27, 1995.   , ,
2. Keith Stenning and Oliver Lemon. Aligning logical and psychological perspectives on Diagrammatic Reasoning. *Artificial Intelligence Review*, 2000. (in press).   ,
3. Oliver Lemon. Comparing the Efficacy of Visual Languages. In Barker-Plummer, Beaver, Scotto di Luzio, and van Benthem, editors, *Logic, Language, and Diagrams*. CSLI Publications, Stanford, 2000. (in press).   , , ,
4. Oliver von Klopp Lemon and Ana von Klopp Lemon. Cognitive issues in GUI design: Constructing website maps. In Andreas Butz, Antonio Kruger, and Patrick Olivier, editors, *AAAI Smart Graphics Symposium*, pages 128 – 132, 2000.   , ,
5. Sinan Si Alhir. *UML in a Nutshell.* O'Reilly, 1998.   ,
6. David Harel. On Visual Formalisms. *Communications of the ACM*, 31(5):514 – 530, 1988.

7. Oliver Lemon and Ian Pratt. Spatial Logic and the Complexity of Diagrammatic Reasoning. *Machine GRAPHICS and VISION*, 6(1):89 – 108, 1997. (Special Issue on Diagrammatic Representation and Reasoning). , , ,

8. Atsushi Shimojima. *On the Efficacy of Representation.* PhD thesis, Indiana University, 1996.

9. Dejuan Wang. *Studies on the Formal Semantics of Pictures.* PhD thesis, Institute for Logic, Language, and Computation, Amsterdam, 1995. , ,

10. Atsushi Shimojima. Constraint-Preserving Representations. In *Logic, Language and Computation, volume 2*, number 96 in Lecture Notes, pages 296 – 317. CSLI, Stanford, 1999.

11. Hector J. Levesque. Logic and the Complexity of Reasoning. *Journal of Philosophical Logic*, 17:355–389, 1988.

12. Volker Haarslev. Formal Semantics of Visual Languages using Spatial Reasoning. In *IEEE symposium on Visual Languages*, pages 156 –163. IEEE Computer Society Press, 1995.

13. John Gooday and Anthony Cohn. Using spatial logic to describe visual languages. *Artificial Intelligence Review*, 10:171–186, 1996.

14. Stephen Casner. Automatic design of efficient visual problem representations. In *AAAI*, pages 157–160, 1992.

15. Jock Mackinlay. *Automatic Design of Graphical Presentations.* PhD thesis, Computer Science Department, Stanford University, 1986.

16. Jacques Bertin. *Graphics and Graphic Information Processing.* Walter de Gruyter, Berlin, New York, 1981.

17. Stephen E. Palmer. Fundamental Aspects of Cognitive Representation. In Elanor Rosch and Barbara B. Lloyd, editors, *Cognition and Categorization*, pages 259 – 303. Lawrence Erlbaum Associates, Hillsdale, N.J., 1978.

18. H. G. Eggleston. *Convexity.* Cambridge University Press, Cambridge, 1969.

19. G. Kuratowski. Sur le probleme des courbes gauches en topologie. *Fund. Math.*, 15:271–283, 1930.

20. George Englebretsen. Linear Diagrams for Syllogisms (with Relationals). *Notre Dame Journal of Formal Logic*, 33(1):37–69, 1992. ,

21. Keith Stenning and Jon Oberlander. A Cognitive Theory of Graphical and Linguistic Reasoning: Logic and Implementation. *Cognitive Science*, 19(1):97 – 140, 1995.

22. Oliver Lemon and Ian Pratt. On the insufficiency of linear diagrams for syllogisms. *Notre Dame Journal of Formal Logic*, 39(4), 1998. (in press, to appear 2000).

23. Mike Scaife and Yvonne Rogers. External cognition: how do graphical representations work? *International Journal of Human-Computer Studies*, 45:185 – 213, 1996.

24. Oliver Lemon and Ian Pratt. Logical and Diagrammatic Reasoning: the complexity of conceptual space. In Michael Shafto and Pat Langley, editors, *19th Annual Conference of the Cognitive Science Society*, pages 430–435, New Jersey, 1997. Lawrence Erlbaum Associates.

Picking Knots from Trees[*]
The Syntactic Structure of Celtic Knotwork

Frank Drewes[1] and Renate Klempien-Hinrichs[2]

[1] Department of Computing Science, Umeå University
S-901 87 Umeå, Sweden
drewes@cs.umu.se

[2] Department of Computer Science, University of Bremen
P.O.Box 33 04 40, D-28334 Bremen, Germany
rena@informatik.uni-bremen.de

Abstract. Interlacing knotwork forms a significant part of celtic art. From the perspective of computer science, it is a visual language following mathematically precise rules of construction. In this paper, we study the syntactic generation of celtic knots using collage grammars. Several syntactic regulation mechanisms are employed in order to ensure that only consistent designs are generated.

1 Introduction

A typical characteristic of visual languages is that the diagrams in such a language are related by a common structure and layout. In other words, the language is defined by a set of syntactic visual rules yielding the acceptable pictures. Formal picture-generating methods help to understand the structure of the languages in question, to classify them, and to generate them automatically by means of computer programs.

Artists from many cultures have been using visual rules since ancient times in order to design diagrams of various sorts. Celtic diagrams, and in particular celtic knotwork, are a famous visual language of this type. Figure shows an example of a celtic knot. Basically, such a diagram is the two-dimensional picture of one or more continuous strands weaved in a particular fashion. If we look at two successive crossings of a particular strand,

Fig. 1. A celtic knot

this strand lies on top of the first strand

---

[*] Research partially supported by the *Deutsche Forschungsgesellschaft* (DFG) under grant no. Kr-964/6-1, the EC TMR Network GETGRATS (General Theory of Graph Transformation Systems), and the ESPRIT Basic Research Working Group APPLIGRAPH (Applications of Graph Transformation).

M. Anderson, P. Cheng, and V. Haarslev (Eds.): Diagrams 2000, LNAI 1889, pp. 89–   , 2000.
© Springer-Verlag Berlin Heidelberg 2000

crossed if (and only if) it passes below the second. Furthermore, the distances and angles which appear are determined by an invisible grid.

Traditional methods to construct celtic knotwork are described in e.g. [,   ,   ]. Mainly, one first draws a grid of squares, triangles, or a similar pattern. This grid is used to obtain a plait, i.e. a knot with regular interlacing, which is subsequently modified by breaking crossings and connecting the loose ends. In the last step, the style of the strands and the background is determined. Although one can identify certain characteristic features of celtic knots as well as typical ways to construct them, the description of celtic knotwork as a visual language remains informal.

In [   ], Sloss presents an algorithmic way to generate knots. An introduction to celtic knotwork from the perspective of computer graphics is given by Glassner in [,   ,   ]. In the present paper, a formal description for celtic knotwork is provided by means of *collage grammars* [,   ,   ], one of the picture-generating devices studied in computer science. Using collage grammars in the form of [   ], we shall describe (a picture of) a knot by a term, i.e. an expression over graphical operations and primitives. Such a term corresponds to a derivation tree in a collage grammar, the value of the term (the result of the derivation tree) being the generated knot. Thus, the tree describes the syntactic structure of the knot, whereas the evaluation of the tree yields the actual knot. In particular, two knots may share their syntactic structure although their visual appearance differs—which is reflected in the formal model by the fact that the underlying trees are identical, but the symbols are given different interpretations as picture operations.

The paper is organised as follows. In Section   , we develop a first grammatical description of plaitwork, which is the basic form of knotwork, in terms of table-driven collage grammars. Sections   and   build upon this in order to generate square knots—designs with breaks in the interlacing—resp. plaits in the so-called carpet-page design. In Section   , we turn to rectangular knots. Section   is devoted to knots composed of triangular primitives. In Section   , variations in the drawing of structurally identical knots are discussed. Finally, Section   contains some concluding remarks.

Due to lack of space, grammars cannot be discussed in detail. In particular, this concerns Sections   – . Interested readers may find the details in the long version [   ]. Furthermore, we invite readers to download the system TREE-BAG, which was used to implement the examples in this paper, from the internet at   .
The distribution contains all examples presented in the following (and many more from other areas), so there is plenty to experiment with.

2   Plaits

Celtic knotwork is based on plaiting, where the strands are interwoven so that they turn only at the border of a design. A square plait is shown in Figure   . Being square, it can be quartered by a horizontal and a vertical line through

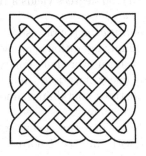

Fig. 2. A square plait

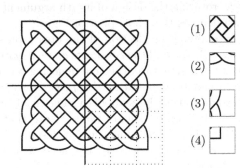

(1)
(2)
(3)
(4)

Fig. 3. Dividing the plait into tiles

its centre, see Figure , and the quarters can be obtained from each other by a rotation about the centre of the whole. This observation offers already a method to construct a description for the whole plait from a description of one of its quarters. Now consider the lower right quarter plait. The dotted lines in Figure indicate that it is made up of four basic designs shown at the right: four copies of design (1) are used in the inner part, designs (2) and (3) yield the edges, and design (4) the corner. As a tiling of the plane can be obtained by seamlessly repeating design (1), we will from now on call a basic design a tile.

Turning the quarter plait counterclockwise by 45 degrees as in Figure makes it easy to see that the tree to its right describes its structure. For this, the symbols occurring in the tree are considered as operation symbols, which turns the tree into an expression that can be evaluated. In order to

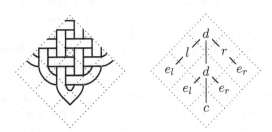

Fig. 4. Constructing a tree for the plait quarter

produce the quarter, the following meaning is given to the symbols. The nullary symbols e_l, e_r, and c stand for the tiles (2), (3), and (4). Their evaluation simply draws the respective tile (with the upper corner placed at the origin, say). The ternary operation d stands for

(a) drawing tile (1),
(b) shifting the design of the first argument to the left square below tile (1) and, respectively, the third argument down to the right square, and
(c) shifting the second argument to the square directly below tile (1).

Similarly, the unary operation l (resp. r) denotes shifting the design of its argument to the lower left (resp. right) square, and drawing tile (1). Plugging four copies of the tree into a 4-ary operation *initial* which denotes, as discussed above, rotating the design of its ith argument by $(i-1)\cdot 90$ degrees yields a tree which denotes the whole plait of Figure .

Formally, the collection of the operation symbols $initial, d, l, r$ of respective arities 4, 3, 1, 1 and the constants e_r, e_l, c (which are simply operation symbols of arity 0) is a signature, the tree denoting our plait is a term over that signature (which can be denoted as $d[l[e_l], d[e_l, c, e_r], r[e_r]]$ in linear notation), and the interpretation of the symbols yields an algebra of pictures containing, among others, the desired square plaits. Note that there are many trees over these symbols which denote inconsistent designs. Clearly, a tree denoting a consistent square plait must at least be balanced, as is the case with the tree denoting the plait of Figure .

Until now, we have discussed a single tree whose evaluation yields a particular plait. But how can we describe *the set of all square plaits* in a formal way? With the signature and the algebra given as above, we only need a way to generate suitable trees over that signature. This can be done by means of a grammar which has three nonterminal symbols C, L, R, the axiom $initial[C, C, C, C]$, and the following six rules:

$$C \to d[L, C, R], \qquad L \to l[L], \qquad R \to r[R],$$
$$C \to c, \qquad L \to e_l, \qquad R \to e_r.$$

Such a rule is applied by plugging the root of the right-hand side tree into the place of the replaced nonterminal. A derivation starts with the axiom and proceeds in a nondeterministic way until there is no nonterminal symbol left. For convenience (and since it does not affect the resulting trees), we shall always consider maximum parallel derivations, i.e., in each step all nonterminals are replaced in parallel. Clearly, using the rules from above, some of the generated trees are balanced, and thus denote square plaits, whereas others are unbalanced.

Unbalanced trees are generated by derivations in which the terminating rules are applied too early resp. too late in some places. In order to exclude such derivations, we have to regulate the generation process in an appropriate way. For this, the notion of a table-driven, or TOL, grammar [,   ,   ] (see also [   ] for the case of collage grammars) can be put to use. In such a grammar, there may be more than one set, called table, of rules (with each table containing at least one rule for every nonterminal), and a derivation step from a tree t to a tree t' consists of choosing one table, and rewriting in parallel every

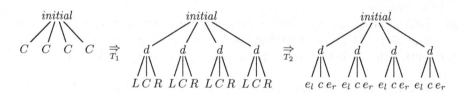

Fig. 5. A derivation with the tables T_1 and T_2

nonterminal in t with an appropriate rule in that table. As an example, we may use two tables T_1 and T_2, where T_1 contains the three nonterminating rules on the left, and T_2 contains the three terminating ones on the right. A derivation using these two tables is shown in Figure . Every application of T_1 expands the tree, whereas T_2 finishes the derivation as it contains only terminating rules. The interpretation of the tree generated by this derivation results in the plait of Figure . Note that there is a square plait whose size lies between this one and that of Figure . It can be produced by adding three constants e'_r, e'_l, c' to the signature, which are interpreted as the tiles

(5)   (6)   (7)

Fig. 6. Interpreting the tree generated in Figure

respectively, and a table T_3 analogous to T_2 to the grammar. The three tiles will reappear in Section , where they will be used to generate knots rather than plaits.

When the grammars get more complicated than the one presented above, it is usually not appropriate to discuss their rules and the operations of their associated algebras separately and in all detail. In fact, once the basic idea of the construction is understood, it tends to be a straightforward (though sometimes time-consuming) programming exercise to devise appropriate rules and define the required operations. Furthermore, while the distinction between trees and their evaluation is a very useful principle from a conceptual point of view, it is normally easier to understand the idea behind a particular grammar when it is presented in terms of pictures. There is an elegant way of doing so: We just have to extend the considered algebra *so that it interprets nonterminal symbols as primitive pictures* (special tiles, for instance). As an immediate result, the nonterminal trees occurring in a derivation can be evaluated as well, yielding a pictorial representation of the derivation. If we use this idea to visualise the derivation of the plait in Figure , we get a sequence of pictures as shown in Figure if nonterminals are interpreted as grey squares with arrows inside, indicating the direction into which a nonterminal is propagated by the rules.

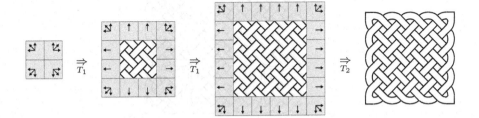

Fig. 7. Deriving the plait in Figure

In a knot, strands turn not only at the border but also in between. The traditional construction of a knot usually starts with a plait (at least conceptually). Afterwards, one repeatedly chooses a crossing of strands s and s' and cuts both strands, yielding loose ends s_1 and s_2 respectively s'_1 and s'_2. Then, s_1 is

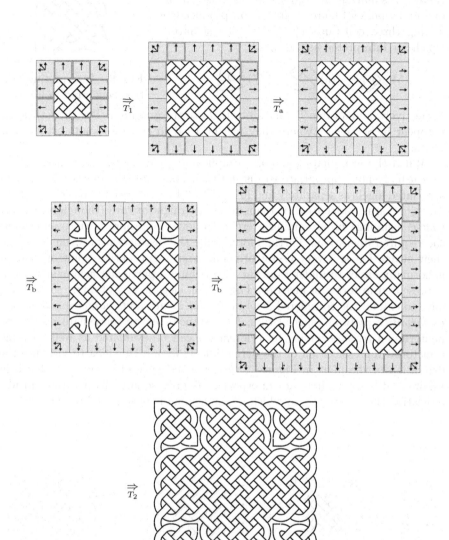

Fig. 8. Deriving a knot

Fig. 9. Two derivable square knots

connected with s_1' and s_2 is connected with s_2'. As a result, the crossing has disappeared and two new turns have emerged.

Let us use this method in order to turn the square plaits of Section into knots. We need five additional tiles, which we obtain by modifying tile (1):

(1a)   (1b)   (1c)   (1d)   (1e)

If we want to use these tiles in order to break some of the crossings in a plait like the one shown in Figure , we have to face two major difficulties. First, in order to obtain a consistent design it must be ensured that, for instance, whenever tile (1a) is used in some place, it is accompanied by tile (1c) on its left. Second, celtic knots are usually symmetric with respect to the placement of breaks, which we therefore wish to ensure as well. Again, tables turn out to be a useful mechanism for dealing with both requirements.

In order to add breaks in a systematic way, we increase the number of tables. The tables T_1 and T_2 introduced in Section remain useful and play the same rôle as before. A few rules must be added to these tables in order to deal with new nonterminals, but these rules can be inferred from the existing ones in a straightforward way. Besides the old ones, there are two new tables T_a and T_b. Intuitively, an application of table T_a 'informs' the four nonterminals C in the corners of the square that they and their descending chains of L's and R's are *allowed* to insert breaks. Figure pictures a sample derivation (where the first picture, which equals the first one in Figure , is left out). Here, the nonterminal symbols C, L, and R are depicted as in Figure , but there are several additional nonterminals. Two of the square knots generated by the discussed grammar are depicted in Figure . For more details, the reader is referred to [   ].

4 Elements of Carpet-Page Designs

In this section, the basic design of a square plait is varied by interrupting the plaitwork with regularly placed holes. The celtic artist would use these holes for further decoration, creating illustrations in the so-called carpet-page design. The simplest form consists of a square hole in the middle of a square plait, resulting in a closed plaitwork border (Figure   ). Placing square or L-shaped holes at the

Fig. 10. A square border

Fig. 11. Two crosslets

four corners of a square plait yields crosslets (Figure   ), or cross panels if the holes are inside the plait (Figure   ). Of course, these three possibilities can be combined in one design, as the examples shown in Figure   illustrate.

Closed borders and crosslets may be seen as special cross panels. The latter can be obtained from square plaits by replacing certain tiles with the empty tile, and the tiles around the resulting hole(s) with appropriate edge tiles. Intuitively, arranging regular holes on the corners is achieved by applying one table which

Fig. 12. Two cross panels

'marks' the four nonterminals C and thereby defines the left and right boundary of a hole to be created. Such a marked C may then be activated at once, producing with its immediate and outermost descendants an edge and with the other descendants the (square!) hole, until a table is applied which either terminates the derivation (the holes go over the boundary) or produces an inner edge (the holes stay inside). An L-shaped hole is obtained if the design is increased for some steps before the marked C is activated. The creation of holes can be iterated, either by using the same marked position again, or by re-marking the Cs

Fig. 13. Panels in the carpet-page design

at some current step; compare the last and first panels in Figure . Overlapping holes—e.g. an L-shaped, long-armed one with a square—can be achieved with two (or more) distinct markers for the left and right boundaries of the respective holes, but these markers have to be controlled independently (which means many additional nonterminals, rules, and tables in the grammar).

5 Rectangular Knots

So far, only square knots have been discussed. Let us now turn to the more general case of rectangular knots. In order to generate such knotwork, one could in fact use the operations from above together with a more sophisticated grammar, but it seems more instructive to consider an-  other sort of tiles. On the right, an alternative division of the interior of a plait into tiles is shown. Two basic types of tiles occur, one of them being a 90-degree rotation of the other. We shall use them in order to generate the desired square plaits (and, afterwards, knots). The basic idea is to split the generation into two phases each of which is implemented by a table. In the first phase the width of the plait is determined. The plait is extended horizontally to either side by one tile in each step. The second phase yields the vertical extension. Thus, the width (height) is determined by the number of applications of the first table (respectively second table), which yields rectangular plaits with arbitray width/height ratios. Depicting nonterminals as grey squares with inscribed arrows of various descriptions, Figure depicts a typical derivation (where we have used tiles with a black background and added a frame to the border tiles). Based on this picture, the necessary operations and rules can be defined in a rather straightforward way. As mentioned in the introduction, the reader is invited to fetch the grammar and the TREEBAG system from the web and make their own experiments. One could, for instance, use two further tables in order to make the plaits grow either to the left, right, top, or bottom in each step (which, however, seems to require numerous additional nonterminals). In this way, rectangular plaits with an even number of rows and/or columns can

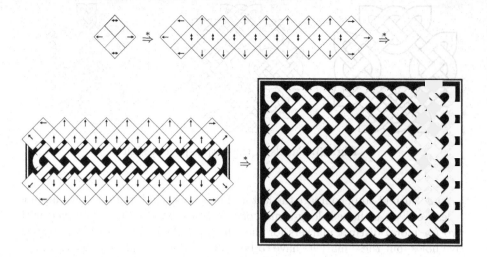

Fig. 14. Derivation of a rectangular plait (as usual, starred arrows denote sequences of derivation steps)

be generated as well (which can, of course, also be achieved with the current grammar if we use more than one axiom).

The new set of tiles lends itself to the generation of knots. Obviously, consistency is always preserved, as any crossing can be broken just by replacing the tile. Thus, no consistency constraints must be observed—the plait grammar can be turned into one that generates rectangular knots by inserting breaks at random. The required modifications of the grammar are almost trivial. It suffices to make two copies of each rule whose right-hand side contains a crossing, and to replace

Fig. 15. A rectangular knot obtained by inserting breaks at random

the crossing by a horizontal resp. vertical break. One of the resulting knots is shown in Figure . The disadvantage of this method is that symmetric knots are encountered only by chance. Aiming at symmetric knots, we have to search for a more controlled approach.

A rather ambitious task is to aim for a device that generates *all* knots that can be obtained from a rectangular plait by replacing some of the crossings with horizontal or vertical breaks in such a way that the pattern of breaks is symmetric with respect to both middle axes. Some knots of this type are pictured in Figure . To generate the set of all these knots, tables do not seem to be

Fig. 16. Symmetric rectangular knots

powerful enough. One needs 'binary tables', a regulation principle described in detail in [   ].

6 Knots Based on Triangular Tiles

Although rectangular tiles like those used in the previous sections are suitable to produce a large number of different classes of knots, other types of tiles can be useful as well. In this section, we concentrate on triangular tiles, which are particularly well suited to generate triangular or hexagonal knots.

We shall briefly mention two examples of this kind. The complete discussion can be found in [   ]. The first type of knot consists of the tiles in Figure   ,

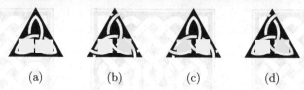

(a)   (b)   (c)   (d)

Fig. 17. The trinity knot (a) and three tiles obtained from it (b)–(d)

Fig. 18. A triangular knot composed of the tiles in Figure

which are derived from the famous *trinity knot*. Assembling these tiles to larger triangles in a suitable way, one obtains knots like the one shown in Figure   .

Our second example of triangular tiles, shown in Figure   (a), is based on a motif mentioned in [, Plate 2] and occurring similarly on the Rosemarkie stone, Rossshire. Using rotated copies of this tile and the variant in Figure   (b), knots like the one depicted in Figure   are obtained. In order to ensure consistency of the design, one has to employ the method of binary tables mentioned at the end of Section   .

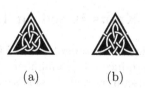

(a)   (b)

Fig. 19. Two further tiles, which can be used to generate hexagonal knots

Fig. 20. A knot obtained from the tiles in Figure

7 The Final Treatment of a Knot

In the preceding sections we have seen how, given an example for some type of knotwork, it is possible to deduce syntactic rules which allow to describe the structure of the sample knot. At the same time, the interpretation of that structural description was already determined by the sample. In this section, two further examples are given to illustrate the flexibility gained by the possibility to choose various algebras for a signature.

Consider the first 'knot signature' in Section . The algebra used there produces one single broad ribbon for each strand of the knot. A lacier effect can be achieved by splitting the ribbon using the so-called swastika method; with this treatment, the knot of Figure   can be obtained. Clearly, all that needs to be done is to exchange the tiles the algebra uses for the operations and constants. For instance, in the interpretation of the symbol c the tile on the right is used. A simpler (but analogous) variant would be to adorn the single ribbon with lines or colouring it in.

When drawing a knot by hand, the particular style of ribbon can be decided upon when the design is already quite advanced. In contrast, the form of the underlying grid has to be fixed in the very first stage of the drawing. Widely popular for knots are the Pictish proportions 3 by 4, resp. 4 by 3. The knot

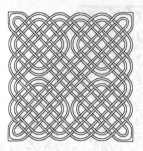

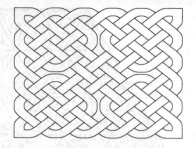

Fig. 21. Knot with split ribbons

Fig. 22. Knot in 4 by 3 proportion

of Figure   implements the second ratio. Its algebra is derived from the first algebra in Section   by interpreting the symbol *initial* so that the rotation of each argument is followed by a scaling of 4 by 3. Alternatively, it is possible to adjust all operations to the rectangular grid, and scaling the tiles accordingly. Then, the tile on the right is the one used in the operation c.

8 Conclusion

In this paper, we have shown that syntactic methods, in particular collage grammars with regulated derivations, can be used to model the construction of celtic knotwork. The formal model has the advantage that structure and representation can be treated individually and independently. The knots which we have generated are as yet of quite basic type, but there are several directions open to further investigation.

Firstly, one may think of using more complex tiles in the generation of knots. For instance, adding a third dimension yields knot models which could practically serve as jewellery designs. Pleasing effects can also be achieved by supplying the knotwork ribbons with colour.

Secondly, there are phenomena not yet provided for in our framework, such as e.g. the wide corner arcs of the knot in Figure   . Moreover, while we have to some extent integrated the method of regularly placed breaklines to untangle crossing strands, the breaklines occurring in a celtic knot can form rather more complex, but still regular patterns, and it is not yet clear how more involved structural properties of this kind can be expressed.

On the other hand, the theory of formal languages offers sophisticated tools to control the generation of objects. Thinking of the table-driven grammars we use, refining the regulation techniques for the application of the tables is one possibility. Furthermore, the grammars admit only a top-down generation of trees, i.e., information cannot be propagated bottom-up. More powerful grammars may prove to be helpful for the generation of knots in the carpet-page design where the holes are not quite so uniform.

A question which naturally arises when designing a knot is the number of strands. The celtic artist usually aimed at one continuous strand for the whole knot; a higher number may be useful e.g. for colouring. A description of a knot in terms of its syntactic structure should allow to compute that number.

The aim of the work reported here was to develop a formal model for celtic knotwork as close to the original as possible. It should nevertheless be noted that such form languages develop in time. Moreover, the methods employed here have proved to be applicable to other visual languages such as fractals or Escher-like pictures, and we believe that combining aspects of these languages may be quite pleasurable, both in results and in the doing.

Acknowledgement

We thank the referees for their useful comments.

References

[Bai51] George Bain. *Celtic Art. The Methods of Construction*. Constable, London, 1951.

[Buz] Cari Buziak. Aon celtic art and illumination. http://www.aon-celtic.com/.

[Dav89] Courtney Davis. *Celtic Iron-On Transfer Patterns*. Dover Publications, New York, 1989.

[DK99] Frank Drewes and Hans-Jörg Kreowski. Picture generation by collage grammars. In H. Ehrig, G. Engels, H.-J. Kreowski, and G. Rozenberg, editors, *Handbook of Graph Grammars and Computing by Graph Transformation, Vol. 2: Applications, Languages, and Tools*, chapter 11, pages 397–457. World Scientific, 1999.

[DK00] Frank Drewes and Renate Klempien-Hinrichs. Picking knots from trees. The syntactic structure of celtic knotwork. Report 4/00, Univ. Bremen, 2000.

[DKK00] Frank Drewes, Renate Klempien-Hinrichs, and Hans-Jörg Kreowski. Table-driven and context-sensitive collage languages. In G. Rozenberg and W. Thomas, editors, *Proc. Developments in Language Theory (DLT'99)*. World Scientific. To appear.

[Dre00] Frank Drewes. Tree-based picture generation. *Theoretical Computer Science*. To appear.

[Gla99a] Andrew Glassner. Andrew Glassner's notebook: Celtic knotwork, part 1. *IEEE Computer Graphics and Applications*, 19(5):78–84, 1999.

[Gla99b] Andrew Glassner. Andrew Glassner's notebook: Celtic knotwork, part 2. *IEEE Computer Graphics and Applications*, 19(6):82–86, 1999.

[Gla00] Andrew Glassner. Andrew Glassner's notebook: Celtic knotwork, part 3. *IEEE Computer Graphics and Applications*, 20(1):70–75, 2000.

[HK91] Annegret Habel and Hans-Jörg Kreowski. Collage grammars. In H. Ehrig, H.-J. Kreowski, and G. Rozenberg, editors, *Proc. Fourth Intl. Workshop on Graph Grammars and Their Application to Comp. Sci.*, volume 532 of *Lecture Notes in Computer Science*, pages 411–429. Springer, 1991.

[HKT93]   Annegret Habel, Hans-Jörg Kreowski, and Stefan Taubenberger. Collages and patterns generated by hyperedge replacement. *Languages of Design*, 1:125–145, 1993.

[KRS97]   Lila Kari, Grzegorz Rozenberg, and Arto Salomaa. L systems. In G. Rozenberg and A. Salomaa, editors, *Handbook of Formal Languages. Vol. I: Word, Language, Grammar*, chapter 5, pages 253–328. Springer, 1997.

[Mee91]   Aidan Meehan. *Knotwork. The Secret Method of the Scribes*. Thames and Hudson, New York, 1991.

[PL90]   Przemyslaw Prusinkiewicz and Aristid Lindenmayer. *The Algorithmic Beauty of Plants*. Springer-Verlag, New York, 1990.

[Roz73]   Grzegorz Rozenberg. T0L systems and languages. *Information and Control*, 23:262–283, 1973.

[Slo95]   Andy Sloss. *How to Draw Celtic Knotwork: A Practical Handbook*. Blandford Press, 1995.

Differentiating Diagrams: A New Approach

Jesse Norman

Department of Philosophy
University College, London
Gower Street, London WC1E 6BT, UK
jesse.norman@dial.pipex.com

Philosophers and practitioners commonly distinguish between descriptions, depictions and diagrams as visual representations. But how are we to understand the differences between these representational types? Many suggestions have been made, all of them unsatisfactory. A common assumption has been that these representational types must be evaluated in terms of the presence or absence of a single property. I argue that this assumption is both questionable and overly restrictive, and advance a two-property analysis in terms of what I call Assimilability and Discretion. I argue that this analysis allows us give a general differentiation of the various types and to understand better what factors could affect changes in classification. This suggests an outline framework for empirical research. Philosophically, it can also be used to capture a core idea of perspicuousness, and to ground an argument for the general perspicuousness of diagrams as a representational type.

1 Introduction

Among the methods often used to represent information visually are descriptions, depictions and diagrams. If I want to show some friends how to reach my house, for example, I can do so via a description (say, a set of instructions), via a depiction (say, a photograph of the region) or via a diagram (say, a sketch map).[1] And there will also be hybrid solutions from among the three.

That we can and do discriminate descriptions, depictions and diagrams as distinct representational types seems, on the face of it, relatively unproblematic, though of course there are hard cases and grey areas. But how in principle are these types to be characterised and distinguished from each other?

---

[1] Classifying a sketch map as a diagram is not intended to be controversial (compare: a map of the London Underground, which is widely regarded as a diagram). Some maps are better classified as depictions, however. I discuss the general question of what factors might affect such classification in section 3 below.

M. Anderson, P. Cheng, and V. Haarslev (Eds.): Diagrams 2000, LNAI 1889, pp. 105-116, 2000.

A satisfactory answer to this question has so far proven elusive. Such an answer would, it is often thought, provide some criterion for distinguishing examples of one type from those of another. Ideally, it might also provide an explanation of some apparently common intuitions: that diagrams are somehow an intermediary case between depictions and descriptions, that there can be central or peripheral cases within each type, and that certain factors can affect how we classify a given representation within each type and, perhaps, between types.

One can in the first place distinguish representational from non-representational differences between types: the former category might include visual differences as well as differences in the relation between the representation and its range or target; the latter might include differences of use or function, for example, such as communication, problem-solving, or conveying aesthetic or linguistic meaning. Sometimes, moreover, non-representational characteristics (for example, the content of a description) can affect the efficacy or perspicuousness of a given representation. It would seem, however, and it is often assumed by writers in this area, that distinguishing one representational type from another is primarily a matter of distinguishing common representational characteristics among their tokens.

This suggests a limited strategy by which to tackle the question raised above. Rather than attempt any general account of descriptive, diagrammatic or depictive representation as such, one may focus on the extent to which a candidate property or criterion meets the intuitive tests mentioned above; that is, allows a successful categorisation or analysis of the representational characteristics of each type. I will adopt this approach.

2   Exploring the "One Property" Assumption

Numerous attempts have in fact been made to characterise and explain the representational differences between the three types. Representations have been classified as, to take a few examples, „graphical/linguistic", „analogical/Fregean", „analogue/propositional", „graphical/sentential", and „diagrammatic/linguistic".[2] Commentators have sought, not to give necessary and sufficient conditions on a representation's being of one type or another, but to identify a single property alleged to be characteristic of a given type, which can serve as a criterion for distinguishing examples of that type. Such alleged properties include, for a given representation: the sequentiality or linearity of its sub-elements; its degree of compactness; whether or not its colour is relevant; whether or not it represents from a certain point of view; whether or not its target or range can be seen; its relative scope to misrepresent; its scope for labelling or annotation; whether it is „analogue" or „digital"; its degree of semantic and syntactic „density", and „relative repletion"; whether or not it includes certain characteristic types of shape or figure; whether it is or requires to be one-, two- or three-dimensional; whether or not it resembles, or is homomorphic or isomorphic to, its range or target; and whether or not it has certain properties

---

[2] By, respectively, Shimojima; Sloman; Palmer and Lindsay; Stenning, Oberlander and Inder; Larkin and Simon, Barwise and Etchemendy, and Hammer. See Shimojima [10] for further discussion.

(permitting free rides, restricted by over-specificity) imposed by the structural constraints on the relation between the representations and its target.[3]

There is little agreement at any level of detail amid this plethora of claims, which reflect differing goals, methods, background disciplines and governing assumptions. As Shimojima [10] has identified, many of these approaches have been motivated by the desire to find a single property-criterion for such a wide diversity of representations. Consider what can, without straining, be considered diagrams: many types of map, product instructions, Venn diagrams, Cartesian (X-Y) graphs, blueprints, Feynman diagrams, chemical valency diagrams, wiring diagrams and the wide variety of diagrams used in business (flowcharts, pie charts, scatter plots, gant charts, bar charts, organisational hierarchies etc.), to take just a few examples.[4] What, one might ask, do these have in common?

2.1   The Diagrammatic/Sentential Distinction Examined

The question here can be sharpened by considering an influential example, the distinction between the diagrammatic and the sentential proposed in Larkin and Simon [7]. Larkin and Simon describe two types of external representation, both using symbolic expressions, as follows:

"In a sentential representation, the expressions form a sequence corresponding, on a one-to-one basis, to the sentences in a natural-language description of the problem... In a diagrammatic representation, the expressions correspond, on a one-to-one basis, to the components of a diagram describing the problem."

A little further on, the authors make a more substantive claim:

"The fundamental difference between our diagrammatic and sentential representations is that the diagrammatic representation preserves explicitly the information about the topographical and geometric relations among the components of the problem, while the sentential representation does not."

This latter suggestion has the merit of analysing the concepts "sentence" and "diagram" in terms of a supposedly more fundamental property, that of preserving geometrical relations; but it is highly problematic.   Some sentences preserve geometrical relations about their ranges (e.g. "aLb" or "Andy is left of Betty"), while many diagrams (e.g. flow diagrams, organisational charts, pie charts, bar charts etc.) do not relate to topological or geometrical ranges at all, and so cannot preserve information about them.

---

[3] Some of these alleged properties are advanced by more than one person; cf. Larkin and Simon [7], Hopkins [5], Eberle [2], Goodman [3], Barwise and Hammer [1], Shimojima [10].
[4] For types of diagram generally, see Harris [4]; for specific types of graphs, see Kosslyn [6].

Later in the article the authors do give explicit definitions. A sentential representation is defined as "a data structure in which elements appear in a single sequence", while a diagrammatic representation is "a data structure in which information is indexed by two-dimensional location". These definitions avoid the above problems: sentences are defined in terms of sequentiality, not the absence of geometrical properties, while diagrams are defined in terms of the indexation of information, which leaves open whether or not the information in the range is itself supposed to be two-dimensional. Moreover, the definitions decompose the two concepts into apparently more basic terms, and in so doing provide what is in principle needed: a single property, that of sequentiality of elements, by which a sentential representation can be distinguished from a non-sentential representation; and a collateral (though not necessarily mutually exclusive) property, that of indexation to two-dimensional location, by which a diagrammatic representation can be distinguished from a non-diagrammatic representation.

These definitions may serve their purpose in the tightly restricted context of the original paper, that of analysing formalised computer or human inference procedures. Taken in a general way, however, they are again insufficient. Some diagrams, such as flow diagrams, are sequential; others, such as that below (a diagrammatic solution to the problem of the sum of the series "$1/2 + 1/4 + 1/8$ ..."), are arguably linear and so not two-dimensional:

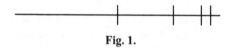

Fig. 1.

Alternatively, take a diagram representing that Andy is left of Betty:

Andy ------------------------- Betty

Fig. 2.

This seems clearly sequential and linear. Conversely, there are sentences in some languages which are two-dimensional: alphabets such as those of Hebrew and Korean utilise non-linear spatial relations between letters in forming words.

3   A Two-Property Approach

As this example illustrates, it has proven difficult to differentiate the various types of representation from one another in terms of the presence or absence of a single property. One might therefore question the need for this highly restrictive and ambitious assumption. Why not explore whether a satisfactory differentiation is possible in terms of two or more properties? After all, in making other visual discriminations we often analyse in terms of the combination of two or more properties: for example, distinguishing a mug from a cup may require us to make a judgement based on a combination of such properties as size, shape or type of material. We might agree that something would qualify as a cup rather than a mug if its shape were altered in a particular way, and its size remained the same; or if its size were altered and its shape remained; or if both size and shape were altered. Clearly, it would not be generally sufficient to appeal to just one property to make the required distinctions; and especially not if that property proved to be "cuppiness" or some equivalent. Given this kind of case, the one-property assumption starts to look, not merely overly restrictive, but less well-motivated than has been thought.

I will explore a two-property approach of my own below; but there are likely to be others to be explored, of course, and concerns about the one-property assumption are logically independent of the success or failure of the two-property alternative described below.

Some preliminary comments are in order, however. I shall assume, here and later, that all such representations are well-formed, appropriately sized and visually salient. Second, it is plausible to believe that while all three representational types use symbols as marks (which must in some sense be interpreted to have meaning), they can all also employ symbolism in a different way.

3.1   „Bare" vs. „Endowed" Symbols

Let me explain this last point with an example.[5] Take the case of a picture which represents the Holy Spirit as a dove. The picture bears a series of marks which we can recognise as a dove; but it also represents the Holy Spirit. It does not, one wants to say, *depict* the Holy Spirit; the fact that the Holy Spirit is connoted by the dove is a further piece of symbolic content, which is supplied by the observer's background information. This content is secondary; if the dove were not represented, the Holy Spirit would not be. One wants to say, following Hopkins 1998, that there are two levels of representation here. The picture uses „bare" symbols; these are enough, when organised in a certain way, for it to convey information as a depiction. It also uses what I shall term „endowed" symbols, however; but these are not intrinsic to depiction, for a description of the same scene could represent the dove in a similar manner (the symbolic or allegorical content could be assigned to the word „dove"). I think this distinction, between what may also be called iconic and symbolic representation, is fundamental to understanding how descriptions, depictions and diagrams represent.

Finally, it seems generally to be a condition on r's being a representation of X that it be subject to some underlying cause or intention. Thus, we consider a blueprint to represent a building in part because someone has caused or intended it to do so; we consider an ECG reading to represent heart activity because the machine has been (intentionally) set up to make it do so, and there is a causal relation between the heart's electrical outputs as measured and the shape of the line on screen. It is not enough that there be some purely accidental relation between the two; we do not consider that an ECG reading represents a graph of UK GNP even if the two are in fact identical. (Note that this underlying intention does not mean that a representation must convey the information it is intended to convey; it may in fact fail to do so, or the intender may not know in advance precisely what information he or she intends to

---

[5] Taken from Hopkins [5].

convey. Nor does it mean that, even where it succeeds in conveying its intended information, it must convey only that information; there are cases in which the reasoner or creator of the representation has no such specific intention, indeed many diagrams are valuable because they give free rides to unintended or unexpected information.) Finally, there will clearly be information which is intended to be part of a representation, is relevant to the representation's conveying the intended information, but is not part of the intended information itself. Thus, in creating the representation that „there is a ball on a box", I intend that it contain seven words, it is relevant to its representing what it does that it contain seven words, but it is not part of the intended information that it do so.[6]

3.2   Discretion and Assimilability

We can now say that one important property of descriptions is this: they need only express their intended information; they can leave unsaid and indeterminate some aspect of what is described; they need not convey more than is required. They are, one might say, *discreet*. A description of a ball on a box may be enormously detailed, or it may simply be the sentence: „there is a ball on a box". Depictions, by contrast, generally represent more information than is needed, in three ways. First, it may be in the nature of the specific type of depiction to display more information than is needed; thus, a photograph of a box will almost invariably contain additional information as to background, while a drawing could leave this information indeterminate while still remaining depictive. Second, though, there are some types of information which depictions must generally represent in order to depict other information. Thus, it is difficult simply to depict information about shape without also depicting information about size and orientation; and similarly, to depict information about texture, colour or intensity without (perhaps only vaguely) depicting information about shape.   Thirdly, while the background may be indeterminate, it is difficult for a depiction to leave its intended information indeterminate, while remaining genuinely depictive: any depiction of a ball on a box is likely to show the ball in some particular location.[7] Depiction is, to this extent, intrinsically indiscreet.

The idea of discretion can be expressed as follows. For representation r and intended information P:

$$\text{Discretion} \quad = \quad \frac{P}{\text{Total information which } r \text{ must convey to convey the information that P}} \tag{1}$$

---

[6] I do not discuss the idea of relevance here, but cf. Sperber and Wilson [11].

[7] Shimojima [9] analyses this overspecificity in the case of diagrams persuasively in terms of structural constraints between representations and their targets.

That is, a representation is discreet to the extent that, in conveying information that P, it need only convey the information that P.[8] The distinction between description and depiction can then be expressed thus: given a certain context, a well-formed description will tend to have discretion of close to 1. A well-formed depiction will typically never have discretion close to 1.

An important general property of depictions, by contrast, seems to be this: that the information they represent is typically easy for the observer to process or assimilate. In many cases, obtaining information from a depiction does not require a conscious process of inference; it is just *seen*, at a glance, in a way which is phenomenologically similar to how one sees the colour of one's shoes, for example. The content of a depiction can, moreover, be grasped in this way by a range of different viewers, speaking different languages and with different cultural backgrounds. Obtaining information thus does not seem, at this level at least, to be heavily dependent on background knowledge; or specifically, on knowledge of language. Descriptions, by contrast, are in natural language; for them to convey information requires an act of linguistic interpretation. Descriptions cannot simply be viewed, one might say; they must be *read*.

The point here does not concern the degree of effort expended by the reasoner as such – one may obtain information from reading a description at a glance, without conscious inference – though it will, I think, in general be true that a (well-formed) depiction of (positive, non-disjunctive) given information will require less effort to process than a comparable description. Nor do I mean to suggest that one cannot also derive information from depictions more slowly, or through an explicit process of inference. Rather, the point concerns the cognitive resources which are brought into play in considering the representation. Descriptions seem to require symbolic interpretation, in the second „endowed symbol" sense described above; their content cannot be grasped at all except in terms of the (endowed) values assigned to their constituent symbols, and this seems to require the exercise of higher level resources of understanding. By contrast, depictions seem to be able to represent at a cognitively lower level, in terms of what I called „bare symbols" above; and this seems principally to require the exercise of perceptual resources. But of course there may also be a further level of representation in a depiction, in which symbolical or

---

[8] Of course, much will depend here on how „information" is defined, and on the role played by context and the observer's background knowledge. In particular, it seems evident that the information conveyed by a representation can differ from one observer to another based on differences in background knowledge. I explore this question, and assess various theories of information in detail, in Norman [8]. The key point to note here is that this phenomenon does not seem problematic for the idea of discretion, which is defined in terms of what information must be conveyed in order to convey (only) the *intended* information P. For example: A may intend to convey information P (it is raining in Thailand) to B by writing „it is raining in Thailand". This will have optimal discretion: simply to understand the words is to understand P, the intended information. B may of course infer (perhaps without being consciously aware of doing so) the information that the streets will be wet, or that the harvest will be damaged, etc.; but this will not affect the discretion (as defined) of the original sentence. But of course, the discretion of the sentence may be lowered to the extent that additional information is supplied by it in conveying P.

allegorical content is brought into play, and this will require a higher-level grasp of endowed symbols in the sense already mentioned.

We can now be more precise about the status of diagrams. Diagrams score highly on the scale of discretion, it seems, though not in general as highly as sentences. A well-drawn diagram need not convey much if any extraneous information in order to convey the intended information that P. In this regard the great variety and multiplicity of types of diagram may come to the user's aid, allowing him or her to select a different type rather than allow indiscretion to occur; and the recognition that it is possible by such choices to avoid conveying extraneous information itself suggests that users generally expect diagrams as a type to be discreet.

On the other hand, however, diagrams also score highly on the scale of ease of processing or assimilability. A diagram may serve as well as a picture, indeed in some contexts perhaps even better than a picture, at representing to an observer that there is a ball on the box.

This gives a fairly clear sense in which diagrams are indeed an intermediary type, between descriptions and depictions. Perhaps appropriately, this set of relationships can be viewed diagrammatically, as in Figure 3 below. (The circles are intended to indicate roughly where the central cases of each type fall; there will clearly be overlaps and borderline cases.)

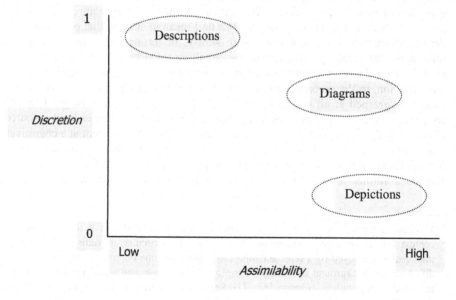

Fig. 3.

3.3   Changes in Classification

The above account suggests that changes in discretion or assimilability could in principle affect the degree to which a given representation would be taken as a representative token of its type. At the limit, enough change might cause us to want to reclassify a representation under a different type. And we can make broad predictions, in terms of discretion or assimilability, as to some of the dynamic factors which could affect the location of a particular type of representation along the two axes.

- Descriptions would lose discretion by conveying unnecessary collateral information; perhaps, through verbosity or periphrasis. They could increase their assimilability in at least two ways: at the level of form, perhaps through a change in font type, or through replicating the shape of their object, in a few rare cases;[9] at the level of content, perhaps through improved syntax, provided this did not affect the information conveyed.
- Depictions would lose assimilability by being overly symbolic or allegorical, as discussed above, or by being cluttered (where „clutter" is defined as *uninformative* detail). Decoration could in principle either increase or decrease assimilability.[10] Depictions would gain discretion by becoming more schematic; that is, by omitting irrelevantly informative detail.
- Diagrams would lose assimilability by being cluttered; and would lose discretion by conveying irrelevant information.

Some of these dynamic factors can also be represented, as in Figure 4:

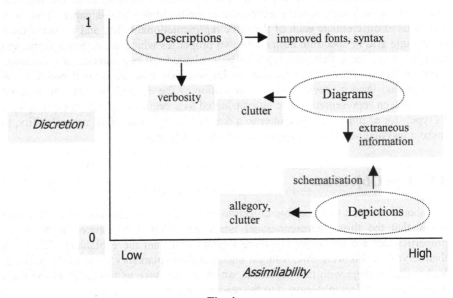

Fig. 4.

---

[9] A similar phenomenon is occasionally seen in works of devotion (such as „the Altar" and other pattern poems of George Herbert); and also in the Mouse's tale in *Alice in Wonderland*.
[10] Contra Tufte [12] and [13], whose data-ink ratio assumes decoration must decrease assimilability.

Inevitably, this latter suggestion is somewhat speculative, though not, I think, implausible. It would be interesting to explore, however, whether the broader framework here could be used empirically, to test some of the suggested outcomes. This might involve recasting certain of the above variances and trade-offs into a series of empirical experiments on various defined sets of representations. Experimental results showing that variances in discretion or assimilability caused subjects to regard given representations as more or less descriptive, diagrammatic or depictive could perhaps help to confirm or disconfirm the analysis given here. The same is in principle true for reclassification of representations from one type to another.

4   Conclusions

This discussion is, of course, far from complete. Nor, of course, does it claim to exhaust our need for explanation of the representational differences in play here. In particular, it is important to note that this two-property account is only one "cut", so to speak, at the issue. As such, it is not necessarily in conflict with other one-, two- or multi-property accounts, nor with more ramified theories of representation such as, for example, Peirce's theory of signs; any more than, for example, the physics of medium-sized objects need be in conflict with physics at the particle level.

The discretion/assimilability account does, however, seem to meet the intuitive tests for a classificatory scheme mentioned at the beginning of this paper. That is, it allows us to differentiate each of the three representational types, and to locate them on a single grid or spectrum defined by two properties which are different from, and arguably more basic than, the types to be explained. It accommodates a common intuition that diagrams are somehow an intermediary case between depictions and descriptions. And it provides some explanation of the dynamic factors which could affect a given representation's being considered as a central or peripheral example of its type: for example, to add clutter to a diagram is, by reducing its assimilability, to make it less diagrammatic and so less exemplary of its type.

4.1   Two Objections

I should mention two possible objections, among many. The first arises from the worry that the three representational types are being differentiated in terms of properties that are not genuinely more fundamental than the explananda, the types themselves. If this were true, it might undercut the explanatory value of the account. Rather than engage with this objection, which would take us too far from the main discussion, I shall simply leave it to the reader to judge.

The second objection is, perhaps, more serious. It arises from the apparent implication here that there is not a deep bifurcation in representational properties at this level between descriptions on the one hand, and diagrams and depictions on the other. This may run counter to some intuitions, to the effect that descriptions are a radically distinct representational type, not to be located even in principle in terms of the same properties as depictions and diagrams. On this view, moreover, the question might arise whether any variance in assimilability and/or discretion could ever cause us to want to reclassify a given description as a diagram. Yet Figure 3 seems to suggest that this is possible.

Of course, intuitions differ here, especially given the range of cases to be considered; and the differentiation of descriptions relative to diagrams and depictions in terms of discretion and assimilability does not of itself suggest that a given reclassification from one type to another is in fact possible. But a better response is simply to note that the contrary view can be accommodated within the picture I have been sketching. There is, I argued above, an important distinction between iconic and symbolic representation; and I have suggested that descriptions (normally) require symbolic interpretation. Conversely, it is often held that iconic representation involves a mapping of some kind between a given representation and its range. These views may or may not be part of a deeper theory of representation; but they do not cut against the present account. What is at issue here is how, if at all, such factors affect the assimilability and discretion of a representation. And it is not, I think, problematic that we can assess different descriptions, depictions and diagrams in the latter terms.

Indeed, we may do so more often than we know. For I want to suggest, finally, that this account can help us to analyse the concept of *perspicuousness*.

4.2   The Concept of Perspicuousness

What is generally meant by perspicuousness, or perspicuity? Something is perspicuous if it is „easily understood" or „clearly expressed", according to the OED. There is a visual metaphor in the Latin root of the word; that of „looking through" something to see something else, of easily discerning the information in a given representation. These senses are captured by the two-property analysis above: perspicuousness can be understood in terms of the combination of discretion and assimilability. Both properties are needed: a representation that P which is discreet but hard to assimilate by the observer will not be a perspicuous representation at all; a representation which is assimilable but contains much additional or irrelevant information will not be a perspicuous representation *that P*.

What is considered perspicuous will of course vary with context and between specific individuals. In general, this analysis suggests that a perspicuous representation – whether it is a description, a depiction or a diagram – does not convey more information than it requires to make its point, so to speak, and it does so in the way most assimilable by the observer. Specifically, a representation will be perspicuous if it avoids clutter, irrelevant information, verbosity, or unnecessary symbolic or allegorical content. Thus it may be possible to compare two descriptions or two depictions on these criteria, and say of one token of a given type that it is more perspicuous than the other. But it may also be possible to compare tokens of different types, and to say that token *a* of type *A* is superior to token *b* of type *B* as a perspicuous representation of information that P. In both cases, some general trade-off will be required between the two criteria in order to make such judgements; but it does not seem problematic that people can and do make such trade-offs in precisely these terms (of clutter, irrelevance, verbosity etc.). Indeed, these criteria are often invoked to explain why, for example, a given diagram may or may not be better than a given description as a means to convey information. Thus, this analysis seems to correspond both to our normal intuitions and to our normal linguistic practices. And it allows us to explain in principle, to at least a first level of approximation, the often-noted special perspicuousness of diagrams as a type.

Acknowledgements

This paper adapts and extends work from Norman 1999. It was prepared under a postgraduate studentship award from the Arts and Humanities Research Board, which I gratefully acknowledge. In addition, I would like to thank the Philosophy Department of University College, London; Marcus Giaquinto and Steven Gross for their comments on earlier drafts of this material; and the anonymous referees of this paper for the Diagrams 2000 conference.

References

1. Barwise, J., Hammer, E.: Diagrams and the Concept of Logical System. In: Allwein, G., Barwise, J.: Logical Reasoning with Diagrams. Oxford University Press (1996) 49-78
2. Eberle, R.: Diagrams and Natural Deduction: Theory and Pedagogy of Hyperproof. PhD thesis, Indiana University (1995)
3. Goodman, N.: Languages of Art. Hackett, Indianapolis (1976)
4. Harris, R.: Information Graphics: A Comprehensive Illustrated Reference. Management Graphics, Atlanta (1996)
5. Hopkins, R.: Picture, Image and Experience. Cambridge University Press (1998)
6. Kosslyn, S.: Elements of Graph Design. Freeman, Oxford (1994)
7. Larkin, J., Simon, H.: Why a Diagram is (Sometimes) Worth Ten Thousand Words. Cognitive Science 11, (1987) 65-100
8. Norman, J.: Diagrammatic Reasoning and Propositional Logic. MPhil Thesis, University College London (1999)
9. Shimojima, A.: Operational Constraints in Diagrammatic Reasoning. In: Allwein, G., Barwise, J.: Logical Reasoning with Diagrams. Oxford University Press (1996) 27-48
10. Shimojima, A.: On the Efficacy of Representation. PhD Thesis, Indiana University (1996)
11. Sperber, D., Wilson, D.: Relevance. Blackwell, Oxford (1986)
12. Tufte, E.: The Visual Display of Quantitative Information. Graphics Press, Connecticut (1983)
13. Tufte, E.: Visual Explanations. Graphics Press, Connecticut (1997)

Logical Systems and Formality*

Patrick Scotto di Luzio

Department of Philosophy, Stanford University
sdiluzio@csli.stanford.edu

Abstract. The question is posed: in which respects and to what extent are logical systems which employ diagrammatic representations "formal"? I propose to characterize "formal" rules to be those which are reducible to simple constructive operations on the representations themselves. Formal systems, then, are those which employ such formal rules. It is argued that "formality" thus characterized underlies a particular strategy for meeting a certain epistemological challenge. Some diagrammatic and heterogeneous logical systems are tested for formality, and it is suggested that any robust heterogeneous system is unlikely to be formal. The analysis of this paper, then, provides a principled account of how some diagrammatic systems differ significantly from linguistic ones.

1 Introduction

Among the various methods and strategies that have emerged for the study of diagrammatic representations is one which involves the formulation of *logical systems* which employ such representations. This particular strategy takes as a starting point the standard template used in the study of linguistic systems (by which the syntax, semantics and rules of inference are precisely specified), and applies it to non-linguistic kinds of representations. For instance, Shin (   ) presents such a logic for Venn diagrams; Hammer (   ) ones for higraphs, Euler circles, and Peirce diagrams; and Luengo (   ) one for a diagrammatic subsystem of Hilbert's geometry. Other logical systems (e.g., Hammer   , Barwise and Etchemendy   ) are heterogeneous in that both sentential and diagrammatic representations are employed.

In broad outline, these systems are all similarly characterized. Rules are given to delineate the class of well-formed representations, a mathematically precise semantics is provided, and proofs are defined to be sequences of well-formed representations according to certain rules of inference. The systems are then shown to be sound and (in some cases) complete. According to Barwise and Etchemendy (, p.214): "The importance of these results is this. They show that there is no principled distinction between inference formalisms that use text and those that use diagrams. One can have rigorous, logically sound (and complete) formal

---

* The author would like to thank Johan van Benthem, John Etchemendy, Krista Lawlor, Keith Stenning, and two anonymous referees for helpful comments on an earlier draft. The author also gratefully acknowledges the financial assistance of the Social Sciences and Humanities Research Council of Canada.

M. Anderson, P. Cheng, and V. Haarslev (Eds.): Diagrams 2000, LNAI 1889, pp. 117–   , 2000.

systems based on diagrams." On the other hand, some have criticized this "logical approach" for not taking adequate account of the distinctly *spatial* features of the representations (Lemon , Lemon and Pratt).

At first glance, then, it may appear that the formulation and study of diagrammatic logics is in certain respects a *conservative* enterprise. There is innovation, of course, in that the representations being studied aren't linguistic. Nonetheless, insofar as the basic techniques and formulations employed appear to be completely analogous to the standard linguistic case, there may be a tendency to view such study as, ultimately, doing "logic as usual", but just on a wider class of representations.

Shin, however, would resist such a tendency. She sees the development of diagrammatic logical systems as a first step towards a broader understanding of representation in general:

> As long as we treat diagrams as secondary or as mere heuristic tools, we cannot make a real comparison between visual and symbolic systems. After being free from this unfair evaluation among different representation systems, we can raise another interesting issue: If a visual system is also a standard representation system with its own syntax and semantics, then what is a genuine difference, if any, between linguistic and visual systems? The answer to this question will be important not just as a theoretical curiosity but for practical purposes as well. (Shin , p.10)

So once we have dispelled the worry that diagrammatic reasoning can't be rigorous, we can begin to address more fine-grained questions about its nature and use. There may turn out to be a principled distinction between linguistic and diagrammatic systems after all, and perhaps the logical approach will help us find it.

In this paper, I endeavour to consider one such question – one which takes seriously the direct employment of diagrammatic representations within logical systems. The question posed is this: in which respects and to what extent are logical systems which employ diagrammatic representations *formal*? I will claim that linguistic systems employ rules of inference which are much more straightforwardly formal than those of certain diagrammatic or heterogeneous systems. Since the "formality" of a system makes it naturally suited to play a central role in a particular kind of justificatory project, an important disanalogy emerges between standard linguistic logical systems and some of the newer diagrammatic or heterogeneous ones. Contrary to initial appearances, then, the incorporation of diagrammatic representations into logical systems cannot be merely seen as doing "logic as usual".

We will proceed as follows. In Sect.2, I present a notion of formality and argue for its naturalness in terms of a particular kind of epistemological strategy. I then apply this notion in Sect.3 to particular diagrammatic systems such as Shin's *Venn-I* and Barwise and Etchemendy's *Hyperproof*. More general issues are addressed in Sect.4, where I consider an argument by Barwise and Etchemendy ( ) against the need for an "interlingua".

2 A Notion of Formality

I propose to characterize "formal" rules to be those which permit the writing of a representation R from $R_1, \ldots, R_n$ (for some n), where R is the result of applying a simple constructive operation on (the form of) $R_1, \ldots, R_n$. It is then natural to call a logical system formal if its rules of inference and well-formedness are formal.

To be sure, more needs to be said to flesh out this notion of a "formal rule". To begin with, the "construction" metaphor should be taken seriously. For instance, in characterizing the set of well-formed representations, it is customary to provide an inductive definition by which more complex representations are literally built-up from simpler ones. This main intuition carries over to the characterization of formal rules of inference, where the "end-representation" (a more neutral term than "end-formula" in this context) is constructed from parts of the premise-representations; that is, such rules take components of the premise-representations and put them together with some other basic building blocks (usually, a representational primitive) to yield the end-representation. In fact, in many cases, such rules are one-to-one; given a set of premise-representations and a rule, there is often only one permissible end-representation.

Some paradigmatic examples may be useful here. Consider, for instance, modus ponens, the main rule in Hilbert-style propositional systems; it is also called $\rightarrow$ - elimination in natural deduction or Fitch-style systems. This is a rule which permits the derivation of ψ from ϕ and $\phi \rightarrow \psi$, where ϕ and ψ are well-formed formulae. Here, quite literally, the end-formula is built up from parts of the premise-formulae. (It is fortuitous that another name for this rule is "detachment".) Similarly, consider $\wedge$-introduction, where the derivation of formula $\phi \wedge \psi$ is licensed from that of ϕ and of ψ. Again, the "constructive" feature of this rule is readily apparent. For a diagrammatic example, consider *Venn-I*'s rule of *erasure of part of an $\otimes$-sequence*. Here, if a diagram is such that part of an $\otimes$-sequence lies in a shaded region, that part of the $\otimes$-sequence can be erased, with the remaining parts reattached.

These examples hopefully give a good idea of what is picked out by this notion of formality. We will consider some complications in the next section. For now, it is helpful to consider reasons for why this conception of formality might pick out something of philosophical interest. One way to motivate such reasons is to ask: why offer a *formal* proof at all? Hammer offers the beginnings of a possible answer:

> For suppose the validity of a proof involving a diagram is challenged. In defense one would claim that each inference in the proof is valid. When asked to defend this claim, one would spell out each of the inferences made in more and more detail. Eventually, perhaps, one would specify the syntax of the statements and their semantics, translating each step

of the proof into this formalized system and showing that each inference in the proof is valid in the system. (Hammer , p.26)

It is worth dwelling upon the argumentative dynamics of the situation described by Hammer, in order to determine more clearly the justificatory role that logical systems might be expected to play. Implicit in the imagined situation is the existence of two "characters", a "prover" who is presenting a proof, and a "skeptic" who is questioning its validity. Notably, the prover's initial reaction to the skeptic is to assert the validity of the steps of the *original* proof. When the skeptic then reiterates his challenge, directing it now to these particular steps, the prover interprets this as a request for a *more detailed* proof in which the inferences in the original proof are replaced by a series of finer-grained ones. The prover does *not* interpret this as a request for some sort of direct justification of the original steps themselves. Even if the prover believes her original steps to be valid (and presumably would be able to furnish independent reasons for this), she nonetheless assumes that it would be inappropriate to defend the original proof "as is", and that the skeptic's challenge (whatever it may be) is better met by providing a proof with a finer grain.

Why might this be so? Perhaps it is generally easier to *see* the validity of finer-grained transitions, and hence the skeptic's challenge might be adequately met by providing a sufficiently detailed proof. Such considerations suggest that what's at issue in this argumentative context isn't just the *actual* validity of a proof's transitions. If validity were all that mattered, then our imagined prover should be able to stick to her guns and assert to the skeptic the acceptability of her original proof (assuming, of course, that it actually employed only valid inferences). We might even imagine our prover indignantly instructing the skeptic to figure out for himself why the original proof is in fact valid.

But this is not how the situation is imagined by Hammer to proceed. So what's at issue here isn't simply whether the inferences of a proof are *in fact* valid, but also whether they are in some sense *recognized* to be valid. In facing the skeptic's challenge, our prover is implicitly taking on the responsibility to provide a valid proof *of a certain kind*, one which "wears its validity on its sleeve" each step of the way. This in turn fuels a "spelling-out" process which (Hammer suggests) might end with the purported proof being "translated" into one within a logical system. Presumably, this latter proof can't legitimately be the target of further challenges on the part of the skeptic, or else our prover's argumentative strategy would have failed, leaving our characters on the brink of a regress. Logical systems are thus seen by Hammer to be frameworks which provide a principled stopping-point at which a certain kind of ongoing skeptical challenge can be met.

But this invites the further query: in virtue of what would a logical system be able to provide such a stopping-point? I contend that its *formality* (in the sense outlined at the beginning of this section) would ground a plausible answer. The use of a formal system faces this skeptical challenge by restricting inferential

---

[1] Hammer's use of this word doesn't coincide with my own. Here he is following common usage in applying this term to any precisely formulated logical system.

moves to simple, safe manipulations of representations. The epistemic privilege of such transitions is secured in part by their characterization solely in terms of formal properties of the representations themselves. Rather than having to ponder the content of any particular step in a proof, the proper use of formal rules, at least in principle, requires little more than an ability to recognize and differentiate the various parts of a simply structured representation. This would involve both an ability to recognize each of the representational primitives for what they are (e.g., to be able to differentiate the various atomic symbols in linguistic systems) as well as an ability to recognize how the symbols are spatially related to each other. The epistemic interest in such abilities is that they are exercised over "concrete" objects (namely, the representations themselves) which are often more immediately accessible than the domain being reasoned about.

As we have seen, the mere validity of a proof – even one whose transitions are characterized solely in terms of the representations' formal properties – is not enough to answer our skeptic's challenge. Hammer suggests that the proof also needs to be of sufficient detail, and that logical systems provide a natural framework for ensuring this. Formality can explain why this should be so, for if the proof is one licensed by a *formal* logical system, then by definition each step of the proof (other than the listing of premises, of course) is the result of a simple, constructive operation on the representations themselves. In other words, only "formally immediate" transitions are licensed by such systems. So once our prover embarks on the process of spelling out her proof in more and more detail, it is reasonable to expect this process to end when the only transitions that are left in the proof are those that correspond to basic formal manipulations, just like those of paradigmatically formal inference rules. Indeed, it's hard to see how much finer the grain could possibly become.

This "fine-grainedness" serves to distinguish formality from another notion that is thought to characterize what is formal; namely, *recursiveness*. Given an effective Gödel numbering of the representations of a given system (by which each representation is effectively assigned a unique natural number), one could say that a recursive rule over representations is one whose image over the natural numbers is recursive (where this latter sense has a precise, mathematical characterization). Through the Church-Turing thesis, recursiveness is almost universally thought to provide a precise characterization of theoretical machine-checkability. What's worth noting here is that machine-checkability and formality do not necessarily address the same skeptical worries, or achieve the same epistemological goals. Not just any machine-checkable rule is likely to reside at the end of a spelling-out process, the point of which is to produce a sufficiently detailed proof. For it may be the case that some machine-checkable rules are too coarse-grained to meet certain argumentative challenges. We will return to this issue in the final section.

It is also worth distinguishing formality from another "formalist" strategy which has been used to address skepticism, albeit of a somewhat different sort. Hilbert's program was explicitly designed to establish a secure foundation for mathematics by means of formalization. The emergence of the set-theoretical paradoxes precipitated a perceived crisis which threatened to dethrone mathematics as the most certain of the sciences. Hilbert saw the paradoxes to be the result of applying what seem to be intuitively correct principles of reasoning over abstract domains "without paying sufficient attention to the preconditions necessary for [their] valid use" (Hilbert , p.192). This is to be contrasted with the domain of formal proofs, each of which is surveyable and whose structure is "immediately clear and recognizable", and where contentual inference is thought by Hilbert to be quite safe and reliable.

The fact that formalist strategies have played an important role in the history and philosophy of mathematics may provide further evidence that formality as presented here picks out a notion of philosophical interest. But it should be noted that there are some non-trivial differences between Hilbert's program and the particular argumentative strategy outlined a few paragraphs back. For instance, they are each addressing different skeptical worries. Hilbert's main goal was to establish the consistency of analysis and set theory, whereas formality as articulated here is seen as providing a way of ending a potential regress, by which the challenge to articulate ever-finer steps in a proof is made to stop. Furthermore, Hilbert's methodology relies on the claim that it is safe to *informally* reason over formal proofs in order to show the consistency of a theory. This is quite distinct from a view which says that reasoning formally is in itself safer than reasoning informally. While it is beyond our present purposes to explore such intricacies, these brief remarks should hold off the temptation to see the almost universally accepted failure of Hilbert's program in light of Gödel's Incompleteness Theorems as an immediate knock-down argument against the notion of formality presented here.

To sum up this section, I have articulated formality to be underlying a plausible strategy for dealing with a particular kind of skeptical challenge. What I haven't done is provide reasons for why it should be deemed central to, say, a theory of general reasoning, nor is it likely that the arguments presented here could be easily adapted towards such an end. Rather, our focus is on some of the abstract worries that philosophers often entertain. To say that a notion is of significant philosophical interest is, of course, not to say that it exhausts all we may be interested in.

3 Applying the Notion

In this section, we test some logical systems for formality, and consider some complications of the notion itself.

3.1 First-Order Logic

We have already seen two paradigmatic linguistic formal rules of inference, modus ponens and ∧-introduction. In fact, pretty much every rule in the standard formulations of first-order logic is straightforwardly formal in the sense presented here. (The reader is invited to verify this.) However, there are two kinds of complications that need to be considered. Firstly, it may appear that non-constructive rules like ∨-introduction (from ψ derive $\phi \vee \psi$) or ⊥-elimination (from ⊥ derive ϕ) break the formal mold, in that a significant chunk of the end-representation isn't found in the premise-representations at all. Secondly, something ought to be said about Hilbert-style systems, where the inferential work is being done more by the axiom schema than by the rules of inference.

As it turns out, "non-constructive" rules like ∨-introduction or even ⊥-elimination do little to threaten the "formality" of a system which contains them. The reason is that the non-constructive aspect of them is of a degenerate kind: *any* well-formed ϕ will do. As long as the rules of well-formedness are also formal (and they in fact are, in standard formulations of first-order logic), the proper application of such non-constructive rules still requires no more than a basic facility with manipulating bits of notation. Similarly, Hilbert-style systems are rather straightforwardly formal, since the particular axiom instances are derived from the axiom schema through a formal substitution operation, by which well-formed formulae (formally constructed) are plugged into the schema.

In the previous section, it was suggested that the philosophical interest in having *formal* rules of inference resides in the relatively unproblematic abilities they require, abilities which correspond to rather elementary operations on the representations. By inspecting the inference rules of first-order logic and most other linguistic systems, we can come up with a rough catalogue of these operations. Well-formedness rules and inference rules like ∧-introduction, for instance, require a concatenation operation, whereas rules like ∧-elimination or modus ponens require a truncation or detachment operator. More complicated rules, like the quantifier rules or the rules that yield particular axioms from axiom schema, require substitution as well.

3.2 Peirce's Alpha-Logic

Peirce's Alpha-logic is a logical system of equal expressive power as propositional logic. It is a diagrammatic system whose representations are "graphs" by which propositional letters are enclosed by closed curves (called "cuts"). I will follow Hammer's presentation of it in Chap.8 of Hammer (   ).

The rules of well-formedness are straightforwardly formal; the graphs are generated from propositional letters and a cut with nothing inside by applying two formal operations: enclosing an entire graph within a closed curve, and juxtaposition. The rules of inference allow for various insertions and deletions of graphs or cuts. For instance, the rule of *Insertion of Odd* allows an arbitrary graph to be inserted within an odd number of cuts. (This rule is similar to first-order logic's ∨-introduction, in that it licenses the derivation of a representation which contains an arbitrary subrepresentation. As argued above, this non-constructivity doesn't seriously threaten the formality of the system, since the rules of well-formedness are themselves formal.) Similarly, *Erasure in Even* allows an evenly enclosed subgraph to be deleted from a graph. The rule *Double Cut* allows one to surround any part of a graph with two closed curves with nothing between them. Given a graph, *Iteration* allows one to copy a subgraph deeper (i.e., inside more cuts) within the original graph, and *Deiteration* allows one to reverse this process (i.e., erase a subgraph that has a less deeply nested copy). Finally, *Juxtaposition* allows independently asserted graphs to be juxtaposed together into a single graph.

The Alpha-logic, then, is a formal logical system. Interestingly, the formal operations that are invoked are somewhat distinct from the ones we encountered in the previous subsection. Juxtaposition is merely the two-dimensional analogue of concatenation, but the Alpha-logic also requires something like an "overlay" or "superposition" operator which allows cuts or subgraphs to be inserted into graphs. Furthermore, the proper application of this operation requires that the user both have the ability to recognize containment (i.e., that a subgraph happens to be enclosed by a cut) as well as a counting ability (since the proper application of the rules of inference depends on the number of cuts in the graph or subgraph). What is *not* required is the recognition of any kind of metric, affine, or even purely projective properties. Merely topological ones will do.

3.3 Shin's *Venn-I*

Venn-I is a logical system of Venn diagrams formulated and studied in Shin (   ), a variant of which is studied in Hammer (, Chap.4). The rules of well-formedness are presented in the usual inductive manner. Just as in Alpha-logic, an "overlay" operator is needed, as diagrammatic primitives (curves, shading, lines, and ⊗'s) are placed over one another to compose a well-formed diagram. However, unlike Alpha-logic, some of the diagrammatic objects of Venn-I are allowed to cross each other in well-formed diagrams. In fact, the closed curves are *required* to intersect so that the resulting diagram has the appropriate number and kind of minimal regions.

Many of the rules of inference are straightforwardly formal as well, usually involving the addition or deletion of a simple diagrammatic object subject to certain formal features holding. Examples include the *rule of erasure of part of an ⊗-sequence* (which was already mentioned in the previous section) and the *rule of spreading ⊗'s* (which allows one to extend a pre-existing ⊗-sequence). The non-constructive *rule of conflicting information* is of the harmless kind, by which *any* well-formed diagram may be derived from a diagram which contains an ⊗-sequence entirely within a shaded region.

The *rule of erasure of a diagrammatic object* is an interesting case. In its original formulation, it allows any curve in a well-formed diagram (wfd) to be erased. This, however, turns out to be too strong, since it would allow the derivation of an ill-formed diagram from a well-formed one. This is illustrated in the next figure. While it is true that the middle wfd is built up from the left wfd, the original rule allows the middle diagram to be transformed into the right diagram (which is ill-formed since there are redundant, disjoint minimal regions). For $n > 3$, it will always be the case that one can transform an n-curve well-formed Venn diagram into an ill-formed one by erasing some curve.

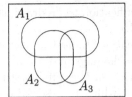 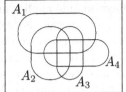 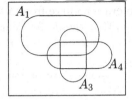

Fortunately, the rule can be modified to block this undesirable result without sacrificing the completeness of the system (Scotto di Luzio   ). The modification requires that curve-erasure be permitted only if what results is well-formed. This may still count as a formal rule; however, it's of a peculiar sort, since it requires that one check the well-formedness of the resulting representation *after* the deletion operation is performed. (Notably, all the other constructive rules we've encountered to this point are such that well-formedness is preserved "for free". The proper application of these rules never required that one check the well-formedness of the result.)

The *rule of unification*, however, resides at the outskirts of formality. This is a rule which essentially allows one to pool together into a single diagram the "information" contained in two. What is notable about this rule is that it does not require that the resulting diagram bear any kind of direct topological resemblance to either of the pair of diagrams that precede it. For instance, in the unification below, notice how the topological relations between curves labeled A_3 and A_4 are not preserved since, e.g., they meet at two points before unification, and four points after.

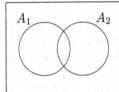 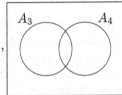 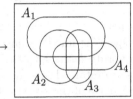

A degenerate case of unification occurs when one of the pair of diagrams being unified is in fact the "empty" diagram (i.e., just a rectangle). This rule then allows for the complete reorganization of the other diagram. For instance, we could relabel the diagram and make the appropriate adjustments to the shadings and ⊗-sequences, as required by the rule. This is illustrated below. (I'll use ● to indicate the shading of a minimal region.)

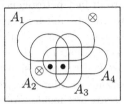 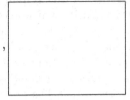 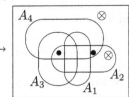

Given the restriction on the erasure of curves which is needed to preserve well-formedness, it turns out that such "reorganizational" uses of the unification rule are necessary to preserve the completeness of the system. (In brief, if one wishes to erase a curve from a diagram, one applies unification to "redraw" the diagram so that the desired curve can be safely erased.)

There is a lot packed-in, then, in the rule of unification. Indeed, in the example just given, it takes a bit of work to see that the shadings and ⊗'s are in fact in the appropriate regions. Unification is of a stripe rather different than either the constructive rules we have encountered (by which representational primitives are added, deleted, or recombined step-by-step according to basic formal operations), or the non-constructive ones which license the derivation of *any* well-formed representation. It is indeed "constructive" in the sense that all the formal building-blocks are provided in the premises (as would be appropriate for a rule named "unification"), but its successful application doesn't appear to correspond to any particular simple formal operation. (Contrast the complexity of this rule with the functionally identical *Juxtaposition* rule of Alpha-logic.) I suggest that it stretches the boundaries of formality as presented here.

By questioning the formality of the rule of unification, I do *not* mean to suggest that the rule isn't completely "syntactic". It's not as if the proper application of the rule requires the sneaky use of some semantic notions. Rather, though completely syntactic, it doesn't seem to correspond to anything like a *basic* formal operation, and as such, doesn't appear to be a good candidate for having the kind of epistemic privilege the notion of formality is supposed to capture.

3.4   Hammer's Heterogeneous Venn System

In Hammer (   ), a heterogeneous system incorporating both *Venn-I* and first-order logic is studied. Adjustments are made to the rules of well-formedness to incorporate *labels* into the diagrams, the rules of inference of each of these systems are retained, and heterogeneous inference rules are added. The "homogeneous" parts of the system thus inherit the formality features (or lack thereof) we've already discussed above.

The well-formed representations of the system include all first-order formulae (defined in the usual manner), as well as Venn diagrams whose curves are labelled by "set terms" of the form $\lambda x \phi$, where ϕ is a well-formed first-order formula with at most variable x free. Also, in addition to ⊗-sequences, diagrams may contain c-sequences, where c is any constant symbol. With the addition of labels and c-sequences, the "diagrams" of this system are in fact heterogeneous representations, and it is this heterogeneity which allows for the *formal* derivation of diagrams from formulae and vice versa. For instance, given a sentence $\exists x \phi$ and a diagram, the ∃-Apply rule allows one to add to the diagram (i.e., copy the diagram and add) an ⊗-sequence occupying at least the region inside the curve labelled $\lambda x \phi$. Similarly, the ∃-Observe rule allows one to infer $\exists x \phi$ from a diagram with a curve labelled $\lambda x \phi$ which has an ⊗-sequence inside. Analogous rules are available for dealing with c-sequences (in which case the relevant formulae are of the form $\phi(c)$), and dual rules are available for the universal quantifier.

3.5 Hyperproof

Hyperproof (Barwise and Etchemendy   ) is a computerized system and textbook designed for elementary logic courses. The system itself is heterogeneous, in that its well-formed representations include first-order formulae (possibly with interpreted predicates like *RightOf*, *Large*, or *Cube*) and "blocks-world" diagrams in which icons of various sizes and shapes are placed on or off a checkerboard. A partial mathematical analysis of *Hyperproof* is presented in Barwise and Etchemendy ( ).

As may be expected, the well-formed formulae are constructed in the usual manner. The well-formed diagrams aren't given an inductive characterization in Barwise and Etchemendy ( ), but rather are directly characterized in terms of the positions of the icons (of various types) on the grid. This isn't a significant difference, as an equivalent "inductive" characterization is easily obtainable. (Similar to the definition of "wfd" in Shin ( ), one could construct the *Hyperproof* diagrams by adding the icon tokens one-by-one onto the grid.) As such, the well-formedness rules are straightforwardly formal.

The rules of inference, however, are a completely different story. There are three ways in which the set of usual first-order natural deduction rules is augmented to get the rules of *Hyperproof*. One is with the addition of the "short-cut" rules Taut Con, Log Con, and Ana Con, which license the derivations of tautological, logical, and analytic consequences respectively. The first two allow the derivation of a formula in one step what would otherwise require many using the usual natural deduction rules. The third can also be seen as a short-cut rule, this time compressing into a single rule the work that would be done by a series of more basic steps plus an axiomatization of the predicates like *RightOf*, *SameShape*, etc. From the formality point of view, these rules are somewhat similar to *Venn-I*'s unification rule – constructive, but not corresponding to a simple formal operation. This is relatively unproblematic, since the application of these rules is in principle reducible to a series of applications of more paradigmatically formal rules in the system.

The second class of additional rules consists of the heterogeneous rules of *Hyperproof*, which license inferences between diagrams and sentences. Now, unlike the Venn diagrams of Hammer's heterogeneous system, the diagrams of *Hyperproof* are hardly at all heterogeneous (their only linguistic elements being the possible use of constant symbols to label icons). This feature virtually guarantees that the heterogeneous rules of *Hyperproof* will fail to be formal, for the rules which license the derivation of diagrams from sentences (and vice versa)

---

[2] These rules are also implemented in Barwise and Etchemendy ( ), a logic textbook and courseware package covering more traditional material than *Hyperproof*. Here it is explicitly stated (p.61) that the Con rules are not "genuine" rules at all, but rather "mechanisms" that are convenient to use from time to time. Interestingly, the authors claim that the main difference between the system's genuine rules and these Con mechanisms is that the former are always checkable by the computer program. It is not obvious how this applies to Taut Con, however, since tautological consequence is of course decidable.

will not be of the kind in which formal components are recombined, simply because this is made impossible by the fact that the premises and conclusion share no formal components. For instance, with the *Observe* rule, one can derive $\exists x\,Cube(x) \wedge \forall y \neg LeftOf(y,x)$ from a diagram in which there are no icons to the left of some cube icon. It should be clear that no amount of formal manipulation of the diagram will ever result in the sentence; the diagram simply isn't made of the right kind of stuff! Nor is this application of the rule anything like that of the other non-constructive rules we've already encountered and which we've seen to be "harmless" from the formality point of view (since they license the derivation of *any* well-formed representation). Here, with *Observe* (like its dual, *Apply*), we have a rule which truly breaks the formal mold.

The same could be said of *Hyperproof*'s third class of non-standard inference rules. These ones allow *declarations* to be asserted as steps in a proof. For instance, the declaration *Assumptions Verified* is warranted through *CTA* (Check Truth of Assumptions), a rule which applies if the active diagram depicts a situation consistent with the active (non-closed) linguistic assumptions. The incorporation of rules like CTA represents a further deviation from the traditional logical template, in that *Hyperproof* proofs are not fully reducible to sequences of representations. We won't dwell on this particular innovation here.

We see, then, that *Hyperproof* is quite decidedly *not* a formal system in the sense of Sect.2.

4 Generalizing from the Data

In the previous section, five logical systems were tested for formality. I will now proceed to argue for some generalizations on the basis of these preliminary findings. Of course, such generalizations can (and should) be subject to further "empirical" testing by which other logical systems are investigated along these lines.

4.1 Purely Linguistic Systems

We looked at first-order logic, *the* paradigmatic logical system, and found it (unsurprisingly) to be formal. It is reasonable to expect that other, purely linguistic logical systems would also be completely formal. (I have in mind the standard modal logics, and even the newer relevance and dynamic ones.) But even here, some finer-grained distinctions may be motivated from the formal point of view. For instance, it may be interesting to distinguish various natural deduction rules of first-order logic in terms of their "complexity", by which one looks more closely at the basic formal operations invoked (like concatenation, detachment, and substitution). Alternatively, principled comparisons *between* linguistic systems may be motivated by looking more closely at the "constructivity" of the rules. (This seems especially relevant for relevance logics.) Such investigations, however, are clearly beyond the scope of this paper.

4.2 Purely Diagrammatic Systems

We got mixed results looking at Hammer's formulation of Peirce's Alpha-logic (which was straightforwardly formal), and Shin's formulation of *Venn-I* (which was less so). As would be expected of two-dimensional diagrammatic systems, their rules of inference invoke distinctly "diagrammatic" formal operations not to be found in linguistic systems.

But some of these rules fare better than others by formalist lights. For instance, we saw that the use of *Venn-I*'s rule of unification doesn't seem to correspond to any simple formal operation. In contrast to the step-by-step nature of other constructive rules, unification licenses a kind of formal "discontinuity" in the proof by which basic topological features of the premises aren't preserved. In fact, given the need to patch up the curve-erasure rule whose "natural" rendering would allow the derivation of ill-formed from well-formed diagrams, such discontinuities are necessary for demonstrating the completeness of the system. Here, then, is a specific example of how topological constraints of the plane interact directly with the naturalness and formality of the rules of a logic. The very structure of Venn diagrams makes it impossible to have a well-behaved, natural, unrestricted rule of curve erasure. The restrictions that are invoked to tame this rule in turn require complicated, non-formal applications of unification in order to preserve a desired meta-theoretic property. The problems are even more pronounced for other diagrammatic schemes, where expressive completeness isn't even possible (Lemon and Pratt   , Lemon   ).

Now, this tension won't be manifest in *all* diagrammatic logical systems, as demonstrated by the Alpha-logic. Indeed, as there are a myriad of ways in which spatial features of representations can be used to represent various aspects of a domain, we should expect variability in this respect as well. But it is quite striking how a system as apparently natural and relatively inexpressive as *Venn-I* has features which don't fit comfortably within a formalist framework.

4.3 Heterogeneous Systems and the "Interlingua" Debate

Recognizing the expressive limitations of any particular diagrammatic scheme, but also wanting to harness their supposed efficacy for particular reasoning tasks, one may see the development of heterogeneous logics as a way of exploiting the strengths of diagrammatic representations without suffering their weaknesses. A question worth asking, then, is: Is this strategy of "heterogeneity" compatible with that of formality?

We've seen how heterogeneous rules of inference may fall on either side of the formality spectrum. On the one hand, the heterogeneous rules of Hammer's Venn system are very much in accord with the formalist strategy. On the other hand, the heterogeneous rules of *Hyperproof* completely abandon it. We can even identify the particular formal feature which accounts for this difference. Hammer's Venn diagrams have a significant linguistic component (namely, set terms labelling the curves) which facilitates formal transitions between the different modalities. In contrast, *Hyperproof* diagrams do not. It appears then, that heterogeneity and formality can be reconciled through the use of some sort of syntactic bridge between representations, an "interlingua" which mediates various systems.

Barwise and Etchemendy (, p.218) argue against the need for an interlingua, and even its desirability. They claim that the *semantic* mappings which link the various diagrammatic schemes to a single, common target domain suffice to allow for their interaction. There is no need to supplement this with some explicit formal analogue; in fact, any attempt to do so is likely to be self-defeating: "To combine these into one system, the mappings from syntax to semantics would have to be complicated to the extent that they lose much of their homomorphic character. The more complicated the mapping becomes, the closer we come to having a notational variant of a linguistic scheme of representation, thereby losing the very representational advantages inherent in a homomorphic system."

It is not hard to imagine that any sort of interlingua for *Hyperproof* would turn out to be quite clumsy. But this clumsiness should not make us forget the skeptical challenge that formality was supposed to answer. Recall that the challenge was partially to justify a stopping-point whereby we no longer need to decompose an inference in a proof into finer-grained ones. On the basis of our investigations in Sections 2 and 3, we see that formality can offer a natural response to this challenge *only in certain cases*; namely, when the representations employed are linguistic, and occasionally when they are diagrammatic. For certain diagrammatic schemes, formality ceases to ground a promising strategy, and appears absolutely hopeless for any robust heterogeneous system.

So where does this leave *Hyperproof*? What features of it can be used to meet the skeptical challenge? I would argue that the fact that its representation schemes are linked to a common domain is at best a necessary condition for there being any prospect at all for supplying an answer to this challenge. But it can't provide the complete answer by itself. One way to see this is to imagine *Hyperproof* without its computer implementation; that is, as a logical system presented only as ink on paper. Would we then see it as a bona fide logical system, as the kind of framework whose proof-steps are in no need of further justification? Recall that this paper-version of *Hyperproof* would still have the Con rules in full strength; its Observe rule would license *any* linguistic consequence of a diagram; its CTA rule would license declarations that a particular diagram is consistent with the set of active assumptions whenever this is indeed the case. It seems clear to me that we would consider such rules to be much too coarse. Cheats even. And yet, the semantics of this system would be identical to that of the original computerized one. So something independent of the semantics needs to be invoked in order to meet the challenge in a satisfying way.

Such considerations suggest that if *Hyperproof* is to be vindicated at all, it's by its *programming*. Having forsaken formality, something else needs to be given a privileged epistemic status if the skeptical challenge is to be met. It appears that this status must be conferred onto the algorithms that underlie the implementation. Unlike the use of formality, this in effect gives epistemic privilege to a process to which the user of the logic does *not* have direct access.

Adapting Hammer's imagined argumentative situation, we could ask ourselves what would have to be the case for the prover to successfully appeal to a logical system like *Hyperproof* in order to face the skeptic's challenge. Here it seems that a *third* character, the "computer", is made to play roughly the role that was initially assigned to a logical system's formality. It is only in virtue of the mechanical work done "behind-the-scenes" by the computer (and the skeptic's trust in that work) that the spelling-out process spurred on by the skeptic's challenge can be made to stop at the level of detail enforced by *Hyperproof*. Notably, in this situation, it's not quite the computer-check*ability* of the rules which is taken to satisfy the skeptic, but rather the fact that particular applications of these rules have actually been check*ed* in the proof.

So here, then, is a principled way in which certain diagrammatic and heterogeneous systems differ from linguistic ones; namely, in the kind of justifications which can be offered in the face of a particular skeptical challenge. Far from reflecting a conservative approach to the study of non-linguistic reasoning, diagrammatic and heterogeneous logical systems help to expose the limitations of a certain kind of epistemic strategy.

References

[1994]Barwise, J., Etchemendy, J.: Hyperproof. CSLI Publications, Stanford (1994)
,
[1995]Barwise, J., Etchemendy, J.: Heterogeneous Logic. In: Glasgow, J., Narayanan, N., Chandrasekaran, B. (eds.): Diagrammatic Reasoning: Cognitive and Computational Perspectives. AAAI/MIT Press (1995) 211–234 , , ,

[1999]Barwise, J., Etchemendy, J.: Language, Proof, and Logic. CSLI Publications, Stanford (1999)

[1994]Hammer, E.: Reasoning with Sentences and Diagrams. Notre Dame Journal of Formal Logic **35** (1994) 73–87 ,

[1995]Hammer, E.: Logic and Visual Information. CSLI Publications and FoLLI, Stanford (1995) , , ,

[1926]Hilbert, D.: On the Infinite. Translated in: Benacerraf, P., Putnam, H. (eds.): Philosophy of Mathematics, Selected Readings. Cambridge University Press (1987) 183–201

[1997]Lemon, O.: Book review of Hammer, E.: Logic and Visual Information. Journal of Logic, Language, and Information **6** (1997) 213–216

[to appear]Lemon, O.: Comparing the Efficacy of Visual Languages. In: Barker-Plummer, D., Beaver, D., van Benthem, J., Scotto di Luzio, P. (eds.): Logic, Language, and Diagrams. CSLI Publications, Stanford (to appear)

[1997]Lemon, O., Pratt, I.: Spatial Logic and the Complexity of Diagrammatic Reasoning. Machine Graphics and Vision **6** (1997) 89–108 ,

[1996]Luengo, I.: A Diagrammatic Subsystem of Hilbert's Geometry. In: Allwein, G., Barwise, J. (eds.): Logical Reasoning with Diagrams. Oxford University Press (1996) 149–176

[to appear]Scotto di Luzio, P.: Patching Up a Logic of Venn Diagrams. In: Copestake, A., Vermeulen, K. (eds.): Selected Papers from the Sixth CSLI Workshop on Logic, Language, and Computation. CSLI Publications, Stanford (to appear)

[1994]Shin, S.-J.: The Logical Status of Diagrams. Cambridge University Press (1994)

Distinctions with Differences: Comparing Criteria for Distinguishing Diagrammatic from Sentential Systems

Keith Stenning

Human Communication Research Centre, Edinburgh University
k.stenning@ed.ac.uk

Abstract. A number of criteria for discriminating diagrammatic from sentential systems of representation by their manner of semantic interpretation have been proposed. Often some sort of spatial homomorphism between diagram and its referent is said to distinguish diagrammatic from sentential systems (e.g. Barwise & Etchemendy 1990). Or the distinction is analysed in terms of Peirce's distinctions between symbol, icon and index (see Shin (forthcoming)). Shimojima (1999) has proposed that the sharing of 'nomic' constraints between representing and represented relations is what distinguishes diagrams. We have proposed that the fundamental distinction is between direct and indirect systems of representation, where indirect systems have an abstract syntax interposed between representation and represented entities (Stenning & Inder 1994; Gurr, Lee & Stenning 1999; Stenning & Lemon (in press)).
The purpose of the present paper is to relate the distinction between directness and indirectness to the other criteria, and to further develop the approach through a comparison Peirce's Existential Graphs both with sentential logics and with diagrammatics ones. Peirce's system is a particularly interesting case because its semantics can be viewed as either direct or indirect according to the level of interpretation. The paper concludes with some remarks on the consequences of sentential vs. diagrammatic modalities for the conduct of proof.

1 Introduction

Several fields have recently identified understanding how diagrams communicate as an important goal: AI, computer science, education, human computer interaction, logic, philosophy, psychology .... To understand how diagrams communicate requires an understanding of how they are assigned their meanings – their semantics – and in particular how their semantics is related to the far more intensively studied semantics of linguistic representations. Any theory of diagrammatic semantics must answer the prior question which representations are diagrammatic, and which sentential. In fact, an insightful answer to this second question is likely to go some way to explaining the nature(s) of diagrammatic (and linguistic) semantics. And insightful semantic theories are a necessary base for the empirical study of the cognitive properties of diagrams.

M. Anderson, P. Cheng, and V. Haarslev (Eds.): Diagrams 2000, LNAI 1889, pp. 132–   , 2000.

We have strong pre-theoretical intuitions about which representations are diagrammatic and which are linguistic. What these intuitions do not make clear is whether this is a discrete boundary, a smooth dimension, or some complex space structured by several dimensions which will allow that actual systems are composite. Here we will argue for the last option, and also that when we have a deep theory of what distinguishes diagrams and languages that it will show that our pre-theoretical intuitions were just misleading about some cases. Theory will reorganise intuitions, as so often happens.

There are various recent accounts of what distinguishes the semantics of diagrammatic representation systems from the semantics of sentential languages. Spatial homomorphisms; type and token referentiality; Peirce's semiotic distinctions between symbols, icons, and indices; nomic vs stipulative constraints; and directness and indirectness of interpretation have all been proposed as dimensions for distinguishing diagrammatic from sentential representations. The first purpose of this paper is to clarify the relations between the various accounts of the semantic differences.

The second purpose is to elaborate the theory that directness vs. indirectness of interpretation is the fundamental distinguishing characteristic, by exploring two pairs of representation systems. Peirce's Existential Graphs (EGs) is an example graphical system equivalent to a well understood propositional system—propositional calculus (PC). Peirce's existential graphs are also compared with Euler's circles (ECs), a well understood, directly interpreted diagrammatic system.

This pair of pairs of contrasting systems illustrate the importance of articulating a theory of 'diagrammaticity'. First it will be argued that EGs are diagrammatic only in a limited sense, and that this sense can be made precise in terms of directness and indirectness of interpretation, combined with a careful consideration of subtly shifting ontologies of interpretation.

The third purpose of the paper is to briefly explore consequences of direct and indirect semantics for the *use* of representations, especially in the justification of reasoning ie. in proof. The comparison of ECs, EGs and PC suggests a distinction in the ways that representations are used in reasoning between *discursive* as opposed to *agglomerative* modes. Only indirectly interpreted representations can be used discursively; and although indirectly interpreted representations can be used agglomeratively to some extent, there are distinct limits to such use. Directly interpreted representations have to be used agglomeratively and this has consequences for their use in proofs.

2 Diagrammatic and Sentential Semantics Compared

2.1 Spatial Homomorphisms

Barwise & Etchemendy 1990 propose that spatial homomorphisms between diagrams and their referents is what distinguishes diagrammatic semantics from sentential semantics. Diagrams consist of of vocabularies of types of icon whose spatial arrangement on the plane of the diagram bears some spatial homomorphism to what is represented. The spatial relations between town icons on a veridical map, for example, bears a homomorphism to the spatial relations between the towns in the real world. This is not obviously true of the spatial relations between the names of the towns occurring in a written description of the spatial relations between the cities. This is the basis of Barwise and Etchemendy's theory of the difference between diagrammatic and sentential semantics.

Etchemendy (personal communication) uses this framework to derive a common distinguishing characteristic he calls type- vs. token-referentiality. In type-referential systems (a paradigm case would be formal languages), repetition of a symbol of the same type (say of the term x or the name 'John') determines sameness of reference by each occurrence of the symbol. Obviously there are complexities such as anaphora and ambiguity overlaid on natural languages, but their design is fundamentally type-referential. Diagrams, in contrast are token-referential. Sameness of reference is determined by identity of symbol *token*. If two tokens of the same type of symbol occur in a single diagram, they refer to distinct entities of the same type (e.g. two different towns on a map). Token referentiality is a particular case of spatial homomorphism: the identity of referents is represented by the spatial identity of symbol tokens: distinctness of referents by spatial distinctness of tokens. We shall see below that the type- token-referential distinction provides a particularly easily applied test of systems.

2.2 Symbols, Icons and Indices

Peirce not only made one of the original inventions of the first-order predicate calculus in both diagrammatic and sentential formalisms, but also defined three classes of signs: symbols, icons and indices . Symbols are conventionally related to their referents. Icons are related though their similarity to their referents. Indices are related to their referents by deixis—pointing. Peirce was quite clear that systems of representation could be composite. He described, for example, elements of sentential languages which were iconic and others which were indexical, even though the dominant mode of signification of sentential languages is symbolic. Diagrams, predominantly iconic, also contain symbolic elements such as words, often indexically related to their reference as labels by spatial deixis.

This framework of kinds of signification has been used to distinguish diagrammatic from sentential semantics. Diagrams are held to be fundamentally iconic. Languages fundamentally symbolic. Both can have indexical features. Most obviously, languages have terms which are deictic—'this' and 'that' have common deictic uses. But diagrammatic representations also have indexical elements. The 'YOU ARE HERE', with accompanying arrow, on the town map displayed at the station, contains a piece of deictic language, but contains it as

---

[1] The relation between Peirce's term and modern uses of 'icon' such as Barwise and Etchemendy's is not transparent. We take up this point below.

a deictic diagrammatic element which is dependent both on the spatial position of the sentence ('you are here'); the arrow; and the spatial position of the map.

Spatial homomorphisms between diagram and represented world would presumably qualify as iconic features of representation systems—likenesses. But deictic relations between representation and referent could also rely on homomorphisms between the two. It is notoriously difficult to make these distinctions work in detail. One pervasive problem is with specifying the similarity which supposedly underlies iconic representations. This problem interacts with vagueness about whether the similarity works to define the meaning of the sign, or merely helps us to remember the meaning of the sign. Etymology, for example, often leads to similarities between the meanings of components of a word, and the meanings of the word itself. These are similarities mediated through conventions, but that is only one problem. They are similarities which may help us remember the meanings of words, but they do not in general define what those meanings are. False etymologies rarely change the meanings of words. Similarly, many diagrammatic road signs have useful mnemonic relations to their referents (a particular curved pattern means a bridge; a pair of the same curves means a bumpy road surface) but are these mnemonic or semantic relations? One might go through life thinking that the common sign for female toilets is an icon for a long-life light bulb, or alternatively having no 'etymology' for the pattern at all, but without being the least unclear about the meaning of the sign, which is, surely, therefore, just as symbolic as the word 'ladies' sometimes also posted on the door.

The overall impression of Peirce's approach to distinguishing diagrams and languages is of insightful distinctions which are hard to weld into a systematic account of the semantics of representation systems. They illuminate a modern understanding of semantics, but they also predate it.

2.3 Nomic and Stipulative Constraints

Shimojima 1999 adopted the goal of synthesising parts of several accounts of the distinction between diagrammatic and linguistic representations in terms of nomic vs stipulative constraints . Nomic constraints are the result of natural laws: topological, geometrical or physical laws. Stipulative constraints are due to stipulative rules such as syntactic well-formedness conditions. The central idea is that diagrammatic representations have nomic constraints (as well as possibly stipulative ones), but linguistic representations have only stipulative constraints.

A paradigmatic example of a nomic constraint is the topological properties of the closed curves that appear in many diagrammatic systems. Containment within a closed curve is a transitive relation. When it is used to represent a transitive relation, a nomic constraint is established.

---

[2] Shimojima actually uses the term 'graphical' rather than 'diagrammatic' and entertains a wider range of representations. I will continue with 'diagrammatic' since this narrowing accurately expresses a sub-class of Shimojima's claims. As we shall see, there is a case for calling EGs graphical but not diagrammatic.

I obviously applaud Shimojima's goal of synthesising an explanatory account integrating the several dimensions of difference that have been pointed out in the literature, but the distinction between nomic and stipulative constraints has its own problems. One is in deciding what is a formal stipulated property, and what a natural topological property. Another is that the distinction between diagrams and languages needs more articulation than a single distinction can provide. Our analysis of EGs and ECs below explores these issues.

Shimojima complains that accounts of limitations on the expressiveness of diagrammatic systems in terms of directness of interpretation cannot *explain* their constraints because computational complexity theory only applies to sentential logics. This seems to be a misunderstanding. Computational complexity theory explains the tractability of reasoning in terms of the size of the search space that a theorem prover has to operate over to find derivations for a class of results. This framework is as applicable to graphical theorem provers as to sentential ones, and explains the same inverse relation between expressiveness and tractability. The fact that the theorem provers for directly interpreted systems are generally integrated with the representation systems themselves should not obscure this point.

2.4 Directness and Indirectness of Interpretation

Stenning and collaborators have proposed elsewhere (Stenning, Inder & Nielson 1995; Gurr, Lee & Stenning 1998; Stenning & Lemon (in press); Stenning (forthcoming)) that the fundamental distinction between diagrammatic and sentential semantics is between direct and indirect interpretation. Sentential languages are interpreted indirectly because an abstract syntax is interposed between representation and the referenced world. An abstract syntax is defined on a single spatial relation—concatenation. In written language, concatenation defines the 'scanning sequence' (e.g. left-to-right-top-to-bottom). In spoken language concatenation is a temporal relation. The interpretation is indirect because the significance of two elements being spatially (or temporally) concatenated cannot be assessed without knowing what abstract syntactic relation holds between them. So written text is interpreted spatially (different spatial arrangements of words have different meanings), but not directly (particular spatial arrangements such as immediate contiguity) do not have uniform meanings.

This semantic property of languages is the fundamental characteristic which underlies all extant accounts of linguistic semantics. It is so fundamental that it is mentioned in the first definition of a language and thereafter taken for granted. The presence or absence of abstract syntax has the virtue of being well defined for sentential presentations of languages, though there are issues about how exactly to extend its application to two-dimensional languages, as we shall see. We will argue that many other important semantic properties flow from it. But as a criterion, it leads to a classification of actual representation systems which goes beneath the surface of what is 'obvious' about which things are diagrams and which language.

Some one dimensional systems which string together tokens of an alphabet are conventionally thought of as languages (finite state or regular languages) but have no abstract syntax. On our classification these systems have direct semantics (the stringing together of elements generally means that the denotata of two immediate neighbouring tokens occur in a particular temporal relation). Notice that here the interpretation of the spatial relation is uniform wherever it occurs, unlike concatenation in a language with abstract syntax. So on our classification these systems are semantically 'diagrammatic'. It is a moot point whether the constraints embodied in finite state languages are nomic or stipulative.

Conversely, there are representation systems conventionally thought of as graphical which our classification would treat as semantically close relatives of sentential languages, with the attendant indirect semantics. That is they have a semantics which interposes an abstract syntax on a two-dimensional version of a spatial concatenation relation between the representations and their interpretation. Examples are semantic networks and conceptual graphs, which derive from Peirce's existential graphs.

2.5 The Criteria Related

How are these different criteria for distinguishing diagrammatic from sentential semantics related? Do they pick out the same classes of system? Can any sense be given to one of them being fundamental? Our main purpose here is to draw out some of the interelations between these ways of distinguishing diagrams from languages.

With regard to homomorphisms, it is well known that sentential systems display homomorphisms between their sentences and the domains on which they are interpreted. This property is generally described as compositionality, and is most evident in the correspondence between syntactic and semantic rules in the definition of formal languages. In natural languages, compositionality is obscured, but is nevertheless generally assumed to hold at some fundamental level.

If this homomorphism exists, then homomorphism is not a good distinguishing characteristic for diagrams? One might rescue homomorphisms as the *differentia* by claiming that the compositionality of sentential languages is not a *spatial* homomorphism. However, even this is problematical. Written languages are interpreted on the basis of a spatial relation—namely concatenation. Concatenation is a spatial relation which gives rise to a homomorphism, presumably a spatial homomorphism?

The proposal to use directness of interpretation as a distinguishing characteristic can be viewed as a way of better defining what is to count as a spatial homomorphism. Direct interpretation precludes an abstract syntax; indirect semantics interposes a layer of abstract syntax between representation and interpretation. It is not that there is any lack of homomorphism in sentential languages, but it's recovery requires access to the syntactic interpretation of concatenation. Once the spatial relation of concatenation has played its part in defining a syntactic interpretation, it plays no further part in determining semantic interpretation.

Type- and token-referentiality are complexly related to directness and indirectness. Token-referentiality is a result of direct interpretation. The token icons of a directly interpreted diagram are placed on a spatial field (the paper) and their identity is an identity of position. Similarly, the identity of token words placed on a spatial field (the paper) is an identity of position. Direct interpretation means that some spatial relation between *all* icons must be interpreted as this identity relation. If any spatial relation between token icons other than identity/distinctness of tokens were chosen to represent identity/distinctness of their referents, the logical properties of the spatial relation would conflict with the logical properties of identity—the represented relation.

Only where there are abstract syntactic relations interposed between icons (here we usually call them words) can this conflict be avoided. The presence of an abstract syntax means that there has to be 'punctuation' of formulae into syntactic units (sentences) prior to interpretation because syntactic units are semantically discontinuous. This punctuation 'insulates' relations between terms that occur in different sentences and clauses. The relation of identity can be named as many times as is required. Identity then becomes just another named relation expressed by token words in the designated syntactic relation. Repetition of tokens of terms across syntactic units is then possible without generating spurious assertions of the non-identity of identicals.

The abstraction properties of diagrammatic and sentential systems discussed by Stenning & Oberlander 1995 are also a consequence of these differences made by abstract syntax. Specificity of directly interpreted systems, and its attendant weakness of expressiveness of abstractions, arises from the fact that all the icons in the diagram's spatial field are automatically related to each other by all the interpreted spatial relations. The insulation of syntactic units one from another ensures that indirect systems can be indefinitely expressive.

The distinction of sentence and discourse levels which comes with abstract syntax has implications for the ways sentential systems are used in reasoning. Their use is necessarily discursive, whereas diagrammatic systems are used agglomeratively. This distinction is defined below in relation to Peirce's EGs. In the last section we will examine its implications for the justification of reasoning in proof.

How do these properties relate to Peirce's iconic, symbolic and indexical ways of signifying? Modern uses of the term 'icon' in discussing diagrams is derived from computer icons and conflicts with Peirce's use. The semantic function of computer icons is most like that of words (the archetypal symbol). The shape and pattern of the symbol may or may not be successful at reminding us of its referent, but that mnemonic value is incidental to the semantic relation between icon and referent. Frequently, such icons consist at least in part of words, and their semantic function is no different whether they are words or patterns. The iconic (in Peirce's sense) aspect of computer icons is quite distinct, and resides in the arrangement of icons with respect to other diagrammatic elements. Archetypally, an icon's being inside a closed curve representing a directory, means that the file denoted by the icon is in the directory denoted by the curve. It is this correspondence of spatial relation in the diagram to the represented relation in the world that makes that part of the system iconic. It is the lack of any concatenation relation defining an abstract syntax that makes the relation direct.

So the computer icon's relation to its referent (the file) is in Peirce's terms, symbolic; its relation to the representation of the directory, and through it to the directory itself, is iconic. It is this relation which also represents a direct spatial homomorphism. Peirce was quite clear that syntactic relations in a language with an abstract syntax could be iconic—one example he cites is of algebraic relations representing transitivity iconically. He was aware that homomorphisms can work through an abstract syntax. But we should appreciate that this limits the usefulness of Peirce's terms for giving articulation to the distinction between diagrammatic and sentential systems.

How is the distinction between nomic and stipulative constraints related to directness of interpretation? Indirect systems with their abstract syntax are assumed by Shimojima to have stipulative constraints: direct systems, nomic constraints. But nomic and stipulative are properties of the *origins* of constraints, and origins are complex matters that take us far beyond semantics. Besides, with indirect systems there are two levels of constraint—constraints on syntactic interpretation of concatenation; and constraints on semantic interpretation of syntax. One or both of these constraints might have natural or nomic origins. Which is to count? Comparing EGs and ECs will provide example quandries.

In summary, directness and indirectness of interpretation, with its attendant absence or presence of an abstract syntax defined on a concatenation relation, gives us a way of understanding the relations between the various proposals for criteria for distinguishing diagrammatic and linguistic representations. As we noted above, this theoretical distinction reveals considerable complexities in classifying real representation systems, and does not always yield the categorisation which pre-theoretical intuition suggests. The next section takes EGs as an illustration of applying the approach.

3   Peirce's Existential Graphs

Peirce's existential graphs (EGs) provide an illuminating case study for these various approaches to distinguishing the modes of semantic interpretation. Comparing Peirce's EGs with the equivalent sentential propositional calculus, and with other diagrammatic representations of fragments of logic clarifies the different accounts. What follows draws heavily on Shin's (1998) recent work which has done much to clarify the status of EGs by providing more perspicuous algorithms for translating between EGs and sentential calculi. We will restrict ourselves to the propositional calculus fragment of EGs (the alpha system), though the same kind of analysis is extensible to the beta system that corresponds to the predicate calculus.

EGs consist of proposition symbols (P, Q, R ...) inscribed on the 'sheet of assertion' (the paper), possibly enclosed within patterns of 'cuts' (closed curves). Unenclosed propositional letters on the sheet of assertion are simply asserted.

Enclosing one of more letters within a single cut, asserts that the enclosed propositions are not all true (the negation of their conjunction in sentential terms). Correspondingly, enclosing a letter already enclosed in a single cut asserts the negation of the already negated proposition, logically equivalent to simply asserting the proposition. Figure   compares an EG with an Euler's Circle representation of the sentences $if P \rightarrow Q, if Q \rightarrow R$.

This EG is perhaps most easily read from inside out. Consider first the inner two circles and their two propositions Q and R—what Peirce called *a scroll* pattern. The outer of the two circles is read as an outer negation, so the sentence equivalent becomes *Its not the case that Q and not R* which is equivalent to $Q \rightarrow R$ (see the remarks on Peirce's scroll readings below).

The Euler diagram is read as asserting that the type corresponding to each minimal sub-region of the diagram *may* exist, but that types not represented by any sub-region do not exist. For example, the ground of the EC diagram between the outer circle and the edge of the square field is a sub-region which represents situations in which neither P nor Q nor R is true. Such situations may or may not exist if the diagram is true. One type of situation which cannot exist if the diagram is true is that in which P is true and Q and R are false, because there is no corresponding region in the diagram—hence the equivalence to the same conditional as the EG.

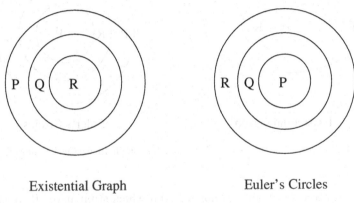

Existential Graph   Euler's Circles

Fig. 1. EG and EC representations of $if P \rightarrow Q, if Q \rightarrow R$

At a glance the two representations are similar except for the arrangement of the letters. One might think that these were rather similar schemes save for the convention for the representation of the direction of transitivity.

The correspondence is in fact completely illusory, as is illustrated in Figure . Peirce's system is at least apparently type referential—note the occurrence of two tokens of 'P'. In fact, letters in the two systems have entirely different semantic functions, and are even formally distinct. The Peirce letters are propositional symbols comparable to the symbols of the propositional calculus formulae. The

Euler letters are labels for circles. 'P' labels the circle enclosing all situations in which the proposition P is true, and excluding all situations in which P is false. which is why the right hand diagram is ill-formed. No two distinct circles can be the same circle, and so they cannot have the same label, nor be labeled with the negation of the same proposition as any other circle. The identity criteria for circles are identity of position—circles are token-referential. Similarly, it is entirely conventional whether Euler circle labels are placed inside, outside, or on their circles, whereas the positioning of letters inside or outside cuts is significant and placing them on cuts is ill-formed.

In EGs, different letters (say P and Q) can represent either the same or different propositions, just as in PC. In ECs distinct labels on different circles represent different propositions. EG cuts can contain any number of letters, whose propositions' conjunction is thereby negated. EG cuts which intersect are ill-formed, whereas Euler circles can intersect freely. EGs are type referential—each occurrence of the same type of propositional letter (say P) denotes the *same* proposition. ECs are token-referential—distinct letters (and circles) denote distinct propositions.

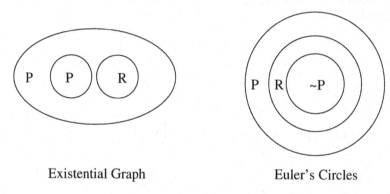

<table>
<tr><td>Existential Graph</td><td>Euler's Circles</td></tr>
</table>

Fig. 2. EG representation of $if P \rightarrow (P \wedge R)$ with ill-formed Euler diagram

Which of these constraints is nomic? And which stipulative? If we stipulate the constraints on Euler circle labeling just mentioned, then it is, for example, a nomic (topological) property of the system that it is self-consistent (see Stenning & Oberlander 1995). If we fail to make this stipulation (and e.g. allow the labeling in Figure) then the system is not self-consistent. Is ECs' semantic constraint of self-consistency nomic or stipulative?

What are we to make of EGs? They are superficially diagrammatic (they are essentially 2-D) yet they are type-referential. Remember EGs do have alternative notations, one of which is sentential (consisting of bracketings of strings of letters: $(P(P)(R))$ in the case of the EG in Figure) and the other is in node-and-link notations of directed acyclic graphs (see Sowa 1984). Something must be wrong if token-referentiality is to be a criterion of diagramhood? And do EGs have an abstract syntax? Is the constraint against intersection of cuts a stipulative constraint or a nomic one? Is their interpretation direct or indirect? And what can the answers to these questions tell us about what is special about EGs? Let us first look more closely at why EGs are interesting.

As Shin (1998; forthcoming) has done much to deepen our understanding of EGs. Structurally, the most striking feature of EGs is that logical particles disappear. There are no connectives in the alpha graphs (shown here) and no quantifiers in the beta graphs. A simple way of thinking about the lack of connectives is to think of EGs as encoding a single-connective version of the sentential calculus. Schaeffer stroke (a single binary connective interpreted as *not both of (P, Q)*) is an adequate connective to express all truth functions so we might try thinking of cuts as equivalent to Schaeffer strokes. As Shin points out, this is a helpful hint, but neither quite right, nor the whole truth. Schaeffer stroke is syntactically a binary operator and Peirce's cuts can contain any number of predicate letters.

More significantly, Shin has shown that EGs are most naturally understood as systematically susceptible of multiple logically equivalent readings in sentential calculi with connectives. These alternatives are often much more comprehensible than the ones derived from the particular outside-in reading scheme promoted by Peirce—his *endoporeutic* reading method. For example, the reading given above for the EG in Fig is in negation normal form; but Peirce also defined a 'scroll' pattern which read the same sub-diagram (a cut enclosing X, itself enclosing a cut enclosing Y is a scroll) directly into equivalent conditional form $(P \rightarrow Q)$. Shin shows that crucial visual properties of EGs (even- and odd-enclosedness) can afford a variety of direct translation algorithms which can start by focusing on various different cuts in an EG. These visual properties also make the inference rules more intuitive. On this basis, she defines an algorithm for multiple equivalent readings of EGs—endoporeutic, negation normal form; conditional normal form etc. So EGs are not merely a calculus with one connective (Schaeffer stroke) but are somehow more radically different.

EGs represent classes of connectiveful sentential readings, in classes of calculi differentiated by their connective vocabularies. This multiplicity offers new possibilities for reasoning. The fact that two sentences are translations of the same EG provides a demonstration that they are logically equivalent.

As well as the presence or absence of connectives, there are also processing differences between EGs and sentential calculi—differences in the ways that we usually envisage using EGs and sentential calculi. These differences can themselves be separated into two kinds: those involving processing at the level of single sentences (or graphs) as opposed to processing at the level of multiple sentence (or graph) discourses.

At the sentence level, we generally envisage sentences being written from left-to-right. Two dimensional EGs have no such fixed order of drawing, so the 2-D representations of EGs obliterate their production history. Logically, we could write sentences in different sequences (and within word-processors may sometimes we do so). In general our production of written sentences bears the heavy mark of the acoustic origins of language, and this carries through to our interpretation algorithms for them, which are left-to-right. But we must remember that Shin's multiple reading algorithms for EGs are as applicable to the EG sentential notation as to the graphs. Shin's multiple reading algorithms are a feature of translation from connectiveless to connectiveful calculi rather than of 2-D notation . The multiple sentential translations of EGs are like the multiple categorial grammar parses of unambiguous natural language sentences—all semantically equivalent (see e.g. Steedman 2000). Multiplicity arises in both cases through the erasure of some history of syntactic generation. We should be wary of assuming that multiple readability is a feature of diagrammatic representation.

Now to the discourse level. Sentential representation systems use sentences discursively. When a conclusion is derived from earlier sentences, it is *rewritten* on a new line of the proof. In contrast, EGs (and their 2-D equivalents, conceptual graphs) are sometimes presented in an *agglomerative* manner . At any time, the assumptions operative in an episode of reasoning are represented by a single EG ie. the whole of what is drawn on the sheet of assertion. If new assumptions are made, they are represented by being added to this single EG. Inferences are made by modifying this single EG—an example would be the erasure of a double cut surrounding a letter. Notice that this operation is equivalent to the rewriting of $\neg\neg X$ as X. But therein lies the difference. Whereas derivation in the sentential system is represented by *rewriting* a new conclusion on a new line of a proof, the derivation in the EG system is represented by a modification of an existing EG—that is to say, not represented at all except in its result. The history of derivation is erased.

EGs do not have to be used in this way in derivations. Shin uses EGs discursively, redrawing the modified diagram that results from inference rule application. A proof then consists of a sequence (or possibly a tree) of diagrams exactly analogous to the sequence or tree of sentences that make up a proof. It is an interesting question how generally the agglomerative mode could be used with EGs or with sentential systems? Instead of using separate representations of the several assumptions, we might conjoin them all into a single sentence; and instead of rewriting each result of inference rule application, a single sentence could be modified by erasures and additions—given the technology we have for erasing and inserting structure. But there is a real semantic problem with such uses. Proof history has to be maintained because the legitimacy of later inferences depends on it. Inference rules invoke their justifications by providing the line numbers of their input propositions. In agglomerative mode, there is no guarantee that the required structure still exists.

---

[3] Strictly speaking, the relation between graphical and sentential EGs is also not one-to-one because there are many ways of linearising an EG diagram. But this is a different level of difference than that between an EG and its multiple readings in sentential calculi with connectives.

[4] Stenning 1992 shows that the concept of an agglomerative representation provides a much needed way of characterising the kernel of empirical content in the theory of mental models by showing how memory and processing differences might set them off from sentential systems, where semantic differences fail to differentiate them.

In contrast, directly interpreted diagrammatic systems *have* to be used in an agglomerative fashion. For example, Euler diagrams are used by adding a new circle to the accumulating diagram each time a new premiss introduces a new predicate. Inferences are made by 'reading off' this agglomerated diagram. All the history of the reasoning that is relevant to future inferences is preserved in this diagram. But no inferences can be made without combining diagrams. Merely writing a sequence of diagrams of each premiss and conclusion gives no method of inference. Inference is by agglomeration and subsequent reading off.

So we have now noted three dimensions of difference between EGs and conventional logical calculi; EGs eliminate connectives; they erase the history of graph production; and they at least invite agglomerative use even though it is not generally feasible. Directly interpreted diagrams, in contrast, *require* agglomerative use. How do these differences bear on directness of interpretation? In order to answer this question we need to go deeper into the processing of EGs, and in particular into differences between the one-dimensional and two-dimensional notations for EGs, and consequences for their processing.

Compare the EG in Figure   and its sentential equivalent repeated here: $(P(P)(R))$. Taking such a simple example formula reveals a striking correspondence. Drawing lines connecting the top of each left parenthesis to its corresponding right parenthesis and the same for the bottoms of each parenthesis, turns the string of symbols back into the EG diagram which it translates. Formally, the sentential language consists, as such languages do, of strings of symbols drawn from a vocabulary—P, R, (,), ...—and, as always, some of these strings are well formed (the one just quoted as equivalent to the EG diagram) and some not (e.g. $PR)(P)$. This language, as all such languages, clearly has an abstract syntax. A phrase structure grammar would be easy to supply. No finite state grammar would be adequate because it could not capture the long distance dependencies between the left and right parentheses. The existence of ill-formed strings suggests stipulative constraints.

What of the 2-D diagram EGs? What is their structure? They are composed of cuts and letters and the formal definition of 2-D EGs is simple enough, but it is made in terms of different elements than the sentential EGs. Cuts are not vocabulary members of a finite alphabet. And the constraints on legitimate arrangements of cuts and letters which form EGs are not statable in the usual grammatical formalisms. Cuts must not intersect each other. Letters must be wholly within or wholly outwith cuts. By simply putting back the 'ligatures' which connect the corresponding pairs of parentheses in the sentential version of EGs, the diagrams provide a concrete representation of their abstract syntax. The transitivity of containment in a closed curve provides a direct representation of the hierarchical syntactic relations in the EG's structure. But observe what the objects of this interpretation are. They are not the propositions which we normally think of as the semantic interpretations of sentences. They are, in fact the sentences and their constituents themselves—linguistic objects. EG diagrams represent directly the syntactic structures of formulae—their trees. EG sentences represent directly only *strings* of symbols—well-formed or not—and an abstract syntax has to be defined on these strings to sift the well-formed from the ill-formed, and to define structures for the well-formed.

So EG graphs directly represent sentences, but only indirectly represent the propositions that those sentences represent. EG sentences indirectly represent propositions, and indirectly represent what those propositions represent. So we can take EG graphs to be diagrams, but only if we understand that they are diagrams of sentences. Euler diagrams, in contrast, represent directly the propositions which they represent.

The topological relation of proper containment by closed curves gives rise to transitive inferences in both EGs and in ECs. Is this a nomic constraint or a stipulative constraint? How can Shimojima's framework do justice to the fact that graphical EGs are diagrams of sentences, but not diagrams of their meanings? It seems that a more structured theory is required, and the concepts of directness and indirectness can supply it.

This more analytical view of levels of objects in distinguishing direct from indirect interpretations also answers the question whether EGs are type- or token-referential. If occurrences of the proposition letters P, Q, R ... in EG diagrams are taken to represent token occurrences of their proposition letters, then the EGs are token-referential. The leftmost 'P' in Figure   is a different thing than the rightmost 'P' in this ontology, and the two distinct occurrences of a 'P-icon' stand for two distinct *occurrences* of the proposition—just like two cities on a map. But at the level at which the two occurrences are understood as representing a single proposition, the EG is type-referential.

Studying Peirce's EG system clarifies some of the relations between alternative ways of distinguishing what is diagrammatic from what is sentential. Approaching the question through the concept of directness and indirectness of interpretation forces us to distinguish several distinct dimensions of difference between EG graphs and conventional sentential calculi, as well as between EG graphs and EG sentential notation: the presence or absence of connectives; the directness of interpretations at several levels; the erasure or not of history of single formula production; and the erasure or not of the history of proof.

On the significance of the presence or absence of connectives, directness/indirectness tells us little. EGs provide both one-dimensional indirect and two dimensional direct notations for connectiveless calculi. Shin's multiple translations of EGs into connectiveful languages are a feature of EGs connectivelessness, not of what diagrammaticity they have. This seems an important dimension of representation which has been almost entirely ignored. The resulting idea of using parsing or translation mechanisms as inference mechanisms might be cognitively significant. But no evidence has been provided that specifically 2-D visual parsing mechanisms are any more useful for inference than 1-D mechanisms such as those evolved for natural language. It seems that presence or absence of connectives is probably an orthogonal dimension to that of diagrammatic or sentential representation. We should assume the dimensions are orthogonal at least until they are shown to interact.

On the second dimension, EG graphs provide directness of interpretation at the level of syntactic structures which EG sentences only provide at the level of strings of symbols. EG graphs' directness should make parsing tractable. However, neither system is directly interpreted at the level what we normally take to be semantics, and at this level we have no reason to expect further differences if ease of use. One theoretical gain from the example is realising that directness and indirectness can be construed at different levels of interpretation. At the level of some linguistic objects (e.g. strings), even sentences are directly interpreted.

The third and fourth dimensions have to do with the erasure of histories—syntactic or proof theoretic. Indirectly interpreted representations must generally be used discursively; directly interpreted systems agglomeratively. EGs seem to invite agglomerative use, even though it cannot be general, presumably because they are apparently diagrammatic. It is moot point whether for some tasks involving only the limited range of inferences where proof-history is not essential, EG graphs may have an advantage over sentential calculi which make agglomerative use unattractive.

Armed with this relation between directness/indirectness and agglomerative/discursive modes of use, I will finish with some necessarily brief remarks about the relation between discursive and agglomerative uses of representations in proof—the discourse of justification of reasoning. Any semantic account of diagrammaticity ought to be able to account for the uses of diagrams and sentences in the business of proof.

Proof makes a 'story' out of a static situation—something sequential out of something simultaneous. A logical consequence (conclusion) of a set of premises is a proposition *already logically contained* in the premises. In order to gain perspective on this conjuring of dynamics from the static, it is helpful to compare conventional sentential proofs with proofs-without-words. In a proof-without-words (e.g. Nelsen 1993) a static diagram unaccompanied by words contains the justification of a theorem. Perhaps the best known are the several diagrammatic proofs of Pythagoras theorem in a single diagram. We can 'see' in this diagram containing a right triangle that 'the area of the square on the hypotenuse' is the same as 'the sum of the areas of the squares on the other two sides'. The diagrammatic system in the figure is directly interpreted and agglomeratively used.

The first obvious problem for proofs-without-words is the statement of the theorem. If we do not know what Pythagoras theorem is, we are unlikely to find it in this diagram. Even if we know that it is about triangles, there are still many triangles in the figure and no indication which is the topic of the theorem. To even state Pythogoras' theorem requires a representation system that can represent only part of the truths about right triangles. In particular, a representation system that can represent the square on the hypotenuse without representing the rest (especially the sum of squares on the other two sides); and *v.v*; and can represent the equality of areas between these two entities. Only indirectly interpreted systems with their syntactic articulation can do this.

On the other hand, proofs without words can supply a kind of insight into proofs which sentential proofs find it hard to match. Proofs-without-words can be sequentialised, after a fashion, by providing a kind of 'comic-strip' animation of a sequence of constructions. But these raise the problems about what inferences and constructions are to count as 'formal' discussed in Scotto di Luzio (this volume).

I have suggested here reasons why indirect interpretation goes with discursive use; and direct interpretation with agglomerative use. If the consequences this theory for the inference rules which connect stages in a proof could be worked out in full, then it should tell us what the limits are of directly interpreted diagrams in the business of proof, for they are neither useless nor adequate. It is intriguing that an interaction between diagrams and language appears to have been what gave rise to the first invention of proof in Greek geometry (see Netz 1999).

4   Acknowledgements

This paper grew from a combination of reading a MS of Shin (forthcoming), and conversations with Patrick Scotto di Luzio about the requirements placed on representation systems by the goal of justifying reasoning—I am grateful to both for their stimulus.

I would also like to thank three anonymous referees for their helpful and tolerant comments on a hasty earlier version of this paper. The paper may still not fully comply with all their suggestions but it has benefited greatly from them. Grant support from the ESRC (UK) through Fellowship Grant # R000271074 and research support from CSLI, Stanford University are gratefully acknowledged.

References

1. R. B. Nelsen. 1993. Proofs without Words: Exercises in Visual Thinking. The Mathematical Association of America.
2. Barwise & Etchemendy Barwise, J. & Etchemendy, J. (1990) Visual information and valid reasoning visualization in mathematics. In W. Zimmerman (ed.) Mathematical Association of America: Washington, D. C.
3. Goodman, N. (1968) *The languages of Art*. Bobs Merrill: Indianapolis.
4. Gurr, C., Lee, J., & Stenning, K. (1998) Theories of diagrammatic reasoning: distinguishing component problems. *Minds and Machines* **8(4)**, pps. 533–557.
5. Netz, R. (1999) *The origins of proof in Greek Geometry*. CUP.
6. Scotto di Luzio, P. (2000) Formality and logical systems. This volume.
7. Shimojima, A. (1999) The graphic linguistic distinction. *Artificial Intelligence Review*, **13(4)**, 313–335.
8. Shin, S. (1998) Reading Peirce's existential graphs. Thinking with Diagrams Conference: Aberystwyth.
9. Shin, S. (forthcoming) *Iconicity in logical systems* MIT Press.
10. Sowa, J. F. (1984) *Conceptual Structures: Information Processing in Mind and Machine* Addison Wesley.

11. Steedman, M. (2000) *The syntactic process: language, speech and communication,* MIT Press: Cambridge, Mass.

12. Stenning, K. (1992) Distinguishing conceptual and empirical issues about mental models. In Rogers, Y., Rutherford, A. and Bibby, P. (eds.) *Models in the Mind.* Academic Press. pps. 29–48.

13. Stenning, K. (forthcoming) *Seeing Reason: image and language in learning to think.* Oxford University Press: Oxford.

14. Stenning, K. & Lemon, O. (in press) Aligning logical and psychological perspectives on diagrammatic reasoning. *Artificial Intelligence Review.*

15. Stenning, K. & Oberlander, J. (1995) A cognitive theory of graphical and linguistic reasoning: logic and implementation. Cognitive Science, **19**, pps. 97–140.

16. Stenning, K., Inder, R., & Neilson, I. (1995) Applying semantic concepts to the media assignment problem in multi-media communication. In Chandrasekaran, B. & J. Glasgow (eds.) Diagrammatic Reasoning: Computational and Cognitive Perspectives on Problem Solving with Diagrams. MIT Press. pps. 303-338.

How People Extract Information from Graphs: Evidence from a Sentence-Graph Verification Paradigm

Aidan Feeney, Ala K.W. Hola, Simon P. Liversedge, John M. Findlay
and Robert Metcalf

Department of Psychology, University of Durham, Science Laboratories, South
Road, Durham DH1 3LE, United Kingdom

Abstract: Graph comprehension is constrained by the goals of the cognitive system that processes the graph and by the context in which the graph appears. In this paper we report the results of a study using a sentence-graph verification paradigm. We recorded participants' reaction times to indicate whether the information contained in a simple bar graph matched a written description of the graph. Aside from the consistency of visual and verbal information, we manipulated whether the graph was ascending or descending, the relational term in the verbal description, and the labels of the bars of the graph. Our results showed that the biggest source of variance in people's reaction times is whether the order in which the referents appear in the graph is the same as the order in which they appear in the sentence. The implications of this finding for contemporary theories of graph comprehension are discussed.

1   Introduction

Graphs are a ubiquitous part of everyday life and their use appears to be still on the increase [1]. Despite the proliferation of graphic communication, comparatively little is known about how people extract information from even the simplest of graphs. A central goal of this paper is to examine the processes involved in comprehending simple bar graphs. In general, the attention paid to graphs by psychologists has been disproportionately small relative to the extent of their use. Moreover, much of the work that exists has been concerned either with people's understanding of the graphical representation of complex concepts [e.g. 2, 3] or with the production of a general, high-level, theory of graph comprehension [e.g. 4, 5]. More recently, however, there has been some interest in the detailed cognitive processes underlying the comprehension of very simple graphs [6, 7]. Below we will discuss this work along with an early sentence-picture verification study that we consider to be highly relevant to our experiment.

Our primary interest in this paper is how people form a representation of sentential and graphical descriptions in order to decide whether they agree. This interest stems from the observation that we rarely see, or produce, graphs outside of a linguistic

M. Anderson, P. Cheng, and V. Haarslev (Eds.): Diagrams 2000, LNAI 1889, pp. 149-161, 2000.

context. That is, most graphs have a title and are accompanied by some text. Furthermore, when a person processes a graph, they very often do this with the objective of verifying that claims made about the graph in the accompanying text are correct. Our starting point, then, is that graph comprehension is not usually an open-ended task. Instead, we view it as goal directed. People comprehend graphs in order to verify claims or to understand some quantitative relationship that is currently the subject of focus.

1.1   The Sentence-Graph Verification Task: A Tool for Studying Goal Directed Comprehension

The approach we adopt to examine goal oriented graphical interpretation is to present participants concurrently with a sentence and graphical display and require them to make a decision as to whether the sentence is an accurate description of the graph. Sentence-picture verification methodology is a paradigm that is highly suited to our needs in this study. In order to perform the task, participants must carry out goal oriented graphical interpretation. Hence, adopting this methodology and carefully manipulating different aspects of both the linguistic statement and the graphical display, will allow us to determine those characteristics of the text and graph that make interpretation more or less difficult.

The most common use of the paradigm has been to investigate how negation is represented and comprehended [e.g. 8, 9, 10]. To our knowledge, the most relevant use of this paradigm has been reported by Clark and Chase [11] who investigated how people verify that a verbal description of the relationship between two referents corresponds to a pictorial representation of the relationship between the same referents. The relationships which Clark and Chase asked their participants to verify were 'above' and 'below' whilst both sentences and pictures referred to 'star' and 'plus' signs. Participants were shown arrays (see Figure 1) where a verbal description such as *star is above plus* was placed to the left or the right of a simple picture of a star above or below a plus sign and were asked to indicate whether the descriptions matched. Clark and Chase proposed an additive model of the processing stages involved in their sentence-picture verification task that accounted for the influence on people's verification times of (i) the presence of negation in the description; (ii) the relational term used; (iii) the order of the referents in the sentence; and (iv) whether the sentence and the picture matched. The first assumption underlying their model was that in order to verify a sentence against a picture both must be mentally represented in the same representational format. Secondly, once a representation has been formed of the first description attended to, a representation of the second description will be constructed using the same relational term as was used in the first representation.

Figure 1: Examples from picture-sentence verification paradigm [11]

The sentence-graph verification paradigm that we adopted in our experiment was analogous to the sentence-picture verification paradigm used by Clark and Chase and others. To illustrate this paradigm, in Figure 2 we present some sample arrays from our experiment. Each array comprised a sentence specifying a relationship between two referents presented below a bar graph representing a relationship between the same two referents.

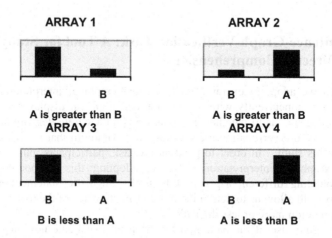

Figure 2: Sample arrays from the sentence-graph comprehension paradigm used in our experiment

The participant's task was simply to decide whether the statement was an accurate description of the graph. In Figure 2 Arrays 1 and 2 constitute matching trials in which the statement is in agreement with the graph while Arrays 3 and 4 are mismatching.

We manipulated four variables: whether the sentence and graph matched; the order of the bar graph labels (alphabetic or non-alphabetic); the slope of the graph (ascending versus descending); the relational term used in the sentence (*greater*, *less*). These variables produce an exhaustive set of possibilities. In combination they determine whether the order of the referents in the sentence is congruent with the order of the labels in the graph. Whilst the order of the terms agree in Arrays 1 and 3 they are in disagreement in Arrays 2 and 4. We will refer to such arrays as being *aligned* (in the former case) or *non-aligned* (in the latter case).

If verification of sentence-graph displays involves similar cognitive processes to those Clark and Chase suggested are involved in sentence-picture verification, then we would expect differences in reaction time suggesting that people represent the second description they encode using the same relational term as that used in the first description. In other words, if a participant reads the sentence *A is greater than B* they should encode the graphical display in terms of the relational term *greater than*. Additionally, Clark and Chase argued that picture-sentence verification involves two separate stages of processing: one of encoding and a second distinct stage during which people compare their representations of the sentence and the picture to check for a match. Consequently, if we can extrapolate their approach to sentence-graph verification, we would also expect our reaction time data to provide similar evidence for two distinct stages of processing.

In this section we have considered Clark and Chase's theoretical account of picture-sentence verification and also the implications this theory might have for an account of sentence-graph verification. However, it is important to point out that the processes involved in graph comprehension may be quite different to those involved in picture comprehension. In the next section we will briefly consider some of these differences and then focus on an account of graph comprehension proposed by Pinker.

1.2   Pinker's Account of Graph Comprehension

Even the most simple of bar graphs contains substantially more information than do the pictures illustrated in Figure 1. This extra information comes in the form of various conventional features of the graph. For example, the X and Y axes, the scale on the Y axis, each of the Bar labels and their positions as well as the physical characteristics of the bars of the graph. Given the additional information, we might expect encoding processes and verification processes in our sentence-graph verification task to be more complex than in the simple picture-verification task used by Clark and Chase. In line with this, Pinker [4] and Gattis and Holyoak [6] have claimed that we possess specific schemas for interacting with graphs and that these schemas appear to emerge at a relatively early stage of development [12].

One way to think about how the graph will be processed is provided by Pinker [4] who suggests that it is the interaction between the graph, the task at hand and the information processor's background knowledge that determines the ease with which information may be extracted from a graph. He argues that the processing of graphical information happens in a number of stages. First, visual processes code the graph into a visual array. Secondly, a 'visual description' or propositional representation consisting of predicates and variables is constructed. This description of the observed graph is compared against graph schema stored in memory in order to decide what kind of graph is being viewed. The activated schema aids the extraction of a conceptual message based on the information present in the visual description. At this point in a sentence-graph verification process the participant's representation of the sentence becomes relevant. If the information required to verify the sentence is not present in the conceptual message, the visual description of the graph is interrogated via the activated schema. Sometimes inferential processes may be carried out on the conceptual message itself in order to extract information required to make a verification decision.

In terms of our task, we expect participants to construct a visual representation of the graph. We would also expect a propositional representation to be constructed after which schematic knowledge about types of graphs would be activated automatically. In fact, in our experiment where participants receive many trials consisting of similar graphical representations, schematic knowledge should be activated to a substantial degree. Pinker also argues that graph readers should be able to translate higher-order perceptual patterns, such as a difference in height between a pair of bars, into an entry in the conceptual message extracted from the visual description via the instantiated schema. Accordingly, it should be a relatively simple task for a participant to interrogate the conceptual message with reference to the relationship between the referents described in the sentence.

Whilst Pinker's account goes some way towards outlining the processes involved in graph comprehension, it has been criticised for its generality [13]. It seems to us best thought of as a description of the macro-processes, rather than micro-processes underlying graph comprehension. Consequently Pinker's account is not sufficiently detailed to allow us to make specific predictions concerning our verification paradigm. We anticipate that our sentence-graph verification paradigm will shed light on some of the detailed processes involved in encoding and extracting conceptual information from graphs. For example, it should be informative as to the flexibility of the encoding processes (Is it equally easy to encode different types of graph?). Our paradigm may also provide insight into how graphical information is represented. We anticipate that whilst it is unlikely that there will be comprehension time differences for each of the sentence forms in the study, it is likely that the information contained in certain sentences will be easier to verify against certain graphical representations than against others. Accordingly, the presence of meaningful patterns in our reaction time data should shed light on the processes involved in interrogating and drawing inferences from [4] the visual description of a graph.

2   Experiment

2.1   Method

Participants: 51 undergraduate students at the University of Durham took part in this experiment.

Materials: Our materials were 64 different visual displays such as the one below. Each display consisted of a simple bar graph specifying the position of two referents on a scale and a verbal description of the relationship between these referents. There were sixteen experimental conditions in this experiment and each will be described with respect to the example above. Figure 3 contains a graph that we classified as descending because the slope of its bars descends left to right. To produce ascending graphs we simply swapped the position of the bars. The relational term used in the verbal description in Figure 3 is 'greater than'. Half of the displays used in this experiment employed 'greater than' whilst the other half employed 'less than'. In addition, the order in which the labels appeared in the graph was manipulated. The labels in Figure 3 appear in alphabetic order but for half of our experimental trials this order was reversed. Finally, the descriptions in Figure 3 match. That is, the sentence provides an accurate description of the graphical display. To achieve a mismatch in Figure 1 we reversed the order of the entities in the verbal description.

We constructed four examples of each experimental condition. These examples differed in terms of the labels used (A&B, C&D, L&M, P&Q), the widths of the bars in the graph, the distance between the bars and the difference between the height of the bars. All factors other than the labels were counterbalanced across conditions so

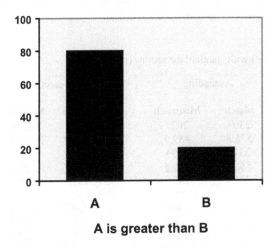

A is greater than B

Figure 3: Sample display from our experiment

we would not expect them to influence our results. Note that the magnitude difference of the bars was always easily discriminable. The order in which the resulting displays were presented was randomised.

Design: The experiment had a 2(Slope) x 2(Label Order) x 2(Relational Term) x 2(Match/Mismatch) within participants design. All participants received 64 trials each requiring them to indicate whether the graph and verbal description of the graph were in agreement.

Procedure: Data was collected from 51 participants in two separate testing sessions (25 participants per session approx.). Each participant was seated in front of a IBM clone computer and monitor and was given a booklet containing instructions and a sample display. Once participants indicated that they understood the instructions they were required to start the experiment. Each trial consisted of an initial display of a fixation cross (duration 1000 ms) followed by the graphical display with the sentence underneath. This display remained on the screen until the participant made a Match/Mismatch decision by pressing one of two buttons. Half the participants were required to make a MATCH response using their dominant hand. The interval between the end of one trial and the onset of the next trial was 1000 ms.

2.2   Results

Reaction Times. For the reaction time (RT) analyses we discarded all trials where participants had made an incorrect response, and all trials with RT's less than 100 ms (none occurred) or greater than two standard deviations from the mean RT for the entire experiment. In total we discarded 8.03% of the data using this procedure, (4.1% errors and 3.9% outliers). We then computed mean RT's for each participant across the remaining trials for each experimental condition.

Table 1: Means RT's (ms) with standard deviations (in italics)

Rel-Term	Label Order	Ascending		Descending		
		Match	Mismatch	Match	Mismatch	
Greater than	Alpha	2307	2077	1902	2394	2170
		575.8	*489.5*	*446.9*	*550.1*	
	Non Alpha	2319	2194	2043	2379	2234
		585.1	*470.8*	*560.2*	*550.0*	
		2313	2136	1973	2387	2202
Less than	Alpha	2035	2440	2298	2083	2214
		555.0	*526.7*	*579.1*	*600.0*	
	Non Alpha	2007	2418	2262	2050	2184
		428.9	*536.8*	*521.5*	*460.0*	
		2021	2429	2280	2067	2199

We carried out a 2x2x2x2 within participants ANOVA on the data. The means and standard deviations from this analysis are presented in Table 1. The analysis revealed two significant results. First, the main effect of whether the linguistic description agreed with the graph ($F(1, 50) = 13.65$, MSE = 173580.5, $p < .001$). The mean RT for matching trials was 2146 ms versus a mean of 2254 ms for mismatching trials. Such a difference between matching and mismatching trials has been observed previously in the literature. For example, Clark and Chase [11] observed that for affirmative sentences, RTs for correct responses to matching trials were shorter than RT's for correct responses to mismatching trials.

More novel is our finding of a highly significant interaction between Slope, Relational Term and the Match variable ($F(1, 50) = 124.08$, MSE = 151295, $p < .0001$). For ease of interpretation, example trials and mean RT's corresponding to each condition involved in this interaction are shown in Table 2. A close inspection reveals that the factors involved in the interaction combine to determine whether the order in which the referents appear in the sentence are aligned with the order in which they appear in the graph. Responses were faster for aligned trials than for non-aligned trials. Post hoc Tukey tests revealed that of the 16 comparisons that could be made to test this interpretation of the interaction, 15 were statistically significant ($p < .05$) in the direction predicted.

Error Data. In order that our analysis of error rates might parallel our RTs analysis, we used the same trimming procedure as for the RT data. For each participant we calculated the mean error rates across conditions. A 2x2x2x2 within participants ANOVA was carried out on errors. Mean error rates are shown in Table 3.

Table 2: Sample arrays and RTs from the interaction between Slope, Relational Term and the Match variable in our experiment

	Ascending		Descending	
	Match	Mismatch	Match	Mismatch

Greater than

	B is greater than A RT = 2313 ms	A is greater than B RT = 2136 ms	A is greater than B RT = 1973 ms	B is greater than A RT = 2387 ms

Less than

	A is less than B RT = 2021 ms	B is less than A RT = 2429 ms	B is less than A RT = 2280 ms	A is less than B RT = 2067 ms

Table 3: Mean percentage errors (standard deviations in italics)

Rel- Term	Label Order	Ascending		Descending		
		Match	Mismatch	Match	Mismatch	
Greater than	Alpha	5.39 *10.38*	3.10 *8.65*	0.49 *3.50*	7.68 *15.71*	4.17
	Non- Alpha	1.47 *5.94*	3.92 *9.18*	1.96 *6.79*	6.21 *14.70*	3.39
		3.43	3.51	1.23	6.95	3.77
Less than	Alpha	5.88 *12.50*	4.58 *10.04*	4.90 *13.25*	3.76 *13.05*	4.78
	Non- Alpha	4.58 *11.22*	3.27 *10.42*	5.39 *13.52*	4.08 *9.63*	4.33
		5.23	3.93	5.15	3.92	4.55

The ANOVA revealed just one significant source of variance - the interaction between Relational term and Match/Mismatch (F (1, 50) = 11.16, MSE = .0079, p < .002). Tests for simple effects revealed that participants made fewer incorrect responses to matching than mismatching trials when the relational term was *greater than* (F(1, 50) = 10.60, MSE = .008, p < .003). However, there was no difference in error rates for matching and mismatching trials when the relational term was *less than* (F(1, 50) = 1.832, MSE = .0089, p > .15). Although we did not predict this interaction *a priori*, differences due to the relational term are predicted by a variety of accounts of how we represent and reason about relationships between objects [for a review see 14].

Given the highly significant three way interaction identified by our analysis of RTs, it is important to note that this interaction does not account for a significant amount of the variance in our error data (F(1, 50) = 2.96, MSE = .0133, p > .05). An examination of Table 3 reveals that the trends present in this interaction do not suggest the existence of a speed-accuracy trade-off in participants' responses. Although non-significant, the trend is in the opposite direction.

3   Discussion

The first thing that is apparent from our results is that there were significant differences in both the RT and error data which presumably reflect differences in cognitive processing during goal oriented graphical interpretation. We believe that these differences demonstrate that our sentence-graph verification paradigm is an appropriate tool for studying graph comprehension.

The second point to note from our results is that the data indicate that participants do not necessarily represent the relationship shown in a graph using the same relational term as that used in the sentence. This finding is in direct contradiction to the central assumption underlying Clark and Chase's [11] account of sentence-picture verification. Recall that Clark and Chase argued that there were two separate stages of processing during picture-sentence verification: an initial stage of encoding and a second distinct verification stage where representations of the sentence and graph are compared. Importantly, Clark and Chase argued that the representations of the picture and the sentence should both employ the same relational term to allow direct comparison.

In our experiment, trials where the order of the referents was aligned produced response latencies that differed from latencies for trials where the referents were not aligned. In simple terms, participants read a sentence like *A is greater than B* and then they processed a graphical display in which the referents are either aligned (i.e. A to the left of B on the abscissa), or non-aligned (i.e. in the reverse direction). Aligned trials resulted in shorter response latencies than non-aligned trials. Clearly, if participants made their verification decisions in the manner advocated by Clark and Chase, then they should construct a representation of the graphical display using the same relational term as that used in the sentence. However, if participants did form representations in this way, then we would have expected shorter response latencies when the non-aligned graph matched the linguistic statement (i.e. the magnitude of bar B being less than that of A) than when it did not match (i.e. the magnitude of bar B being greater than that of A). We observed no such difference in response latencies. Consequently, we consider the current data to provide strong evidence against Clark and Chase's suggestion that linguistic and graphical representations are constructed using the same relational term in order to allow a direct comparison.

An alternative explanation that we favour, is that participants employed a strategy whereby they checked the aligned representations for matching relational terms. When the order of the referents was not aligned between descriptions, participants were forced to transform their representation of either the graph or the sentence to allow direct comparison. This transformation could account for the additional cost in processing time for non-aligned trials compared with aligned trials.

Note that Clark and Chase explicitly manipulated the spatial layout of the displays used in their experiments so that participants would first encode the sentence and then verify this description against the picture or vice versa. However, in our study participants were simply told to "verify that the statement is an accurate description of the graphical display". Given this instruction and simultaneous presentation of the descriptions, participants need not necessarily have formed a representation of the sentence prior to verifying it against that of the graph. Although direct comparisons between the current study and the work of Clark and Chase do not require us to be sure that participants in our study encoded the sentence before comprehending the graph, we have recently completed a separate experiment in our laboratory that provides insight into this question. In this experiment participant's eye movements were recorded as they carried out exactly the same sentence-graph verification task [for a similar approach see 15 & 16]. Preliminary analyses of the data from this experiment reveal that on the vast majority of trials participants immediately make a saccade to the sentence in order to read it prior to fixating different portions of the graphical display. We, therefore, assume that in the current RT experiment participants did the same. That is, they constructed a representation of the linguistic statement and subsequently constructed a representation of the graph before making a verification decision.

3.1 Relevance to Pinker's Account of Graph Comprehension

We will now consider our findings in relation to the account of graph comprehension developed by Pinker [4]. In Pinker's terms our results suggest that the process of interrogating the conceptual message derived from a graph is insensitive to the predicate used to encode the relationship between the referents in that graph. Instead it is driven by the order of the arguments in the proposition.

As mentioned previously, trials where the order of referents in the sentence and the graph were aligned displayed a significant RT advantage over non-aligned trials. This finding is consistent with the claim that the proposition GREATER THAN (A B) in the conceptual message is verified against the proposition LESS THAN (A B) derived from the sentence more quickly than is the proposition GREATER THAN (B A). This effect of alignment further suggests that the graph encoding process is relatively inflexible in that information contained in the sentence does not seem to affect how the graph is initially represented. For example, it appears that a participant who reads the sentence *A is greater than B* and then inspects a non-aligned, descending graph, is unable to prevent themselves constructing a representation in which the relationship is specified GREATER THAN (B A) even though the representation LESS THAN (A B) would be easier to verify. Thus, it appears that certain aspects of the graph encoding process are automatic.

Our results also illustrate that inferential processes are involved during verification. Pinker's use of the term *inference* includes the performance of arithmetic calculations on the quantitative information listed in the conceptual message as well as inferring from the accompanying text what is to be extracted from the graph. The graphical inference for which our experiment provides evidence consists of transformations carried out on representations. As aligned trials have an advantage over non-aligned trials (regardless of relational term, slope or whether the descriptions match), one might assume that the purpose of these transformations is to represent the information from the graph and the sentence so that their referents are in alignment. Once this has been achieved the relationship between the referents specified by each representation may be checked.

Whilst we agree with Pinker that inferential processes are of interest to cognitive psychologists generally, we nevertheless feel that the nature of those inferential processes and the factors which affect them are of considerable interest to any theory of graph comprehension. For example, our finding of an alignment effect is consistent with the claim that participants constructed an analogical representation, such as a mental model [see 17], rather than a propositional description of the graph. They may have then compared this against their representation of the premises (which may also be represented analogically). Although Clark and Chase [11] argued against such an account of their sentence-picture verification paradigm, we have seen that there are good reasons why graph comprehension may differ from picture comprehension. It is conceivable that graphical information may be represented analogically for certain tasks and in certain situations. In the literature on human reasoning [see 14] one tactic used to discriminate between model based [18] and rule based [e.g. 19] theories of deductive inference, has been for each theory to predict what kinds of inferences would be easy under its own representational assumptions. Applying such a tactic to our results, we feel that unless Pinker's propositional account is supplemented with the assumption that people have a preference to build their propositional representation of the graph from left to right, his account cannot easily predict our alignment effect.

3.2   Relation to Other Approaches to Graph Comprehension

Recently, Carpenter and Shah [16] have proposed a model of graph comprehension involving interpretative and integrative processes as well as pattern-recognition. In their model, pattern recognition processes operate on the graph in order to identify its components whilst interpretative processes assign meaning to those component parts (e.g. remembering that an upwardly curving line represents an increasing function). Integrative processes identify referents and associate them with interpreted functions. In a series of experiments where participants' eye movements were recorded whilst they were examining graphs showing complex interactions, Carpenter and Shah claim to have demonstrated that graph comprehension is incremental. That is, people use processes for pattern recognition, interpretation and integration to encode *chunks* of the graph. As the complexity of the graph increases so the cycle of processes is scaled up. Whilst the graphs which Carpenter and Shah investigated were more complex than ours and accordingly were more suitable for their research purposes, their focus is also different from ours. Their interest is in open-ended graph comprehension where we are interested in graph comprehension under constraints. Whilst both methodologies have their strengths and weaknesses, we would argue that graph comprehension is almost always goal oriented.

3.3   Directions for Future Work

Although our work on graphs is in its early stages, possible directions for future work are clear. First, we hope to investigate in more detail the nature of the inferential processes used in graph comprehension. Specifically, we are interested in the transformations performed upon graph representations on non-aligned trials and what those transformations can tell us about graph encoding and comprehension in general. Secondly, we hope to investigate the consequences of inferential processes for graph memory. That is, are trials that require inference prior to verification better remembered than trials that do not? Finally, as we have mentioned above, we have run some experiments where records have been taken of participants' eye movements. These experiments should tell us about the sequence in which processes occur during the extraction of information from graphs. It is hoped that the study of participants' eye movements will shed light on the nature of the inferential processes involved when the referents in the sentence and the graph do not occur in the same order.

4   Conclusions

In this paper we have presented the details of a novel sentence-graph verification technique for the study of how people extract information from graphs. We have argued that this paradigm is analogous to everyday graph comprehension in that it involves goal-oriented rather than open-ended interpretation of graphs. We have described an experiment using this paradigm where it has been demonstrated that the single greatest factor in determining the speed of people's verification response is whether the order of the referents in the graph is the same as the order of the referents in the sentence. We have suggested that this result illuminates some of the micro-processes involved in extracting information from graphs and may be consistent with either an analogical or a propositional explanation of how people represent graphs. Finally, we have outlined some possible directions which future research on this topic might take.

References

1.   Zacks, J., Levy, E., Tversky, B. & Schiano, D.: Graphs in Print. In P. Olivier, M. Anderson, B. Meyer (eds.): Diagrammatic Representation and Reasoning. Springer-Verlag, London (In press)
2.   Shah, P. & Carpenter, P. A.: Conceptual Limitations in Comprehending Line Graphs. Journal of Experimental Psychology: General (1995) Vol. 124 43-61
3.   Lohse, G. L.: The Role of Working Memory on Graphical Information Processing. Behaviour and Information Technology (1997) Vol. 16 297-308

4. Pinker, S.: A Theory of Graph Comprehension. In R. Freedle (ed.): Artificial Intelligence and the Future of Testing. Lawrence Erlbaum Associates Ltd, Hillsdale, NJ (1990) 73-126
5. Kosslyn, S. M.: Understanding Charts and Graphs. Applied Cognitive Psychology (1989) Vol. 3 185-226
6. Gattis, M. & Holyoak, K. J.: Mapping Conceptual to Spatial Relations in Visual Reasoning. Journal of Experimental Psychology: Learning, Memory and Cognition (1996) Vol. 22 231-239
7. Zacks, J. & Tversky, B.: Bars and Lines: A Study of Graphic Communication. Memory and Cognition (1999) Vol. 27 1073-1079
8. Slobin, D. I.: Grammatical Transformations and Sentence Comprehension in Childhood and Adulthood. Journal of Verbal Learning and Verbal Behaviour (1966) Vol. 5 219-227
9. Wason, P. C.: Response to Affirmative and Negative Binary Statements. British Journal of Psychology (1961) Vol. 52, 133-142
10. Greene, J. M.: Syntactic form and Semantic Function. Quarterly Journal of Experimental Psychology (1970) Vol. 22 14-27
11. Clark, H.H & Chase, W.G.: On the Process of Comparing Sentences Against Pictures. Cognitive Psychology (1972) Vol. 3 472-517
12. Gattis, M.: Spatial Metaphor and the Logic of Visual Representation. In AAAI Fall Symposium on Embodied Cognition and Action, Technical Report FS-96-02. AAAI Press, Menlo Park, California (1996) 56-59
13. Lewandowsky, S & Behrens, J. T.: Statistical Graphs and Maps. In F. T. Durso, R. S. Nickerson, R. W. Schvaneveldt, S. T. Dumais, D. S. Lindsay & M. T. H. Chi (eds.): Handbook of Applied Cognition. John Wiley & Sons Ltd, London (1999) 513-549
14. Evans, J. St. B. T., Newstead, S. N. & Byrne, R. M. J.: Human Reasoning: The Psychology of Deduction. Lawrence Erlbaum Associates Ltd, Hove, UK (1993)
15. Loehse, G. L.: Eye Movement-Based Analyses of Graphs and Text: The Next Generation. In Proceedings of the International Conference on Information Systems. Edition 14, ACM (1993)
16. Carpenter, P. A. & Shah, P.: A Model of the Perceptual and Conceptual Processes in Graph Comprehension. Journal of Experimental Psychology: Applied (1998) Vol. 4 75-100
17. Johnson-Laird, P. N.: Mental Models. Cambridge University Press, Cambridge, UK (1983)
18. Johnson-Laird, P. N. & Byrne, R. M. J. Deduction. Lawrence Erlbaum Associates Ltd, Hove, UK (1991)
19. Rips, L. J.: The Psychology of Proof. MIT Press, Cambridge, MA (1994)

Restricted Focus Viewer: A Tool for Tracking Visual Attention

Alan F. Blackwell[1], Anthony R. Jansen[2], and Kim Marriott[2]

[1] Computer Laboratory, University of Cambridge
Cambridge CB2 3QG, UK
Alan.Blackwell@cl.cam.ac.uk
[2] School of Computer Science and Software Engineering, Monash University
Clayton, Victoria 3800 Australia
{tonyj,marriott}@csse.monash.edu.au

Abstract. Eye-tracking equipment has proven useful in examining the cognitive processes people use when understanding and reasoning with diagrams. However, eye-tracking has several drawbacks: accurate eye-tracking equipment is expensive, often awkward for participants, requires frequent re-calibration and the data can be difficult to interpret. We introduce an alternative tool for diagram research: the Restricted Focus Viewer (RFV). This is a computer program which takes an image, blurs it and displays it on a computer monitor, allowing the participant to see only a small region of the image in focus at any time. The region in focus can be moved using the computer mouse. The RFV records what the participant is focusing on at any point in time. It is cheap, non-intrusive, does not require calibration and provides accurate data about which region is being focused upon. We describe this tool, and also provide an experimental comparison with eye-tracking. We show that the RFV gives similar results to those obtained by Hegarty (1992) when using eye-tracking equipment to investigate reasoning about mechanical diagrams.

1 Introduction

Visual attention is an important component in understanding how humans reason with diagrams. Complex diagrams can rarely be taken in at a single glance, and thus following the focus of visual attention can provide important insight into the strategies used in diagrammatic reasoning. For this reason, eye-tracking equipment has been of great benefit to researchers in examining these strategies. This applies not just in the area of diagrammatic reasoning, but also reading [ ], cartography [ ], scene perception [ ] and cognitive processes in general [,].

Yet despite the benefits that traditional eye-tracking provides, it also has significant drawbacks. First is expense: typically, the better the resolution accuracy and sampling rate, the more expensive the system is, and high quality systems are very expensive indeed. Second, many systems are awkward for participants, using head mounted gear, or requiring chin rests or bite bars to suppress head movements. Most systems require frequent re-calibration, and often they cannot

M. Anderson, P. Cheng, and V. Haarslev (Eds.): Diagrams 2000, LNAI 1889, pp. 162–   , 2000.
© Springer-Verlag Berlin Heidelberg 2000

be used with participants who wear glasses. Also, blinks or glances away from the stimulus can cause spurious trajectories in the eye movement data, and finally, and perhaps most important, it can be impossible to determine if a participant is taking in a broad overview of a stimulus, or focusing on a specific region. It is not surprising then that in some papers that discuss eye-tracking data, the results of some participants can not be used due to an inability to accurately record their eye fixations (for example, see [,]).

Here we describe an alternative computer based tool for tracking visual attention: the Restricted Focus Viewer (RFV). This allows visual attention directed towards an image presented on a computer monitor to be tracked. This tool is a cheap, easy to set up system, that has flexibility allowing it to be tailored to specific diagrammatic elements of interest. It is non-intrusive and requires no calibration. We believe it will have wide appeal in research on cognitive aspects of diagrammatic understanding and reasoning. It is important to understand that the RFV is not intended to be a replacement for eye-tracking. Rather it is a complementary technique with its own advantages and disadvantages.

The RFV uses image blurring to restrict how much of the image can be clearly seen, with only a small region in focus. The region of focus can be moved around using a computer mouse. The idea behind using restrictions in the visual field is not new. In research on reading for example, the number of characters that can be processed in one fixation has been examined by using visual restrictions [,]. Image blurring has been used to understand how people take in information from software manuals [ ]. Studies have also been conducted which involve the use of artificial scotomas to disrupt visual processing [ ], and the notion of a movable window has been used before in examining visual search [ ].

The main contribution of this paper is threefold. First, we describe our implementation of the RFV in detail. It is a generic computer based tool, explicitly designed for research into diagrammatic reasoning. Features include graded blurring, motion blur and a data replay program (see Section 2). Second, we provide an empirical validation of the RFV. Results obtained from a mental animation experiment using eye-tracking equipment are compared with results obtained using the RFV instead (Section 3). Our third contribution is a detailed discussion of the relative advantages and disadvantages of the RFV over eye-tracking (Section 4).

2 The Restricted Focus Viewer

The Restricted Focus Viewer (RFV) is a computer program which takes a diagram, blurs it and displays it on a computer monitor, allowing the participant to see only a small region of the diagram in focus at any time. The region in focus can be moved using the computer mouse. The RFV records what the participant is focusing on at any point in time, and the data can be played back using a replayer. We now describe the RFV and the replayer in more detail.

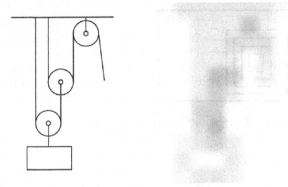

Fig. 1. Example of a visual stimulus and its corresponding blurred image

2.1 Description of the RFV

The human visual system can only focus on objects at the centre of the visual field. The region surrounding this area of sharp focus is still perceived, but the further from the centre of the visual field an object is, the more coarse is the perception of it [ ]. Most of the time, visual attention is directed to the centre of the visual field, although this is not always the case as it is possible to covertly attend to other locations [ ].

The RFV has been designed to reflect these aspects of the human visual system. The key idea is that only the part of the image under the focus window is in focus, with the remainder of the image out of focus. To keep most of the diagram out of focus, the RFV uses image blurring. Figure   gives an example of this with a pulley system diagram as the stimulus. The original system is shown on the left, with a blurred representation of it on the right. The blurred image still allows the general form of the diagram to be perceived, thus allowing the user to move directly from one region of the image to another. However, it does not reveal the finer details of diagram, with the individual components being indiscernible unless the user specifically focuses on them with the focus window. (Note that the blurred images in this paper have been made slightly darker so that they are clearer when printed. Some detail has been lost due to a reduced greyscale when printing.)

It is clear from the example figure that large structural features of the diagram are suggested in the blurred image. Our objective is that broad structure should be perceived as easily in the blurred image as by someone glancing momentarily at the original diagram. The degree of blurring required for a specific type of diagram depends on the size and visual characteristics of the diagram elements of interest. This is discussed in more detail later in this section.

The *focus window* of the RFV is the region in which the stimulus is visible in full detail. Two important issues concerning the focus window for the RFV are firstly, how 'natural' it will look, and secondly, how big it should be.

Initial experience with the RFV led to a clear conclusion about the focus window: it is not sufficient to simply have a box on the screen in which the stimulus

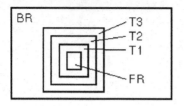

FR – Focus Region
T1 – Transition Region 1
T2 – Transition Region 2
T3 – Transition Region 3
BR – Blurred Region

Fig. 2. Regions of the stimulus used to achieve the graded blurring effect

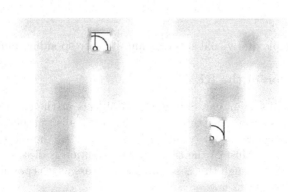

Fig. 3. Two examples of the focus window on different regions of the stimulus

is in focus, while the rest of the image is blurred. The boundary between the two regions is too distinct, leading to a very unnatural effect. Also, a sharp cut-off will enable the participant to guess about neighbouring parts of the image from Gestalt continuation effects. A graded blurring effect, such that the transition from blurred to focus appears smooth and seamless, is needed to prevent this.

A graded blurring effect is achieved by the technique illustrated in Figure . The outer rectangle defines the stimulus area which is fully blurred. The innermost box is the region of focus. Surrounding this focus region are three transition regions. Each transition region is slightly more blurred than the last, so that there is only a subtle difference between neighbouring regions. The overall result is the appearance of a smooth transition from the region of the image in focus, to the region which is fully blurred. Using the mouse to move the focus window therefore moves not only the focus region, but also the transition regions.

Figure gives two examples of the focus window positioned over different regions of the stimulus shown in Figure . The size of the focus window is determined not only by the dimensions of the focus region, but also by those of the three transition regions. However, the two most important sets of values to be considered are the size of the focus region box, and the size of outermost transition region box. For the experiment discussed in this paper, the focus region and transition region boxes are all square, and centred around the current mouse co-ordinates, although this need not be the case. At the centre of the focus

Table 1. Guidelines for setting RFV parameters

Focus Region Size	Goal	Should be slightly smaller than bounding box of typical element of diagram
	Lower Limit	Must allow recognition of any one element of the diagram when region is centred over the element
	Upper Limit	Should prevent simultaneous recognition of two neighbouring elements when placed between them
Transition Region Size	Goal	Should indicate direction of neighbouring connected elements
Level of Blurring	Lower Limit	Should be sufficient that any two elements are indistinguishable and that overall connectivity cannot be established
	Upper Limit	Should allow identification of diagram boundaries (at least convex hull)

window there is also a small grey dot, to allow users to keep track of the focus window location when it is centred on an empty region of the image.

The degree of blurring and the size of the focus region should be adjusted in accordance with the size of syntactic elements in the diagram. In our experience with the RFV to date, the rules of thumb described in Table should be applied in determining these factors. Note that although these parameters can be adjusted easily for a given experiment, the RFV program does not allow them to be changed during an individual trial. This is to prevent inconsistencies in the way that the participant views the stimulus.

Another feature that was implemented so that the RFV would more accurately mimic the way humans perceive visual stimuli is *motion blur*. This is based on the fact that during saccadic eye movements, visual information is not processed [ ]. Thus, if the user of the RFV moves the mouse at high speed (that is, over a large distance on the screen in a small amount of time), the focus window will not achieve full focus. Once the user reduces the speed of the mouse motion back to below a certain threshold, or stops moving the mouse completely, full focus in the focus window will return. This feature helps in defining the temporal boundary between fixations and movements.

When the focus window is stationary or moving slowly, all of the regions listed in Figure are present. During motion blur however, only the outermost transition region is added to the blurred stimulus. Because this region has less blurring than the rest of the image, the user is still able to track the location of the focus window on the stimulus. However, it is not possible to determine the finer details of that location without slowing or stopping the mouse. Only then will full focus be available. Table describes guidelines for appropriate motion blur settings. That is, how fast the mouse needs to move before motion blur occurs.

Table 2. Guidelines for the RFV motion blur setting

Motion Blur Onset	Goal	Should allow separation between fixation and movement
	Lower Threshold	Should not allow "brass rubbing" strategy, that is, identification of diagram by waving window rapidly over it
	Upper Threshold	Should allow slow navigation with continuous focus over a connected diagram when the task requires it

For each stimulus, the RFV outputs a data file that records the motion of the focus window. The file consists of a brief header to specify which trial the data pertains to. For the remainder of the file, each line contains the details for each updated mouse movement. The lines are composed of four data. First is the time elapsed (in milliseconds) since the RFV was initiated for that particular trial. The next two values are the x and y co-ordinates of the centre of the focus window with respect to the top left corner of the stimulus image. The final piece of information is a flag to indicate whether or not the focus window was motion blurred or not. This data allows the experimenter to exactly replicate the state of the RFV while the participant was performing the task.

2.2 Data Replayer

The data replayer is a companion program to the RFV that can read in a data file generated by the RFV, and replay the way that the RFV participant moved the focus window over the stimulus. This has two important benefits.

Firstly, it can be used as an experimental analysis tool by the experimenters when reviewing participant actions. For example, it can play back the focus trace at faster than real-time. The data replayer also has another function that is useful to experimenters. It can draw a scan-path line over the original stimulus based on the locations of the centre of the focus box. Figure   gives an example of this for the pulley system diagram. It can be seen that in this instance, the person started at the free end of the rope on the right of the diagram. From there they moved to the top pulley, continued to the middle pulley, and finally moved down to the bottom pulley. At each pulley, time was spent examining that region of the diagram.

The second important benefit is that the data replayer can be used during experimental sessions to elicit retrospective verbal reports from participants, about their strategy and actions during an experimental trial. As noted by Ericsson and Simon [ ], it can be very difficult for participants to verbalise concurrently while carrying out a complex problem solving task. We have used the data replayer to play back focus movements slower than real-time, so that the participant can record a verbal protocol describing their actions after completing the main task. Where Ericsson and Simon report that it is difficult to report retrospectively on a problem taking longer than 10 seconds to solve, the use of the data replayer

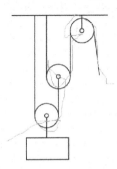

Fig. 4. Example of a data replayer scan-path

with the RFV reminds the participant of his or her actions, and allows more extensive protocols than can usually be obtained.

3 Validation of the RFV

The key question before we can consider using the RFV in experimental work is to determine if it interferes with the strategies used by humans who perform diagrammatic reasoning. Clearly, there is some overhead since the participant must use a computer mouse rather than simple eye and head movements to change their focus. However, it is important to the validity of the RFV technique that this should not affect the strategies used. To test this, we have replicated a classic experiment on diagram interpretation [ ], originally done with eye-tracking equipment, and instead used the RFV. Our hypothesis is that participants will use the same strategy with the RFV as with eye-tracking.

3.1 Method

Participants Eleven participants successfully completed the experiment. All were graduate or undergraduate students from the Computer Science and Psychology Departments. All participants were volunteers with normal or corrected-to-normal vision. Data from one additional participant was excluded due to excessive error rates, however the errors were due to a misinterpretation of the instructions, and were not attributable to the RFV.

Materials and Design Two diagrams of pulley systems were constructed based on those used in Experiment 1 in Hegarty [ ] (see Figure). They each consisted of three pulleys, a weight, braces from the ceiling to support some pulleys and sections of rope that were attached to the ceiling or weight, and went over or under the pulleys. In each system there was also a free end to the rope, and participants were required to infer the motion of the system components when the free end of the rope was pulled. The mirror images of these pulley systems were also used giving a total of four pulley system images.

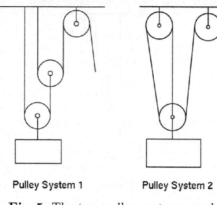

Pulley System 1   Pulley System 2

Fig. 5. The two pulley systems used

For each pulley image there were twelve statements, each about the motion of one of the systems components. Six of these statements were true and six were false. For this experiment only kinematic statements were used (referring to the system in motion), with none of the statements addressing static aspects of the pulley system.

When the free end of the rope is pulled in any pulley system, a causal chain of inferences can be made about the motion of the components. For example, in pulley system 1, pulling the rope causes the rope to move right over the upper pulley, turning it clockwise. From this knowledge we can infer the motion of the middle pulley and so on. In this way, we can define each pulley in the pulley systems as being at the beginning, middle or end of the causal chain of events. The statements about the motion of the pulley system components are equally divided among the pulleys at each of these three locations in the causal chain. In pulley system 2, "the rope moves to the right under the lower pulley" is an example of a true statement about a kinematic event at the middle of the causal chain, and "the upper right pulley turns counterclockwise" is an example of a false statement about a kinematic event at the beginning of the causal chain.

Each statement was presented as a single line of text. A stimulus was composed of a statement appearing on the left, with the diagram of a pulley system on the right. This gave a total of 48 stimuli. Each participant was shown all 48 stimuli twice, with a rest between the two blocks. In each block, the stimuli were presented in a different pseudo-random order.

Procedure Participants were seated comfortably in an isolated booth. Items were displayed in black on a white background on a 17" monitor at a resolution of 1024×768, controlled by an IBM compatible computer running a version of the RFV program whose settings had been tailored to this experiment. The original eye-tracking experiment was conducted using an Iscan corneal-reflectance and pupil-center eye-tracker (Model RK-426), that has a resolution of less than one degree of visual angle, and samples the participants' gaze every 16 milliseconds.

The size of the images were 200×293 pixels for pulley system 1 and 160×300 pixels for pulley system 2. The text statements were on average 338 pixels across (range 244–417) and 16 pixels high. The RFV focus box had an edge length of 36 pixels, while the outermost transition box had an edge length of 50 pixels. The motion blur was set to a high tolerance, allowing for full focus to be maintained even during moderately fast movements of the mouse.

Participants were given a brief statement of instructions before the experiment began, and were shown diagrams which labelled all of the pulley components referred to in the statements. They were then given some practice items involving a very simple (only two pulleys) pulley system. The practice items allowed the participants to become familiar with the RFV program.

The only interface mechanism used by the participants was a standard computer mouse. Progress was self-paced, with each trial initiated by a single click with the mouse on a button at the bottom of a blank screen containing the prompt "Press the button to continue". This action started the timer (to provide a reference time for the rest of that trial), and a blurred image of the stimulus appeared on the screen. On the left was a statement and on the right was a diagram of a pulley system. The participant could move the window of focus over different regions of the stimulus by moving the mouse. By doing this, the participant could then read the text of the statement and look at the attributes of the diagram.

The participants were required to determine if the statement was true or false with respect to the pulley system presented. At the bottom of the screen were two buttons, one labelled "TRUE" and the other "FALSE". When the participant had decided on the validity of the statement, they single clicked on the appropriate button. This stopped the timer, and the program recorded the response given and the time taken. No feedback was given to the participant.

Participants were instructed to try and respond as quickly as possible, while still trying not to make too many errors. The experiment consisted of two blocks of 48, with a brief rest period in between. The full set of stimuli was shown in each block, resulting in two repetitions for each item in the experiment. The experiment took approximately forty minutes to complete.

Data Treatment To reduce the unwanted effects of outlying data points, an absolute upper cut-off was applied to response latencies, such that responses longer than 30 seconds were excluded from the response time data analysis and designated as an error. So as to be consistent with the original eye-tracking experiment, the data analysis was only conducted over the items that contained true statements, with the false items only acting as fillers in the experiment.

For the one participant whose results were excluded, this was due to getting no correct responses at a particular causal chain location in one of the pulley systems, and thus not allowing the calculation of a mean response time at that position.

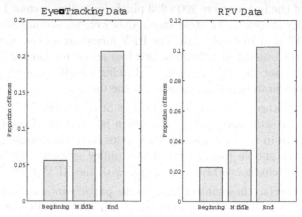

Fig. 6. Comparison of eye-tracking data from Hegarty's original experiment and RFV data from our replicated experiment, examining the proportion of errors at different causal chain positions

3.2   Results and Discussion

The aim of this experiment is to compare the results obtained using the RFV with eye-tracking results, to determine if the RFV affects the strategies used by humans who are reasoning with diagrams, in a manner different to the way that eye-tracking equipment might affect strategy. The main focus of this analysis is therefore to examine key data trends and significant results obtained in the original eye-tracking experiment, and see if the RFV results correlate.

Errors The overall error rate was only 5.3%. This is much lower than the error rate of the participants in the original eye-tracking experiment. However, this is not surprising given that the participants in the original experiment were all psychology undergraduates, and many of the participants in this experiment had more experience in dealing with technical diagrams.

The data comparisons were conducted using two-way ANOVAs (causal chain position × repetition), carried out over participant data. In the original eye-tracking experiment, the position in the causal chain of the component referred to in the statement had a significant effect on error rates. This effect was also present in the RFV data ($F(2, 20) = 5.33$, $p < 0.05$). This can be clearly seen in Figure , where the data from this experiment on the right is compared with the data from the original eye-tracking experiment on the left. There was also a trend for participants to make fewer errors on the second repetition of the stimuli (3.0%, sd = 5.4) than on the first repetition (7.6%, sd = 12.5), however this trend was not statistically significant ($F(1, 10) = 4.52$, $p = 0.059$). This trend was also apparent in the eye-tracking experiment.

Response Times Figure   shows the mean reaction times (overall height of the bars) for each pulley system, for statements referring to components at dif-

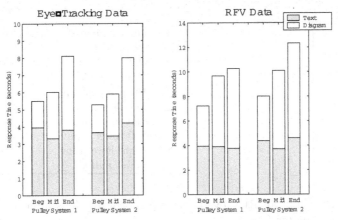

Fig. 7. Comparison of eye-tracking data from Hegarty's original experiment and RFV data from our replicated experiment, examining the mean response times for different trial types

ferent positions in the causal chain. As with the error graphs, the data from this experiment is shown on the right, with the eye-tracking data on the left for comparison. The times have been divided into the time spent reading the statement, and the time spent inspecting the diagram.

Response times for the two pulley systems were analysed separately, to allow for differences in the configurations. In the original eye-tracking experiment, the repetition caused a practice effect which resulted in participants responding significantly faster on the second repetition of the stimuli than on the first. The same effect was seen in the RFV data. The response advantage was 3.65 seconds for pulley system 1 ($F(1, 10) = 44.40$, $p < 0.01$), and 4.05 seconds for pulley system 2 ($F(1, 10) = 38.07$, $p < 0.01$). The original experiment also showed that the position in the causal chain of the component referred to in the statement had a significant effect on response times. Again, the RFV data corresponds to the data from the original experiment, with position in the causal chain significantly influencing response times ($F(2, 20) = 18.87$, $p < 0.01$ for pulley system 1; $F(2, 20) = 24.15$, $p < 0.01$ for pulley system 2). This effect can be seen in Figure .

These results are clearly in agreement with the eye-tracking experiment. However, the participants using the RFV take approximately 50% longer to respond. Response time is likely to be affected by the fact that the participants interface with the stimulus using a computer mouse. This is due to the difference in the ballistic speed of human motor control of the arm and the eye. Also, the blurring of the image with only a small region of focus means that one would expect participants to take longer in moving from one area of the diagram to another area. Despite the extra time taken, the overall trend in the data is very similar.

Further data analysis that was done in the eye-tracking experiment involved examining how long participants were inspecting different components of the pulley systems. In particular, for each statement, the components in the diagram

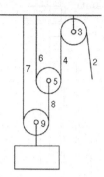

Gaze	Object Fixated	Gaze Duration
1	Statement	1884 ms
2	Pull rope	500
3	Upper pulley	595
4	Right upper rope	345
5	Middle pulley	1160
6	Left upper rope	180
7	Left lower rope	150
8	Right lower rope	61
9	Lower pulley	5917

Fig. 8. Example of an RFV eye-fixation protocol for the statement *The lower pulley turns counterclockwise*

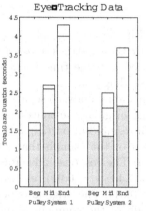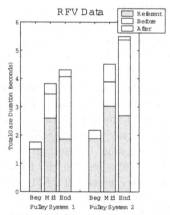

Fig. 9. Comparison of eye-tracking data from Hegarty's original experiment and RFV data from our replicated experiment, examining the breakdown of gaze duration on different components of the pulley systems

were divided into those whose motion occurred before the component referred to in the statement, the referent itself, and those components whose motion occurred after the referent. This allows for a further breakdown of response time spent viewing the diagram. Rectangular bounding boxes were used to enclose regions of the diagrams containing pulleys, rope strands, the ceiling and the weight, just as in the original eye-tracking experiment. This allowed the order of fixations on the components of the diagrams to be determined, along with how long those fixations were. Figure   gives an example of an eye-fixation protocol taken from the RFV data.

The gaze duration is defined as the total time spent fixating on components, in certain locations in the causal chain with respect to the referent in the statement. The graphs of the gaze duration data are shown in Figure   . Again, the data from this experiment is on the right, with the eye-tracking data on the left. The original eye-tracking experiment showed that when looking at the pulley system, participants spend most of their time inspecting the referent and the components whose motion precedes that of the referent in the causal chain of events. The RFV data shows the same result for both pulley system 1 (91%, sd = 4.9) and pulley system 2 (92%, sd = 4.6).

Due to the fact many participants in our experiment had more experience with technical diagrams than the participants in the original eye-tracking experiment, we would expect some minor differences in strategy. However, the results of this experiment indicate that the response time, accuracy and gaze duration trends obtained from the original experiment using eye-tracking techniques, are the same as those obtained using the RFV. There were no key significant results from the original eye-tracking experiment that the RFV failed to obtain.

4 Comparison of the RFV with Eye-Tracking

The Restricted Focus Viewer (RFV) is designed to collect data about visual attention, as are eye-trackers. However, the RFV is not intended as a replacement for eye-tracking techniques. Rather it is an alternative experimental technique and apparatus that adds to the toolbox available to cognitive scientists as they try to understand the processes of diagrammatic reasoning.

Here, we explain the relative advantages and disadvantages of the RFV and eye-tracking. One primary concern is whether the RFV affects the high level strategy used by participants. We note that this issue is not just confined to the RFV: in eye-tracking also, repeated calibrations and head gear (such as head mounted cameras and infra-red reflectance devices) may influence participant strategy. The previous experiment suggests that high level strategy is not changed by using the RFV instead of eye-tracking equipment. However, this depends on the task and the choice of RFV parameters. In particular, there are two important issues to consider.

The main issue is that with the RFV, the experimenter can modify the size of the focus window and the amount of blurring, so as to ensure that the participant must explicitly focus on those components of the diagram they are interested in. Such changes however, may change the strategy used by the participant.

Consider for example, a mathematical equation as the visual stimulus. It is preferable that the size of the equation not be excessively large, since this would make any task involving it seem less natural to the participant. Trying to determine the specific symbols that are being focused on would be practically impossible using eye-tracking, since each eye fixation would take in a large collection of symbols. With the RFV, the size of the focus window can be reduced so that only a few or even only one symbol can be viewed at a time. This would allow visual attention to be recorded at a level of detail not available using eye-tracking techniques. However, the reduction in focus window size also reduces the amount of information available to the participant at any given moment. This could affect the way a given task is approached. It appears that the more accurately you record the focus of the participants attention (by reducing the size of the focus window), the more likely you are to affect the strategy that they would normally use. However, this is not a defect of the RFV, but rather an aspect that experimenters need to be aware of in experiment design.

The second issue is that a computer mouse is used rather than eye and head movements to change the direction of attention. Thus, participants in the experiments need to be confident in the use of a mouse. Also, response times will be slower due to the use of the arm and hand. For some tasks that require fast responses, this may present a problem. However, we believe that for most tasks this should not affect data trends or the significance levels of experimental results. One difference between the RFV and eye-tracking is what happens when the participant is not explicitly focusing on any part of the diagram, because their attention is directed towards internal processing. At this time, their gaze may drift across the stimulus. With eye-tracking this means that spurious fixations may appear in the data, while with the RFV the object last in focus will have a longer gaze duration.

By comparison, we have experienced far more significant difficulties when using eye-tracking equipment (ASL 5000) for experimental tasks to which we had easily applied the RFV. These problems will be familiar to experienced users of eye-tracking equipment, but are nevertheless significant obstacles to new researchers. Eye-trackers are expensive. They require substantial expertise in calibration and adjustment. They do not work reliably in strong daylight conditions. They can be unreliable with participants who have shiny skin, watery eyes or contact lenses. Output data is often subject to positional drift, in addition to local nonlinear uncertainty. Vertical resolution is poor compared to horizontal resolution, making them less useful with detailed two-dimensional diagrams. Analysis of data requires subjective classification of fixation and saccade thresholds. Fixations are often at a point between two display elements, leaving it unclear whether the participant is defocused or viewing both elements as a unit. The tracker can lose stable gaze identification during the experiment, leading to invalid trials. By comparison to experiments conducted with the RFV using the same stimuli, eye-tracking results provided very little useful data.

Overall, the RFV has several advantages over traditional eye-tracking techniques. The system is cheap and easy to set up, providing accurate data about the region that is being focused upon. It is non-intrusive, requiring no special gear to be worn or restrictions on the movement of participants. It does not require any calibration and can be used by participants who wear glasses. The RFV data is not corrupted by blinks or glances away from the stimulus, and the replayer provides a useful tool for immediate feedback on participant performance. Finally, the RFV has flexibility in its parameter settings, allowing it to be tailored to meet specific goals.

5 Future Work

We have described a new tool, the RFV, for tracking visual attention during diagrammatic reasoning. We have also provided an experimental comparison with eye-tracking equipment. The primary direction for future work is to expand this experimental comparison by considering other classes of diagrams. We also intend to extend the RFV and data replayer by adding extra features. Such features include modifying the data replayer to allow it to replay multiple scanpaths at once. This would be useful in allowing for a direct comparison of the strategies used by different participants. The RFV is available in the public domain for other researchers to use at

http://www.csse.monash.edu.au/~tonyj/RFV/

Acknowledgements

The authors wish to thank Mary Hegarty for her helpful information and discussions on the original work with the pulley systems using eye-tracking equipment. We also wish to thank Greg Yelland for his suggestions during the development of the RFV program. Alan Blackwell's research is funded by the Engineering and Physical Sciences Research Council under EPSRC grant GR/M16924 "New paradigms for visual interaction".

References

1. Patricia A. Carpenter and Priti Shah. A model of the perceptual and conceptual processes in graph comprehension. *Journal of Experimental Psychology: Applied*, 4(2):75–100, 1998.
2. Stanley Coren, Lawrence M. Ward, and James T. Enns. *Sensation and Perception*. Harcourt Brace and Co., fourth edition, 1994.
3. K. Anders Ericsson and Herbert A. Simon. *Protocol Analysis: Verbal Reports as Data*. MIT Press, Cambridge, Massachusetts, revised edition, 1993.
4. Mary Hegarty. Mental animation: Inferring motion from static diagrams of mechanical systems. *Journal of Experimental Psychology: Learning, Memory and Cognition*, 18(5):1084–1102, 1992.
5. John M. Henderson, Karen K. McClure, Steven Pierce, and Gary Schrock. Object identification without foveal vision: Evidence from an artificial scotoma paradigm. *Perception & Psychophysics*, 59(3):323–346, 1997.
6. Marcel Adam Just and Patricia A. Carpenter. Eye fixations and cognitive processes. *Cognitive Psychology*, 8:441–480, 1976.
7. Naoyuki Osaka and Koichi Oda. Moving window generator for reading experiments. *Behavior Research Methods, Instruments & Computers*, 26(1):49–53, 1994.
8. Keith Rayner. Eye movements in reading and information processing: 20 years of research. *Psychological Bulletin*, 124(3):372–422, 1998.
9. Keith Rayner and Alexander Pollatsek. *The Psychology of Reading*. Prentice-Hall, Englewood Cliffs, New Jersey, 1989.   ,
10. Keith Rayner and Alexander Pollatsek. Eye movements and scene perception. *Canadian Journal of Psychology*, 46:342–376, 1992.

11. Susan K. Schnipke and Marc W. Todd. Trials and tribulations of using an eye-tracking system. In *CHI 2000 Extended Abstracts*, pages 273–274, 2000.

12. L. W. Stark, K. Ezumi, T. Nguyen, R. Paul, G. Tharp, and H. I. Yamashita. Visual search in virtual environments. *Proceedings of the SPIE: Human Vision, Visual Processing, and Digital Display III*, 1666:577–589, 1992.

13. Theodore R. Steinke. Eye movement studies in cartography and related fields. *Cartographica*, 24(2):40–73, 1987. Studies in Cartography, Monograph 37.

14. Martin J. Tovée. *An Introduction to the Visual System*. Cambridge University Press, 1996.

15. Nicole Ummelen. *Procedural and declarative information in software manuals*. PhD thesis, Universiteit Twente, Utrecht, 1997.

16. Alfred L. Yarbus. *Eye Movements and Vision*. Plenum Press, New York, 1967.

Communicating Dynamic Behaviors:
Are Interactive Multimedia Presentations Better than
Static Mixed-Mode Presentations?

N. Hari Narayanan[1] and Mary Hegarty[2]

[1]Intelligent & Interactive Systems Laboratory
Department of Computer Science & Software Engineering
Auburn University, Auburn, AL 36849, USA
narayan@eng.auburn.edu
http://www.eng.auburn.edu/~narayan
[2]Department of Psychology
University of California
Santa Barbara, CA 93106, USA
hegarty@psych.ucsb.edu
http://psych.ucsb.edu/~hegarty

Abstract. Static mixed-mode presentations consisting of verbal explanations illustrated with diagrams have long been used to communicate information. With the advent of multimedia, such presentations have become dynamic, by migrating from paper to the computer and by adding interactivity and animation. The conventional wisdom is that computer-based multimedia presentations are better than printed presentations. However, does the communicative power of mixed-mode representations stem from their careful design to match cognitive processes involved in comprehension or from their interactive and animated nature? This is an important issue that has never been investigated. This paper first presents a cognitive model of comprehension of mixed-mode representations. We describe how this model generates design guidelines for mixed-mode representations that present expository material in two domains - the concrete domain of mechanical systems and the abstract domain of computer algorithms. We then report on a series of studies that compared computer-based interactive multimedia presentations and their paper-based counterparts. Both were designed in accordance with the comprehension model and were compared against each other and against competing representational forms such as books, CD-ROMs, and animations. These studies indicate that the effectiveness of mixed-mode presentations has more to do with their match with comprehension processes than the medium of presentation. In other words, benefits of interactivity and animation are likely being overstated in the current milieu of fascination with multimedia.

M. Anderson, P. Cheng, and V. Haarslev (Eds.): Diagrams 2000, LNAI 1889, pp. 178-193, 2000.
© Springer-Verlag Berlin Heidelberg 2000

1   Introduction

Mixed-mode presentations consisting of verbal explanations illustrated with diagrams have long been used to communicate technical information. With the advent of multimedia, such representations have become dynamic and highly interactive, and have migrated from paper to the computer. Instead of the traditional combination of text and pictures on a static medium like paper, a designer can now choose from among a variety of static and dynamic verbal and visual media: e.g. static text, animated text, aural narratives, static diagrams, pictures, photographs, animations, and video. Furthermore, interactivity that was limited to scanning, rereading, annotating and page turning (back and forth) with the medium of paper expands to interactive control over the presentation of textual, aural and visual material on a computer. The conventional wisdom is that computer-based interactive multimedia presentations are more effective than printed presentations. However, does the communicative power of multimedia presentations stem from the careful design of their content and structure to match cognitive processes involved in comprehension or from their interactive and animated nature? One possibility is that content and structure are much more important than interactivity and dynamism. This will imply that the cost-benefit ratio of enhancing a static mixed-mode presentation with interactive and dynamic features is likely to be quite high. Another possibility is that interactive and dynamic features significantly increase the effectiveness of presentations. This will imply that even static mixed-mode presentations that have been carefully designed and proven to be effective can further benefit from the addition of interactive and dynamic features that today's interactive multimedia systems provide. This is an important issue that until now has not been seriously investigated.

This paper describes a research program designed to address this issue. We start with the thesis that mixed-mode presentations, static or dynamic, are more likely to be effective communicators of information when they are designed to be congruent with cognitive processes involved in comprehending information from external representations. Such a comprehension model, derived from a wealth of prior research on text comprehension, diagrammatic reasoning and mental animation, is presented first. This model highlights potential sources of comprehension error, which in turn, indicate guidelines for the design of mixed-mode presentations to ameliorate these. We show how the guidelines were used in designing interactive and animated multimedia presentations for teaching novices about dynamic systems in two domains - the concrete domain of mechanical systems and the abstract domain of computer algorithms. We then report on a series of four experimental studies in these two domains that compared interactive multimedia representations and their paper-based counterparts. Both were designed in accordance with the comprehension model and were compared against each other and against competing representational forms such as books, CD-ROMs, and animations. These studies indicate that the effectiveness of textual and diagrammatic representations has more to do with their match with comprehension processes than the medium of presentation. In other words, benefits of interactivity and animation are likely being overstated in the current milieu of fascination with multimedia.

2   Multimodal Information Integration and Comprehension: A Cognitive Model and Its Design Implications

We have developed a cognitive model of how people comprehend multimedia presentations, including text, static diagrams, animations, and aural commentaries, that explain how systems work. This model has been developed and generalized based on our research in the domains of mechanics and algorithms [3,7,19] and we are currently applying it to the domain of meteorology. The model views comprehension as a process in which the comprehender uses his or her prior knowledge of the domain and integrates it with the presented information to construct a mental model of the object or situation described. In addition to text comprehension skills, our model proposes that comprehension is dependent on spatial skills for encoding and inferring information from graphic displays, and integrating information in text and graphics [8].

According to this model, people construct a mental model of a dynamic system by first decomposing it into simpler components, retrieving relevant background knowledge about these components, and mentally encoding the relations (spatial and semantic) between components to construct a static mental model of the situation. They then mentally animate this static mental model, beginning with some initial conditions and inferring the behaviors of components one by one in order of the chain of causality or logic. This mental animation process depends on prior knowledge (e.g. rules that govern the behavior of the system in question) and spatial visualization processes. Thus, mental model construction under these circumstances requires the following stages. Although we list them sequentially, we acknowledge that they are not always accomplished in this linear order.

2.1   Stage 1: Decomposition

The system to be explained typically consists of individual components or variables. Pictorial representations employed to illustrate such systems contain several diagrammatic elements such as geometric shapes and icons. These represent elements of the target system. For example, a geometric shape might represent a mechanical component in a diagram of a machine or an icon might represent a measurement of atmospheric pressure in a meteorological chart. The first step in comprehension is to parse the graphical displays into units that correspond to meaningful elements of the domain (which may be composed of other elements in a hierarchical structure). This process might be considered to be analogous to identifying discrete words in a continuous speech sound.

In the mechanical domain, we have found that decomposition is often a hierarchical process. That is, complex visual units, representing complex elements of the domain, are often made up of sub-units. In the domain of algorithms, decomposition is the process of understanding the individual steps or operations from a technical verbal description, usually called the pseudocode. Decomposition of graphical representations can cause comprehension problems because diagrams are often under-specified in that they do not contain enough information for a user to identify whether two or more connected units represent separate elements or parts of a single element from the target domain [2]. Therefore, interactive displays ought to be designed to explicitly support accurate and early parsing of the graphical and symbolic representations they present.

2.2   Stage 2: Constructing a Static Mental Model by Making Representational Connections

A second major stage in comprehension involves making memory connections among the visual and verbal units identified during decomposition. This stage involves making two types of connections: (a) connections to prior knowledge and (b) connections between the representations of different domain elements.

Connections to Prior Knowledge. First, the comprehender must make connections between the visual and verbal units identified in the previous stage and their referents. These referents may be real-world objects, as in the case when a rectangle refers to a cylinder, or abstract objects, as in the case when the symbol A[I] refers to the Ith data element in an array named A. This process will be partially based on prior knowledge of both elements of the domain and the conventions used to portray the elements visually or verbally. Therefore, a mixed-mode presentation should contain realistic depictions of real-world components to facilitate accurate identification and recall of prior knowledge. If target users are not expected to have sufficient prior knowledge (this may be the case with abstract notations and icons), the relevant knowledge needs to be explicitly provided as part of the presentation. One way to accomplish this is by linking graphical components to explanations of the entities that they depict. For example, in the case of a meteorological display, moving the cursor over any part of an isobar might highlight the entire contour and pop up an explanation that the contour represents a region of equal pressure.

Connections to the Representations of Other Components. Second, the comprehender must represent the relations between different elements of the system. In a mechanical system, this involves understanding the physical interactions (connections and contact) between components. For more abstract causal domains such as meteorology, this involves noticing how variables depicted by different graphical elements co-vary in space or time and how they interact to cause some weather pattern. For a non-causal domain such as algorithms in which logic determines component interactions, this involves understanding how operations on data elements can have repercussions on other data elements. Mixed-mode presentations should be designed to make explicit the spatial, causal or logical connections among elements of the domain.

2.3   Stage 3: Making Referential Connections

An important component of integrating information from different representations is co-reference resolution, i.e., making referential links between elements of the different representations that refer to the same entity. For example, when diagrams are accompanied by text, the reader needs to make referential links between noun phrases in the text and the diagrammatic units that depict their referents [16]. Visual and verbal information about a common referent must be in working memory at the same time in order to be integrated [15]. Previous research indicates that people who see a text and-diagram description benefit if the text is presented close to the part of the diagram to which it refers [15]. If the visual representation is an animation, people benefit if the corresponding verbal representation is presented as an auditory commentary simultaneously with the animation [14,16]. These results imply that a mixed-mode presentation should be designed so that visual and verbal information are presented close together in space and in time.

By monitoring people's eye fixations as they read text accompanied by diagrams, Hegarty and Just [5] observed how people coordinated their processing of a text and diagram to integrate the information in the two media. Interpretation of eye-fixations in this research is based on the empirically validated eye-mind assumption [9], that when a person is performing a cognitive task while looking at a visual display the location of his or her gaze on the display corresponds to the symbol that he/she is currently processing in working memory. People began by reading the text and frequently interrupted their reading to inspect the diagram. On each diagram inspection, they tended to inspect those components in the diagram that they had just read about in the text. This research indicates that information in a text and diagram describing a machine is integrated at the level of individual machine components and that the text plays an important role in directing processing of the diagram and integration of the two media. It suggests the importance of providing text accompanying a diagram that directs people to the important information in the diagram.

2.4   Stage 4: Hypothesizing Lines of Action

The next stage of comprehension involves identifying the chains of events in the system, which are determined by causality or logic. Previous studies [4,20,21] have shown that people tend to infer the events in the operation of a system along the direction of causal propagation In complex physical systems with cyclical operations or branching and merging causal chains, finding the lines of action requires knowledge of both the spatial structure of the system and the temporal duration and ordering of events in its operation. This can introduce errors both in hypothesizing lines of action and in integrating lines of action of interacting subsystems. For example, the flushing cistern device (Figure 2) has two subsystems – one that brings water into the cistern and another that flushes water out. Lines of action (i.e., causal chains) in the operation of these two interacting subsystems are temporarily dependent on each other. In our studies of comprehension of this device from mixed-mode presentations we found that subjects were able to infer the motions of components within each causal chain, but had much more difficulty integrating information between the two causal chains. This suggests that mixed-mode presentations should make the lines of action in the various subsystems explicit. In addition, any temporal or spatial interactions between these subsystems should also be made explicit, in order to facilitate integration.

Even in the case of simple physical systems, providing animations of the events that occur in the system's operation may not be sufficient for accurate comprehension of lines of action. In complex or abstract domains, there is the additional problem that cause-effect relationships between observable behaviors of components may not be immediately evident from animations. Meteorology is a case in point. Lowe [11] found that novices erroneously ascribe cause-effect relationships to weather phenomena visible in typical meteorological animations based on their temporal relationships alone. In the domain of computer algorithms, the underlying logic, and not laws of physics, determines cause-effect relationships among variables. This strongly indicates a need for designing novel visualization techniques that make lines of action explicit in static and animated diagrams.

2.5   Stage 5: Constructing a Dynamic Mental Model by Mental Animation and Inference

The final stage of comprehension is that of constructing a dynamic mental model by integrating the dynamic behaviors of individual components of the system being explained. Cognitive and computational modeling [4,18] suggest that this is often accomplished by considering components individually, inferring their behaviors due to influences from other connected or causally related components, and then inferring how these behaviors will in turn affect other components. This incremental reasoning process causes the static mental model constructed in earlier stages of comprehension to be progressively transformed into a dynamic one. This stage can involve both rule-based inferences that utilize prior conceptual knowledge [28] and visualization processes for mentally simulating component behaviors [20,21,27,29].

The incremental reasoning process implies that comprehension or reasoning errors that occur early in this reasoning process can compound. Thus, when asked to predict the motion of a machine component from that of another component in the system, people make more errors and take more time if the components are farther apart in the causal chain [1,8]. A mixed-mode presentation should include information to prevent these kinds of errors. One way to accomplish this is to show snapshots of the system's operation at various points in the causal chain and encourage the user to mentally animate or simulate events that occur between snapshots prior to showing a complete animation. We have empirically validated the design guideline that learners should be induced to mentally animate a mechanical device before seeing a computer animation of its operation – they learn more from viewing the animation when this is done than if they passively view the animation [6].

3   Structure of Expository Mixed-Mode Presentations in Two Domains

Based on design guidelines derived from the comprehension model, we developed mixed-mode presentations that use multimedia to explain the inner workings of systems in two domains – the concrete domain of mechanical systems and the abstract domain of computer algorithms. Figure 1 shows the structure of the presentation in the domain of machines, which explains how a flushing cistern works. It is structured as two main modules and two supporting modules. The first module is designed to help the user parse the diagram, and construct representational and referential connections. It presents labeled cross-sectional diagrams of the machine and its subsystems in which users can click on individual components to highlight them. Users can also interactively explode and re-combine the machine diagrams, to further help them decompose the system. Figure 2 contains the image of a screen from this first module. The second module illustrates the actual operation of the machine using a narrated animation. The commentary explains the operation of subsystems in sequential order, accompanied by a flashing red arrow in the animation highlighting the corresponding causal order or lines of action. The user can also view a silent animation without the commentary and the red arrow. Two supporting modules provide access to explanations of fundamental principles involved in the operation of the machine and questions intended as a self-assessment tool for the user.

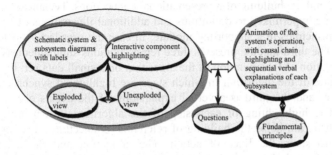

Fig. 1. Structure of a mixed mode presentation of machines

Sub-Systems

The purpose of the toilet tank is to flush water into the toilet bowl and to refill the tank with water. There are two basic subsystems in a toilet tank: a water output system and a water inlet system. You can see the Exploded view of the inlet and output systems.

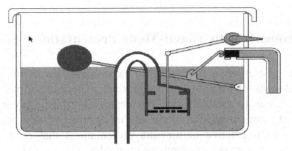

Fig. 2. One view of the flushing cistern manual

The algorithm domain differs from the mechanical domain in several important respects. Algorithms are abstract entities with no physical form. Components of algorithms are not physical objects; rather they are operations on data. An algorithm is like a recipe in that it is made up of a finite number of steps designed to solve a specific problem. The specification of an algorithm, analogous to the cross-sectional diagram of a machine, consists of a description of these steps in a semi-mathematical language called "pseudocode". Thus, the cognitive process corresponding to parsing the illustration of a machine into its components is parsing the pseudocode into the sequence of operations it represents. The structure of the mixed-mode presentation explaining computer algorithms (Figure 3) reflects these differences.

It consists of four main modules and two supporting modules. The first module explains the core operations of an algorithm and illustrates them using a familiar, real-world analogy. For instance, the analogy for the MergeSort algorithm shows animated playing cards, which are divided and merged to create a sorted sequence of cards. The second module is designed to aid the parsing process by presenting, side-by-side, the pseudocode description and a textual explanation of the algorithm. The explanation also serves to aid the building of representational connections. Technical terms in this explanation are hyperlinked to definitions and additional illustrations of fundamental algorithmic principles in a supporting module, in case the user lacks sufficient prior knowledge. The third module presents three representations simultaneously. One is a detailed animation of the operation of the algorithm on a small data set; the second is the pseudocode of the algorithm in which steps are highlighted synchronously with the animation, and the third is a verbal explanation of the events occurring in the animation. The highlighting of each step of the algorithm in synchrony with its graphical illustration helps the building of referential connections. This highlighting also serves to show the lines of action. The animation and the corresponding explanation help with the construction of a dynamic mental model of the algorithm's behavior. The fourth module is intended to reinforce this mental model by presenting an animation of the algorithm's operation on a much larger data set, without the other representations. It also allows the user to make predictions about parameters of the algorithm's behavior and compare these against actual values. Two supporting modules explain basic algorithmic principles and provide questions. Hypermedia manuals for several algorithms were built in accordance with this structure. Figure 4 contains a frame of an animation from the third module of this system.

4   Experiments with Mixed-Mode Presentations

This section describes a series of four experiments involving interactive mixed-mode presentations of machines and algorithms. These experiments had two goals. The first goal was to establish that interactive multimedia presentations designed according to the comprehension model are more effective than traditional means of presentation *and* interactive multimedia presentations that do not confirm to the guidelines derived from our comprehension model. The second goal was to compare the effectiveness of static and interactive mixed-mode presentations when both are designed according to our comprehension model. Thus, in each domain, the computer-based interactive presentation was first compared against competing representational forms such as books, CD-ROMs and other animations, and then compared against a static printed version of itself.

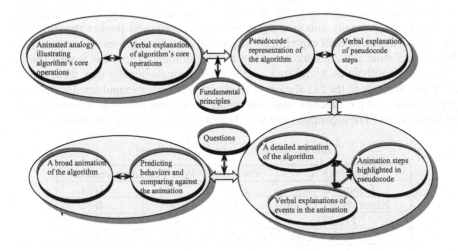

Fig. 3. Structure of a mixed mode presentation of algorithms

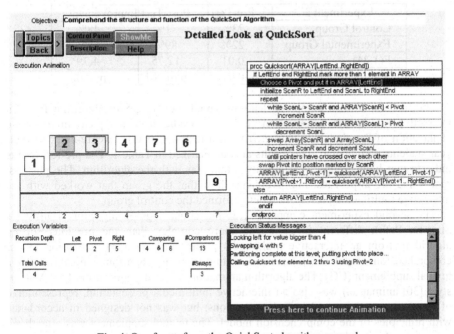

Fig. 4. One frame from the QuickSort algorithm manual

Three experiments in the algorithm domain used the following procedure. Subjects were second or third year undergraduate students majoring in computer science, who volunteered to participate. GPA and SAT/ACT scores were used to create two matched groups: a control group (referred to as CG) and an experimental group (referred to as EG). Their learning was measured by means of a pre-test/post-test combination that probed conceptual and procedural knowledge about algorithms. Students were tested on their ability to recognize and reorder pseudocode descriptions of algorithms, mentally simulate algorithmic operations, and predict resulting data structure changes. In each experiment, subjects began with a pre-test to measure prior knowledge about the algorithms used in that experiment. Then, the groups learned about the algorithms using different presentation methods. Finally they were given a post-test. Students in the EG always worked with an interactive multimedia algorithm presentation with the structure shown in Figure 3. No time limit was imposed for any of the activities. The algorithms that were taught were new to all participants.

Table 1. Statistical summary of Experiment 1

Experiment 1	Pre-Test	Post-Test	Improvement
Control Group	27%	43%	16%
Experimental Group	28%	74%	46%
F(1,27)	0.01	10.9	6.7
Significance level	p= 0.93	p<0.01	p<0.015

Table 2. Statistical summary of Experiment 2

Experiment 2	Pre-Test	Post-Test	Improvement
Control Group	23%	71%	48%
Experimental Group	22%	89%	68%
F(1,24)	0.01	12.75	4.79
Significance level	p=0.91	p<0.001	p<0.05

The first experiment compared learning about the MergeSort algorithm from an interactive multimedia presentation (EG) to learning from a typical mixed-mode explanation of the algorithm extracted from a textbook (CG). Twenty-eight second-year students participated. The results are summarized in Table 1. The pre-test results indicate that both groups were equally unfamiliar with the algorithm. The post-test scores and pre-to-post-test improvements show that the group that worked with the interactive multimedia presentation outperformed the control group.

The second experiment compared learning about a graph algorithm (Dijkstra's Shortest Path), different in style and more complex than the previous sorting algorithm, from an interactive multimedia presentation designed according to our comprehension model (EG) against learning from an algorithm animation and a textual supplement (CG). The algorithm animation we used (replication of a TANGO-style [30] animation) was also an interactive multimedia presentation, representative of previous research on algorithm animations, but was not designed in accordance with principles of the comprehension model. The textual supplement was a textbook extract similar to the one used in experiment 1. Twenty-five third-year students participated. The results are summarized in Table 2. The post-test scores and pre-to-post-test improvements show that the group that worked with our interactive multimedia presentation significantly outperformed the control group.

The third experiment compared learning about the QuickSort algorithm from an interactive multimedia presentation (EG) against learning from a printed version of the same presentation (CG). The interactive presentation was designed in accordance with the comprehension model. The static printed version was produced from the computer-based version, preserving the same sequence of screens, verbal information, and diagrams. When the original had an animation, the printed counterpart contained a sequence of diagrams showing the initial state of the animation, the final state, and several intermediate states. Hyperlinks to additional explanations in the original were replaced by parenthetical references to pages of an appendix containing the same explanations. Thus, interactive facilities were replaced by static counterparts wherever possible. The only differences between the interactive and printed versions were lack of dynamism (i.e., smooth animations were replaced with a series of pictures) and lack of interactive controls that could not be substituted on paper (e.g. animation speed controls, controls for changing the data input to animations). Thirty-eight third-year students participated. The results, summarized in Table 3, indicate that the groups learned and performed at similar levels on both the pre-test and post-test.

Table 3. Statistical summary of Experiment 3

Experiment 3	Pre-Test	Post-Test	Improvement
Control Group	7.048	11.095	4.048
Experimental Group	5.444	10	4.5
F(1,37)	0.629	0.6	0.163
Significance level	p= 0.3289	p= 0.283	p=0.689

Table 4. Statistical summary of Experiment 4

Experiment 4	Causal Chain	Function	Troubleshooting
HMM vs WTW-Printed	4.19, p <.01	0.732, p <.05	0.223, n.s.
HMM vs WTW-CD-ROM	4.64, p <.01	0.571, n.s.	0.929, n.s.
HMM vs HMM-Printed	-0.25, n.s.	-0.214 n.s.	-0.83, n.s.

A parallel experiment in the domain of machines made similar comparisons to those made in the three experiments on computer algorithms. In this experiment 15 undergraduate students studied the hypermedia flushing cistern manual described above (HMM) and 16 students studied a paper printout of the information in this manual (HMM-Printed). A further 16 students studied the description of a flushing cistern from the book "The Way Thing Work" [12] (WTW-Printed) and finally, 16 students studied the corresponding materials from "The New Way Things Work" CD-ROM (WTW-CD-ROM) [13]. Afterwards, all students wrote a description of how the device worked. They were instructed to imagine that they push down on the handle of the flushing cistern and to describe, step by step, what happens to each of the other components of the cistern as it flushes. Their score on this measure was the number of correct steps in the causal chain that they included in their description. Then they answered questions about the functions of different components of the device, e.g. "What is the function of the float and float arm?" Finally, they answered troubleshooting questions, which described faulty behavior of the system and asked what components might be responsible, e.g. "Suppose that after flushing the toilet, you notice that water is continuously running into the tank. What could be wrong? (list all possible answers)."

There was a significant effect of instruction on students' descriptions of how the machine worked ($\underline{F}$ (3, 52) = 9.59 $\underline{p}$ < .01). By analogy to the experiments in the domain of computer algorithms we make three comparisons. First we compare learning from the hypermedia manual designed according to our comprehension model (HMM) to learning from a typical mixed-mode explanation of the mechanical system extracted from a book (WTW-Printed). Table 4 shows difference scores for the two conditions being compared and the significance level of post-hoc tests (Fisher's PLSD). Students who learned from the HMM included more causal steps in their descriptions of how the device worked (mean 12.00) compared to those who read the WTW-Printed mean (7.81), and were better able to answer function questions. There were no differences between the groups in answering troubleshooting questions. For the other two comprehension measures (causal chain and function), the results were comparable to those in Experiment 1 in the computer algorithm domain.

Next we compare learning from the interactive hypermedia manual designed according to our comprehension model (HMM) to learning from an award-winning commercially available interactive presentation that was not informed by our model (WTW-CD-ROM). As shown in Table 4, students who learned from our HMM outperformed students who learned from the CD-ROM on the measure of describing how the system worked (12.00 causal steps vs 7.36 described). There were no differences between the groups on function or troubleshooting questions. The differences in ability to describe how the system worked can be compared to those in Experiment 2 in the computer algorithm domain.

Finally we compare learning from our interactive hypermedia manual (HMM) to learning from the same information presented on paper (HMM-Printed). In this case we find no differences on any of the comprehension measures (12.25 causal steps described by the HMM-Printed group). These results are similar to those of Experiment 3 in the algorithm domain.

5   Discussion

Figure 5 shows the research roadmap we have pursued simultaneously in two domains – simple machines and computer algorithms – that differ in a number of ways. In the domain of simple machines, systems are made up of components that are familiar and have a physical manifestation, and operate according to laws of physics and causality, of which even novices have intuitive knowledge. In the domain of algorithms, systems consist of components that are unfamiliar and do not have physical form, and operate according to laws of mathematics and logic, of which novices do not have an intuitive knowledge. We have discovered similar results in both domains – that mixed-mode presentations designed according to a cognitive model outperform both traditional and computer-based presentations. However, this effectiveness appears to stem primarily from structure and content that can be replicated in static mixed-mode presentations on paper, and not from the interactivity and animation afforded by new media.

Although we replicated the same results in the two domains, the advantages of the presentations designed according to our cognitive model were not as strong in the mechanics domain, particularly on the measure of troubleshooting. It is possible that participants already had prior knowledge of possible faults in the operation of a flushing cistern, based on their everyday experience or intuitive knowledge, and this reduced the possible effects of our instruction. Alternatively, our design guidelines might be more effective in abstract domains, in which the diagrams and animations are *true* visualizations, rather than in the domain of machines, in which animations show processes that are visible in the real world. Future research, including our current research in the meteorology domain, will address these issues further.

In this paper, we have treated the issues of static vs. dynamic media and match with a comprehension model as separate and orthogonal hypotheses. However it is possible that in some situations a dynamic presentation might be more compatible with a cognitive model than a static presentation. Engaging animations and interactivity may have an impact on motivation. However, static presentations may also have benefits, such as promoting active learning and engaging cognitive processes of comprehension. There is also a real risk of the "hands-on, minds-off" problem, in which the sensory stimulation and interactivity of multimedia and its externalization of information distract from, and discourage, the cognitive processing that deep learning requires.

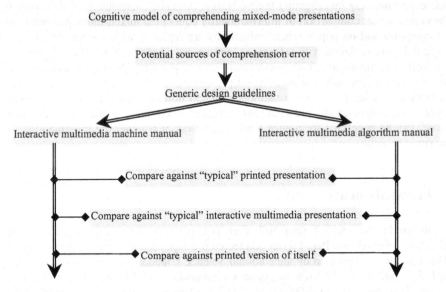

Fig. 5. Research roadmap

Our results can be related to a growing literature on the possible benefits, or otherwise, of animations. One research program [24, 25] showed advantages of animations over static graphics in teaching Newtonian Mechanics to fourth and fifth grade students, but found no such difference in teaching the same content to university students. Another study showed an advantage of an animation over written instructions to perform a task using a Graphical User Interface, but this difference was eliminated after practice on the task [22]. The researchers speculated that animated demonstrations of procedural tasks may encourage processing at a superficial level (a form of mimicry), which does not lead to long-term retention and transfer. Experiments in the domain of meteorology [10, 11] indicate that novices have poor comprehension of both static and animated weather maps. In both cases, the information they extract is perceptually salient rather than thematically relevant, and inappropriate causal attributions are made to changing weather phenomena. A recent review concluded that in all previous studies that showed advantages of animations over static media, the materials were not informationally equivalent, i.e. the animations presented more information than the static graphics [17]. An exception to this is a study [23] in which static informationally equivalent versions of multimedia lessons on dynamic processes in biology were systematically constructed and compared against multimedia. The researchers found no significant evidence that the multimedia presentations further enhanced student understanding of declarative information in the lessons when compared to static presentations.

Research reported here addresses both a theoretical gap in our understanding of how external graphical representations work and a lack of practical design guidance for such representations [26] by developing a cognitive model of multimodal comprehension, designing mixed-mode presentations in accordance with the model, and experimentally investigating their efficacy. A major contribution of this work is the experimental comparison of informationally equivalent mixed-mode presentations on computer and on paper, where animations are replaced with a series of pictures, hyperlinks are replaced with references, and so on. There is very little literature on this sort of comparison, which is essential to prove purported benefits of interactivity and animation that only computer-based presentations can provide. Our experiments indicate that at least in two substantially different domains, mechanics and computer algorithms, mixed-mode presentations designed according to the comprehension model can be beneficial for novice adults regardless of whether the presentations are dynamic or static.

Acknowledgments

Steven Hansen and Teresa Hübscher-Younger conducted experiments on the algorithm manual, and Pam Freitas conducted the experiment on the machine manual. This research is supported by the National Science Foundation under contracts CDA-9616513 and REC-9815016 to Auburn University, and by the Office of Naval Research under contract N00014-96-11187 to Auburn University and N00014-96-10525 to the University of California.

References

1. Baggett, W. B., Graesser, A. C.: Question Answering in the Context of Illustrated Expository Text. In: Proc. 17th Annual Conference of the Cognitive Science Society. Lawrence Erlbaum Associates, Hillsdale NJ (1995) 334-339
2. Ferguson, E. L., Hegarty, M.: Learning with Real Machines or Diagrams: Application of Knowledge to Real-World Problems. Cognition and Instruction **13** (1995) 129-160
3. Hansen, S. R., Narayanan, N. H., Schrimpsher, D., Hegarty, M.: Empirical Studies of Animation-Embedded Hypermedia Algorithm Visualizations. Tech. Rep. CSE98-06. Computer Science and Software Engineering Dept. Auburn University (1998)
4. Hegarty, M.: Mental Animation: Inferring Motion from Static Diagrams of Mechanical Systems. JEP: Learning, Memory and Cognition **18**(5) (1992) 1084-1102
5. Hegarty, M., Just, M. A.: Constructing Mental Models of Machines from Text and Diagrams. J. Memory and Language **32** (1993) 717-742
6. Hegarty, M., Narayanan, N. H., Freitas, P: Understanding Machines from Multimedia and Hypermedia Presentations. In: Otero, J., Leon, J. A., Graesser, A. (eds.): The Psychology of Science Text Comprehension. Lawrence Erlbaum Associates, Hillsdale (to appear)
7. Hegarty, M., Quilici, J., Narayanan, N. H., Holmquist, S., Moreno, R:. Multimedia Instruction: Lessons from Evaluation of a Theory Based Design. J. Edu. Multimedia and Hypermedia **8**(2) (1999) 119-150
8. Hegarty, M., Sims, V. K.: Individual Differences in Mental Animation During Mechanical Reasoning. Memory and Cognition **22** (1994) 411-430
9. Just, M. A., Carpenter, P. A.: Eye Fixations and Cognitive Processes. Cog. Psy. **8** (1976) 441-480
10. Lowe, R. K.: Selectivity in Diagrams: Reading Beyond the Lines. Edu. Psy. **14**(4) (1994) 467-491
11. Lowe, R. K.: Extracting Information from an Animation during Complex Visual Learning. European J. Psychology of Education **14**(2) (1999) 225-244
12. Macaulay, D.: The Way Things Work. Houghton Mifflin Company, Boston (1988)
13. Macaulay, D.: The New Way Things Work. CD-ROM. DK Interactive Learning, New York (1998)
14. Mayer, R. E., Anderson, R. B.: The Instructive Animation: Helping Students Build Connections between Words and Pictures in Multimedia Learning. J. Edu. Psy. **84**(4) (1992) 444-452
15. Mayer, R. E., Gallini, J.: When is an Illustration worth Ten Thousand Words? J. Edu. Psych. **83** (1990) 715-726
16. Mayer, R. E., Sims, V. K.: For whom is a Picture worth a Thousand Words? Extensions of a Dual-Coding Theory of Multimedia Learning. J. Edu. Psy. **86**(3) (1994) 389-401
17. Morrison, J. B., Tversky, B., Betrancourt, M.: Animation: Does It Facilitate Learning? In: Smart Graphics: Papers from the AAAI Spring Symposium. Technical Report SS-00-04. AAAI Press, Menlo Park CA (2000) 53-60

18. Narayanan, N. H., Chandrasekaran, B.: Reasoning Visually about Spatial Interactions. In: Proc. 12th Int. Joint Conference on Artificial Intelligence. Morgan Kaufmann Publishers, Menlo Park CA (1991) 360-365

19. Narayanan, N. H., Hegarty, M.: On Designing Comprehensible Interactive Hypermedia Manuals. Int. J. Human-Computer Studies **48** (1998) 267-301

20. Narayanan, N. H., Suwa, M., Motoda, H.: A Study of Diagrammatic Reasoning from Verbal and Gestural Data. In: Proc. 16th Annual Conference of the Cognitive Science Society. Lawrence Erlbaum Associates, Hillsdale NJ (1994) 652-657

21. Narayanan, N. H., Suwa, M., Motoda, H.: Diagram-Based Problem Solving: The Case of an Impossible Problem. In: Proc. 17th Annual Conference of the Cognitive Science Society. Lawrence Erlbaum Associates, Hillsdale NJ (1995) 206-211

22. Palmiter, S., Elkerton, J.: Animated Demonstrations for Learning Procedural Computer-Based Tasks. Human-Computer Interaction **8** (1993) 193-216

23. Pane, J. F., Corbett, A. T., John, B. E.: Assessing Dynamics in Computer-Based Instruction. In: Proc. ACM Human Factors in Computing Systems Conference. ACM Press, New York (1996) 797-804

24. Rieber, L. P.: Using Computer Animated Graphics in Science Instruction with Children. J. Edu. Psy. **82**(1) (1990) 135-140

25. Rieber, L. P., Boyce, M. J., Assad, C.: The Effects of Computer Animation on Adult Learning and Retrieval Tasks. J. Computer-Based Instruction **17**(2) (1990) 46-52

26. Scaife, M., Rogers, Y.: External Cognition: How Do Graphical Representations Work? Int. J. Human-Computer Studies **45** (1996) 185-213

27. Schwartz, D. L., Black, J. B.: Analog Imagery in Mental Model Reasoning: Depictive Models. Cognitive Psychology **30** (1996) 154-219

28. Schwartz, D. L., Hegarty, M.: Coordinating Multiple Representations for Reasoning about Mechanical Devices. In: Cognitive and Computational Models of Spatial Representation: AAAI Spring Symposium Papers. Tech. Rep. SS-96-03. AAAI Press, Menlo Park (1996)

29. Sims, V. K., Hegarty, M.: Mental Animation in the Visual-Spatial Sketchpad: Evidence from Dual-Task Studies. Memory and Cognition **25**(3) (1997) 321-332

30. Stasko, J.: TANGO: A Framework and System for Algorithm Animation. IEEE Computer **23**(9) (1990) 27-39

Capacity Limits in Diagrammatic Reasoning

Mary Hegarty

Department of Psychology, University of California,
Santa Barbara, CA 93106
hegarty@psych.ucsb.edu

Abstract. This paper examines capacity limits in mental animation of static diagrams of mechanical systems and interprets these limits within current theories of working memory. I review empirical studies of mental animation that examined (1) the relation of spatial ability to mental animation (2) the effects of working memory loads on mental animation, (3) use of external memory in mental animation and (4) strategies for task decomposition that enable complex mental animation problems to be accomplished within the limited capacity of working memory. The effects of capacity limits on mental animation are explored by implementing a simple production system model of mental animation in the 3CAPS production system architecture, limiting the working memory resources available to the model, and implementing strategies for managing scarce working memory resources. It is proposed that mental animation involves the visual-spatial and executive components of working memory and that individual differences in mental animation reflect the operation of these working memory components.

1 Introduction

Studies of visual-spatial cognition have focused primarily on relatively simple tasks, such as scanning visual-spatial images [17], mental rotation [26] and perspective taking [10]. Diagrammatic reasoning tasks are also examples of visual-spatial cognition in that they are at least partially based on an external visual-spatial representation, i.e. a diagram. However they are more complex in that they typically involve making a series of inferences or spatial transformations based on a diagram, and these visual-spatial cognitive processes are embedded in the larger context of reasoning or solving a problem. In diagrammatic reasoning, as in any complex cognitive task, it is important to consider how the task is carried out within the capacity limits of human working memory.

In this paper I will consider a diagrammatic reasoning task that involves inferring the behavior of a mechanical system from a visual-spatial representation (a static diagram). I refer to this process as <u>mental animation</u> [5]. Mental animation is an

M. Anderson, P. Cheng, and V. Haarslev (Eds.): Diagrams 2000, LNAI 1889, pp. 194-206, 2000.

example of reasoning in that it involves inferring new information from the information given in the diagram. It is an example of spatial cognition in that the input to the inference process is an external spatial representation, i.e. a static diagram.

Consider the sample mental animation problems in Figure 1. These problems are more complex than mental scanning [17] in that they involve inferring motion rather than inspecting a static spatial representation. They are more complex than mental rotation [26] in that they involve imagining the motion of several interacting objects that move simultaneously in different ways, rather than the motion of a single object.

Mental animation tasks are therefore examples of complex visual-spatial inference. In this paper I argue that capacity limits in visual-spatial working memory are a major limiting factor in mental animation. However these limits can be compensated for by strategic processes for decomposing the task and managing limited resources, allowing for quite complex spatial inferences within the limited capacity of spatial working memory. Taking mental animation of mechanical systems as an example, this paper will therefore examine capacity limits in reasoning from diagrammatic representations, and interpret these limits within current theories of working memory.

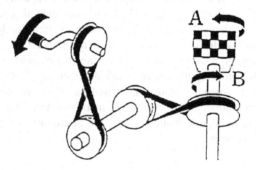

When the handle is turned in the direction shown, in which direction (A or B) will the box turn? .

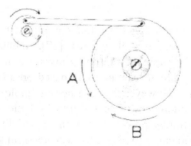

When the little wheel turns around, the big wheel will turn? Direction A, direction B or both directions?

Fig. 1. Examples of mental animation items

2   Capacity Limits in Visual-Spatial Cognition

The notion of capacity limits has been central to theories of imagery. In the dominant model of mental imagery [16], these limits are conceptualized as limitations of the extent and resolution of a spatial buffer. This model can account for the processing of static pictorial images, in which the amount of detail in the image is proportional to the size of the image in the limited resolution buffer, and for simple transformations of static images, such as scanning and zooming. In contrast, mechanical reasoning involves tasks in which the spatial aspects of objects (especially their connectivity) are more important than their visual details, and in which the motions of several interacting objects must be represented.

Another way of conceptualizing limitations of mental imagery is based on theories of working memory [1, 19]. A dominant theory of working memory [1] distinguishes three components of the working memory system, a central control structure called the central executive and two more peripheral, domain-specific systems, the visual-spatial sketchpad and the phonological loop, specialized for maintaining visual-spatial and verbal information respectively. The visual-spatial sketchpad might be seen as the site of mental imagery processes [18]. It is proposed that the visual-spatial and verbal systems have limited resources. Although there has been a lot of research elucidating the verbal component of working memory [14] there has been much less work on the spatial component.

Early models of the visual-spatial component of working memory conceptualized it as a buffer that is merely responsible for maintaining information in memory. However, more recent evidence suggests that the dissociation between verbal and spatial working memory is at the level of mental processes, and not just maintenance of information [25]. In this view, maintenance and processing compete for common resources, often conceptualized as a limited amount of activation [15]. Therefore, one might be able to imagine a complex image, but lose part of the image when one tries to transform it, or one might be able to imagine a simple transformation but not a complex transformation of an image at a particular level of complexity.

Individual differences in spatial ability have been related to the operation of the spatial working memory system. Just and Carpenter [13, 4] have modeled individual differences in mental rotation tasks in terms of differences in working memory capacity. Shah & Miyake [25] developed spatial working memory span tasks analogous to the word span and reading span tests used in studies of verbal working memory. These tasks were highly correlated with paper-and-pencil tests of spatial ability and dissociated from verbal span tests and tests of verbal ability. More recent research has confirmed that spatial ability tests are dependent on the spatial component of working memory, and has also shown that more complex spatial tests are also somewhat dependent on the executive component of working memory [20, 7].

3   A Brief Review of Research on Mental Animation.

3.1   Mental Animation Depends on Visual-Spatial Working Memory

Previous research has provided two types of evidence that mental animation depends on visual-spatial working memory resources [8, 27, 6]. In these studies, the task was to infer how different components of a pulley system move when the free end of the rope was pulled (see Figure 2). First, Hegarty & Sims [8] found that ability to mentally animate mechanical systems is highly related to spatial visualization ability, as measured by psychometric tests, but was not related to verbal ability. Specifically, low-spatial individuals made more errors on mental animation problems but did not differ in their reaction time to solve problems. The difference in performance between low-spatial and high-spatial individuals was greater on more complex mental animation problems. If we conceptualize individual differences in spatial ability as differences in operation of the visual-spatial working memory system, then mental animation seems to depend primarily on this system.

When the free end of the rope is pulled,
the middle pulley turns counterclockwise.

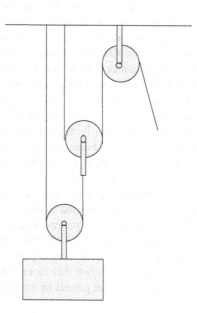

Fig. 2. Sample mental animation item. The task is to verify whether the sentence is true or false

Another type of evidence, often used in studies of working memory, is based on the dual task methodology [cf. 1]. In dual task studies we measure the extent to which a primary task of interest (in this case mental animation) interferes with different secondary tasks, assumed to depend on different components of the working memory system. Sims & Hegarty [27] measured the interference between mental animation tasks and verbal reasoning tasks (primary tasks) and secondary tasks involving maintenance of a visual-spatial working memory load (assumed to tap the visual-spatial component of working memory) and a verbal working memory load (assumed to tap the verbal component of working memory). They found that a visual-spatial working memory load interferes more with mental animation than a verbal working memory load and that mental animation interferes more with a visual-spatial working memory load than does a verbal reasoning task that takes approximately the same amount of time.

It should be noted that not all mental animation tasks are necessarily dependent on spatial processing resources. In relatively simple tasks, such as inferring the motion of interlocking gears, with repeated trials, people infer a simple verbal rule, e.g., that every other gear in the chain turns the same direction [23]. Performance on these tasks is relatively less related to spatial ability and somewhat related to verbal ability [6]. This paper will be primarily concerned with mental animation of pulley systems, which has been shown to be highly dependent on spatial processing resources.

3.2 Mental Animation is Piecemeal

Other research has suggested that people do not mentally animate the motion of all components of a mechanical system in parallel, but decompose the system into components and infer the motion of components one by one, following the causal chain of events [5]. I refer to this as the "piecemeal" strategy of mental animation. Consistent with this strategy, when people infer the motion of components in a pulley system, they take more time to infer the motion of components later in the causal chain compared to components earlier in the causal chain. Second, people's eye fixations reveal that when asked to verify how a component will move, they look at the component in question and components earlier in the causal chain, but not components later in the causal chain. Third, the errors made by low-spatial individuals are primarily in inferring the motion of components later in the causal chain.

The piecemeal strategy requires people to store intermediate results of their inferences. That is, people who make an inference about a sub-component of a device and do not immediately propagate the result to the next sub-component, must store the results for later inference. Given the trade-off between storage and processing in working memory, this might also interfere with the inference process, causing the pattern of errors observed for low-spatial individuals. In this situation, a useful strategy is to offload information onto the external display, for example by drawing an arrow on each mechanical component in a diagram as its direction of motion is inferred. This relieves people of the necessity to maintain this information in working memory, so that working memory can be freed up for mental animation. Hegarty &

Steinhoff [9] allowed some people to make written notes on diagrams of mechanical systems in a mental animation experiment. Although only about half of those who were allowed to make notes actually did so, making notes improved performance as predicted – low-spatial subjects who made notes had fewer errors than those who did not make notes.

4   A Model of Mental Animation

Hegarty [5] proposed a production system model of mental animation of pulley systems that models the piecemeal strategy described above. In this model it was assumed that a pulley system is represented as a set of connected components (pulleys, a weight, rope strands etc.). The model begins with knowledge of how the system components are connected (a structural description) and knowledge that the free end of the pull rope is being pulled. The model then infers how each component of the pulley system moves by applying rules of mechanical reasoning, such that each rule infers the motion of a single component of the pulley system and this inference is made from knowledge of the motion of a component that is adjacent to it (i.e. touches it).

In this model, the application of production rules is not meant as a literal model of mental animation. There is evidence that when inferring the motion of two interacting mechanical components, people actually use an analog imagery process, such that the time it takes them to infer the motion of two interlocking gears is proportional to the angle of rotation [24]. Furthermore, in another mental animation task, Schwartz [22] found that people can make the correct inference only through dynamic analog imagery and not through rule-based reasoning. Although production rules do not model the analog nature of mental animation, they are intended to model the following two aspects:

1.   The motion of a component can only be simulated from that of an adjacent component.
2.   The number of components whose motions can be simultaneously animated is limited.

Hegarty's [5] model provided a good fit to the reaction time data. For example, it accounted for 90% of the variance in time spent inspecting the diagram while inferring how a component of a pulley system would move. However it was limited in that it was a pencil-and-paper model, did not account for the errors of low-spatial individuals, and did not address the precise working memory demands of the task, i.e., the tradeoff between storing intermediate results and inferring the motion of new components. To model these demands, the model was implemented in 3CAPS [15]. 3CAPS is a production-system architecture that has been developed largely in the context of theories of working memory. It has the following distinct characteristics.

• In the 3CAPS architecture elements are not considered to be merely in or not in working memory, but rather they are assigned different levels of activation. The activation of a given element can range from 0 to 1.

- 3CAPS allows one to create different pools of working memory resources to model, for example, the spatial and verbal working memory systems.
- 3CAPS allows one to limit or "cap" the total amount of activation available to any given pool of resources.

In the 3CAPS model, the mental animation process was modeled as taking place in a spatial working memory store. The production system contained three productions that successively inspected and read in information from an unlimited "external" memory to spatial working memory, and 6 productions that inferred the motion of components. The external memory was intended to represent the diagram that is continuously visible to participants during a mental animation trial. Mental animation of a simple two pulley system was modeled (see Figure 3). When no limits were placed on spatial working memory, the production system took 12 cycles (listed in Table 1) to successively inspect and mentally animate all components of the pulley system. The components were inspected and animated one by one and in the order of the causal chain of events in operation of the pulley system.

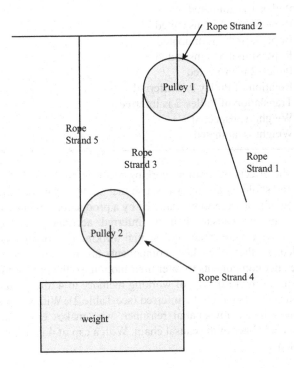

Fig. 3. Diagram of the simple pulley system mentally animated by the 3CAPS simulation. The rope is divided into separate strands that move differently when the pulley system is animated. Robe strand 2 is the section of rope that lies above Pulley 2 and Rope Strand 4 is the section of rope that lies under Pulley 2

Table 1. Description of the action taken on each cycle of the 3CAPS production system trace. "Inspection" of a component means that information about the configuration of that component (what it is attached to etc.) is read into spatial working memory. Note that in the 3CAPS architecture, productions can fire in parallel if more than one production is matched on a given cycle

Cycle	Action Taken
1	Rope Strand 1 is inspected (simulation is given knowledge that this rope strand is being pulled)
2	Rope Strand 1 is animated (i.e. its direction of motion is inferred)
3	Rope Strand 2 is inspected
4	Rope Strand 2 is animated
5	Rope Strand 3 is inspected
	Pulley 1 is inspecte
6	Rope Strand 3 is animated
	Pulley 1 is animated
7	Rope Strand 4 is inspected
8	Rope Strand 4 is animated
9	Rope Strand 5 is inspected
	Pulley 2 is inspected
10	Rotation of Pulley 2 is inferred
	Translation of Pulley 2 is inferred
11	Weight is inspected
12	Weight is animated

To examine the effects of limiting working memory, the spatial store was capped at 3 items and a threshold parameter was set so that an item in working memory had to have an activation of at least .5 to be matched by a production. In this case, the motion of earlier items in the causal chain is inferred accurately. However as more components of the system are "read into" spatial working memory, the activation of all items is degraded, so that when later components are read in, there is not enough activation of the later components to infer their motion, so the production system halts at cycle 8. Increasing the resources of working memory to 4 allows the motion of all components in the causal chain to be inferred (see Table 2). With a cap of 3 items, the simulation behaves like a low-spatial reasoner, who makes errors in inferring the motion of components later in the causal chain. With a cap of 4 items, it behaves like a high-spatial reasoner.

Table 2. Activation of elements in working memory for each cycle of the simulation with spatial working memory capped at 4 items. Items must have an activation of at least .5 to be matched by a production rule. Although the activation of some elements falls below this level in cycles 11 and 12, knowledge of the motion of these items is not needed to infer the motion of the weight, so all components are mentally animated

Cycle	RS1	RS2	P1	RS3	RS4	P2	RS5	W	Tot
1	1								1
2	1								1
3	1	1							2
4	1	1							2
5	1	1	1	1					4
6	1	1	1	1					4
7	.8	.8	.8	.8	.8				4
8	.8	.8	.8	.8	.8				4
9	.53	.53	.53	.53	.53	.66	.66		4
10	.53	.53	.53	.53	.53	.66	.66		4
11	.43	.43	.43	.43	.43	.53	.53	.8	4
12	.43	.43	.43	.43	.43	.53	.53	.8	4

One way of conceptualizing individual differences in performance on mental animation is therefore to assume that high- and low-spatial individuals have different spatial working memory capacities. However another possible difference is that high- and low-spatial individuals differ in how they manage scarce working memory resources. In a further version of the production system model, I added three productions that "cleaned up" the contents of working memory when they were no longer needed to make a further inference. These productions deleted information about a component from spatial working memory if it had already been mentally animated and it was not connected to a component that was yet to be animated. For example, in the case of the pulley system in Figure 3, once the motion or Rope Strand 2 and Pulley 1 had been inferred, information about Rope Strand 1 would be deleted from spatial working memory. When these productions were added, the motion of all components in the simple pulley system could be inferred with a cap of 3 items in spatial working memory.

5   Discussion

In summary, mental animation is conceptualized as occurring in a limited-capacity spatial working memory system. Implementing a production system of a simple mental animation task in the 3CAPS architecture allows us to precisely specify the working memory demands of a mental animation task. The model produces behavior that is similar to human mental animation performance in that time (number of cycles taken) to mentally animate a component is related to its position in the causal chain. When a limit is placed on the spatial working memory resources available to the simulation, it models the behavior of low-spatial participants in that the simulation is able to mentally animate earlier components in the causal chain, but not later components. Differences between capacity limited and unlimited versions of the simulation are in accuracy and not in reaction time, similar to the differences between high- and low-spatial individuals observed by Hegarty & Sims [8].

The model suggests two possible differences between people who are successful and unsuccessful at mental animation. One possible difference is that high-spatial reasoners have more spatial working memory resources than low-spatial reasoners. Another possibility is that high-spatial reasoners have the same working memory resources, but differ in their strategies for managing these resources. Such a strategy difference was implemented in the simulation by deleting elements from spatial working memory once they have been animated. A more robust strategy would be to alter the external display of the problem to reflect earlier inferences [9]. Research to date suggests that both resource differences and strategy differences are responsible for individual differences in mental animation. Hegarty & Steinhoff [9] found that when given the opportunity to make notes on diagrams, only some students did so. These students were able to use a strategy to compensate for limited spatial working memory resources. Other students were not able to use this strategy and were constrained by their limited resources, such that they were unable to mentally animate components later in the causal chain of a mechanical system. Furthermore, strategies for managing limited resources are not equally applicable to all mental animation problems because not all mechanical systems can be decomposed and mentally animated piecemeal [6]. Mental animation of problems that cannot be decomposed is particularly highly related to spatial ability, because these problems are more dependent on spatial working memory resources.

The two possible differences between high- and low-spatial people reflect differences in different components of the proposed working memory system. Differences in spatial working memory resources clearly involve the proposed visual-spatial component of working memory. Differences in task decomposition, scheduling and coordinating of task specific goals and suppression of irrelevant information are ascribed to the operation of the central executive [19]. The analysis presented in this paper therefore suggests that for complex visual-spatial tasks, both the spatial and executive components of working memory play an important role in performance. In an analysis of spatial abilities tasks, Miyake et al [20] found that this was indeed the case – more complex tests that load on the spatial visualization factor (e.g. paper-folding and form-board tests) were more related to executive tasks than simpler spatial tasks (e.g. speeded mental rotation). Hegarty & Kozhevnikov [6], in turn, showed that more complex mental animation tasks are more highly correlated with spatial visualization ability than with simpler spatial abilities.

More generally, this research shows that precisely specifying the working memory demands of a mental animation task gives us insight into which types of problems are most difficult for people, especially those with low spatial ability. This information is of practical value in the design of both multimedia instruction about how machines work [6] and in the design of Computer Aided Design systems, because it specifies when people are able to mentally animate a mechanical system on their own, and when an interactive system should give them additional support. Furthermore, there is currently much interest in the possibility of improving the spatial skills of individuals, especially in domains such as engineering, where spatial skills can be important for success [e.g. 11]. This research provides a precise account of a strategy that might be taught to low-spatial individuals to improve their performance on mental animation tasks.

Further research is needed to examine the effects of working memory limits on other diagrammatic reasoning tasks. For example, mental models theories have characterized difficulties in both syllogistic reasoning and spatial reasoning as limitations in the number of situations consistent with a set of premises that can be maintained in working memory at a time [3, 12]. Other theories have been concerned with the similarities and differences between reasoning based on graphical versus linguistic theories [28]. A precise specification of the working memory demands of these reasoning tasks, and the extent to which they involve different components of the working memory system might provide insights into both human reasoning and the extent to which it employs the representations and processes of different working memory components.

In conclusion, this work combines insights from research on spatial abilities and working memory to provide an account of performance on a diagrammatic reasoning task. It suggests that capacity limits in visual-spatial working memory are a major limiting factor in mental animation. However these limits can sometimes be compensated for by strategic processes that decompose the task and manage limited resources, allowing for quite complex spatial inferences within the limited capacity of spatial working memory.

Acknowledgments

I thank Sashank Varma for fruitful discussions and technical assistance in developing the 3CAPS simulation. This research was supported in part by contract N00014-96-10525 from the Office of Naval Research.

References

1. Baddeley, A. D.: Working Memory. Oxford University Press, New York (1996).
2. Bennett, C. K.: Bennett Mechanical Comprehension Test, The Psychological Corporation, San Antonio (1969).
3. Byrne, R. M. & Johnson-Laird, P. N.: Spatial reasoning. Journal of memory and Language, 28 (1989) 564-575..
4. Carpenter, P. A., Just, M. A., Keller, T. A., Eddy, W. & Thulborn, K.: Graded functional activation in the visuospatial system with the amount of task demand. Journal of Cognitive Neuroscience, 11(1999) 9-24.

5. Hegarty, M.: Mental animation: Inferring motion from static diagrams of mechanical systems. Journal of Experimental Psychology: Learning, Memory & Cognition, 18 (1992) 1084-1102.
6. Hegarty, M. & Kozhevnikov, M.: Spatial abilities, working memory and mechanical reasoning. In J. Gero & B. Tversky (Eds.) Visual and Spatial Reasoning in Design. Key Centre for Design and Cognition, Sydney, Australia. (1999).
7. Hegarty, M., Shah, P & Miyake, A.: Constraints on using the dual-task methodology to specify the degree of central executive involvement in cognitive tasks. Memory & Cognition (in press).
8. Hegarty, M. & Sims, V. K.: Individual differences in mental animation during mechanical reasoning. Memory & Cognition, 22 (1994) 411-430.
9. Hegarty, M., & Steinhoff, K.: Individual differences in use of diagrams as external memory in mechanical reasoning. Learning and Individual Differences, 9 (1997) 19-42.
10. Hintzman, D. L., O'Dell, C. S., & Arndt, D. R.: Orientation in cognitive maps. Cognitive Psychology, 13 (1981) 149-206.
11. Hsi, S., Linn, M. C., & Bell, J. E.: The role of spatial reasoning in engineering and the design of spatial instruction. Journal of Engineering Education, 86 (1997) 151-158.
12. Johnson-Laird, P. N.: Mental Models. Cambridge, MA: Harvard University Press (1983).
13. Just, M. A. & Carpenter, P. A.: Cognitive coordinate processes: Accounts of mental rotation and individual differences in spatial ability. Psychological Review, 8 (1985) 441-480.
14. Just M. A. & Carpenter, P. A.: A capacity theory of comprehension: Individual differences in working memory. Psychological Review, 99 (1992) 122-149.
15. Just, M. A., Carpenter, P. A. & Hemphill, D. D.: Constraints on processing capacity: Architectural or implementational. In D. M. Steier & T. M., Mitchell, (Eds.) Mind matters: A tribute to Allen Newell. Mahwah, N.J.: Erlbaum (1996).
16. Kosslyn, S. M.: Image and Mind, Cambridge, MA: Harvard University Press (1980).
17. Kosslyn, S. M. Ball, W. & Reiser, B.: Visual images preserve metrical spatial information: Evidence from studies of image scanning. Journal of Experimental Psychology: Human Perception and Performance, 4 (1978) 47-60.
18. Logie, R. H.: Visuo-spatial working memory. Hove, UK: Lawrence Erlbaum Associates (1995).
19. Miyake, A. & Shah, P.: Models of working memory: Mechanisms of active maintenance and executive control. New York, NY: Cambridge University Press (1999).
20. Miyake, A., Rettinger, D. A., Friedman, N. P., Shah, P. & Hegarty, M.: Visuospatial working memory, executive functioning and spatial abilities. How are they related? Manuscript submitted for publication (1999).
21. Narayanan, N. H. & Hegarty, M.: On designing comprehensible interactive hypermedia manuals. International Journal of Human-Computer Studies, 48 (1998) 267-301.

22. Schwartz, D. L.: Physical imagery: Kinematic vs. Dynamic models. Cognitive Psychology, 38 (1999) 433-464.
23. Schwartz, D. L. & Black, J. B.: Shuttling between depictive models and abstract rules: Induction and fall-back. Cognitive Science, 20 (1996) 457-497.
24. Schwartz, D. L. & Black, J. B.: Analog imagery in mental model reasoning: Depictive models. Cognitive Psychology, 30 (1996) 154-219.
25. Shah, P. & Miyake, A.: The separability of working memory resources for spatial thinking and language processing: An individual differences approach. Journal of Experimental Psychology: General, 125 (1996) 4-27.
26. Shepard, R. N. & Cooper, L. A.: Mental images and their transformations. Cambridge, MA: The MIT Press (1982)
27. Sims, V. K & Hegarty, M.: Mental animation in the visual-spatial sketchpad: Evidence from dual-task studies. Memory & Cognition, 25 (1997) 321-332.
28. Stenning, K. & Oberlander, J.: A cognitive theory of graphical and linguistic reasoning: Logic and implementation. Cognitive Science, 19 (1995) 97-140.

Recording the Future:
Some Diagrammatic Aspects of Time Management

Stuart Clink[1] and Julian Newman[2]

[1] Caledonian Business School
Glasgow Caledonian University, Cowcaddens Road,
Glasgow G4 0BA, United Kingdom
s.clink@gcal.ac.uk
[2] Department of Computing,
Glasgow Caledonian University, Cowcaddens Road,
Glasgow G4 0BA, United Kingdom
j.newman@gcal.ac.uk

Abstract. Management of time and commitments is a central problem for high-discretion employees in the information society. A variety of conventions have evolved for the representation of time in calendars, diaries, and project management packages. Yet current time management products remain very close to paper-based conventions with respect to their support for visualisation of scheduling problems; indeed their displays may be even more restrictive than the paper diary. We report an exploratory study aiming at "thinking outside the box" of current computerised diaries by an empirical investigation in which a heterogeneous sample of white-collar workers generated diagrammatical representations of their time and commitments. Design issues are raised for diagrammatic representations that can empower the user in such an environment.

1   Introduction

Keeping a diary[1] for business or social purposes is an almost universal tool used to supplement memory and intuition in planning time and future commitments. Diary methods grow along with the individual, who picks up techniques from here and there, with little or no training in operating a diary, though most companies would expect their employees to keep one. "*Few managers have been trained to use a diary. Instead they adopt a combination of daily, weekly, monthly and annual systems to suit*

---

1 In this paper, the classical (British English) meanings are used whereby a DIARY refers to a book for making daily records, noting engagements, etc. whereas a CALENDAR is a table of days, months and seasons.

M. Anderson, P. Cheng, and V. Haarslev (Eds.): Diagrams 2000, LNAI 1889, pp. 207-220, 2000.
© Springer-Verlag Berlin Heidelberg 2000

themselves. Unfortunately, these seldom correlate in practice because updating is carried out in isolation." [1].

Any form of time planning that requires interaction between individuals, or as a group, will be hindered by a disparity of skills and methods. While the methods used by an individual may not be all that effective, it would be considered impolite for anyone to make criticism of another's method. This attitude is closely allied to the premise that polite society considers any entry in a diary to be private and confidential.

Paper and electronic diaries are derived from the need to record engagements in order to remember them. However, individuals do not solely undertake to be in specific places at specific times, and then list these appointments as facts in a diary. They carry out complex tasks, often many at the same time. Diary entries can diagram required outputs for specific moments and locations, but Ehrlich [2] reports that from 15 to 80% of scheduled meetings are, in any case, changed. The most frequent cause of a change being a schedule conflict resulting from a shift in priorities for key meeting attendees. So the diary provides only limited help in managing the time and the resources necessary to arrive at the output schedule.

The human brain can store a limited number of items for a limited time in short-term working memory. The process for retrieval from long-term memory by way of linear associations can be long and painful. Diary entries relieve short-term memory of the need to provide the starting point for these associations.

However, a diary functions as more than a simple list of memory triggers. With areas of a page designated to represent windows in time, the action of entering an item in a diary is also a diagrammatic means of scheduling it relative to other items and linear time. The search for space for a new entry, or general browsing, can trigger prospective remembering. *"These advantages result from the "browsability" of diaries - the readiness with which existing intentions can be scanned. A key aspect of diary use is that, in the course of mundane use, such as entering new intentions, old ones get opportunistically rehearsed."* [3].

2   Diary Diagrams

A diary is a list constructed on a diagrammatic framework. The format is referenced in terms of the period of time that can be viewed in one observation, without turning a page or scrolling a screen; hence a "fortnight-to-view", "week-to-view", or "day-to-view". The size of the time increment is inversely proportional to the duration covered by a single diagram. As the span of time shown in a single diagram increases, the space devoted to a single day becomes smaller, and diaries begin to merge into time planners.

The nature of project management (primarily Gantt and PERT charts) demonstrates the limit of complexity of these diary diagrams; around 100 tasks and 30 resources is about the level of complexity at which the use of pen and paper or simple spreadsheets becomes unworkable. The complexity of these diagrams when produced by computer software makes them difficult for the ordinary diary user to understand

and, as their use is not intuitive, the users do not see implications of situations illustrated in the diagrams.

2.1 Diary Content

Previous research in this area by Kelley, and Chapanis [4], Kincaid *et al*[5] and Payne [3] has generally been slanted towards the use of diaries, but has indicated that people make attempts to use diaries as diagrammatic representations of more information than mere appointments. Nearly 150 styles of diary were identified in use, some dating back as far as the end of second world war, with layout and design not generally based on customer feedback but on traditional good sellers and judgements of publishers' design committees.

Most diary entries consisted of from two to 10 words and were made during a concurrent conversation, often by telephone [4]. Individuals made between two and 60 appointments a week, on average no further in advance than 53 days, although half had routine appointments that ran ahead up to a year. Some subjects also tracked appointments mentally for "*spur-of-the-moment things*".

Changes occur almost as frequently as entries, with both the studies by Kelley & Chapanis [4] and Ehrlich [2] reporting figures as high as 80% of appointments being changed. Ehrlich points out that with many changes to planned meetings occurring close in time to the actual meeting, contingency information on the context of the appointment needs to be available to prioritise when conflicts occur.

The surveys discussed here were mainly concerned with the physical use of diaries, with no reference to individuals' perceptions of the structure of future time, or how they retain and recall information about what they have planned for it. Only a small part of this information comes from diaries, which suggest that diaries may bear little relation to the diagrammatic images used by individuals to manage their time.

3   Investigation of Diary Diagrams and Time Visualisation

Our study originated within the broader research programme of Computer Supported Collaborative Work (CSCW). Work in the CSCW AND DISTRIBUTED APPLICATIONS GROUP at Glasgow Caledonian University has approached the management of time and commitments from a number of perspectives including field studies of professional time management, logical modelling of commitments [6], and the present research aimed at forming a better understanding of the ways in which individuals work in order better to understand the limitations within which group work functions. The present study looks at the issue from the perspective of how individuals plan commitments[2] using the concepts of the mental images of future time used by individuals.

---

2 A COMMITMENT is an agreement to carry out some long-term task, which will require the use of various periods of the agent's time on a number of occasions, to produce some deliverable performance at some point(s) in the future.

Earlier work on CSCW conducted within this Department had established that individuals did not enter certain major recurring activities, such as their teaching times, into their diary. Instead they identified committed time by recalling the individual teaching timetables which had been issued to them[3]. Thus, their diaries looked empty, but this did not really represent free time

3.1 Method

A small-scale study along the lines of those of Kelley & Chapanis, Ehrlich, and Payne (which had all used small sample sizes and produced worthwhile results) was undertaken to investigate, *inter alia*, the basic images used by subjects when thinking of the future. The validity of a small sample size is not a problem when a feel for ideas is required rather than a statistical account of the population.

However, the effect of sample size was minimised in this study by selecting the sample group from a subset of the population, to minimise the uncontrolled variables. This was done by drawing the sample from business and professional persons known to be busy individuals working under pressure of time. They were also selected for having little or no line management responsibility and therefore were neither required to spend time with junior staff, or able to delegate their tasks to others.

A sample size of twenty was selected as similar to those of the previous diary surveys by Kelley & Chapanis (23); Payne (20); and Kincaid *et al* (30 - divided between two groups). All the subjects worked within the central belt of Scotland at the time of the survey, though not all were born, brought up, or had spent all their working life within that area. There may be methods commonly employed in other areas or countries that were not used by any of this geographically local sample group. The group comprised six females and fourteen males, aged between 20 and 55, thirteen of which were qualified to first degree level or higher. The sample group was broadly similar in composition to that used by Kelley & Chapanis and much more cosmopolitan than the group investigated by Payne.

We refrain from reporting the results of this survey in the form of percentages, to discourage the inference of statistical significance to the population as a whole. A sample of this size cannot support such inference. Rather, we are concerned with reporting the existence of the phenomena that we have observed.

Direct observation would have been difficult to carry out for a reasonable number of subjects, and would not have shown the mental process that they were using. The open-ended-question type of interview favoured by Kelley & Chapanis, and again by Payne, was likely to require considerable effort to record and transcribe, and it was also likely that the interview with different informants would tend to develop in different directions. The answers to any "imagination" type of questions would be liable to contamination by an interviewer. It was important that as much as possible of the thoughts of the respondents was put down directly, by way of drawings and

---

[3] In her unpublished master's thesis, a case study of academic time management, Ronghui Jiang found that in an academic department diaries were mostly used for recording meetings, holidays and deadlines, and sometimes to record semester and examination dates.

diagrams. If the respondents were being asked to draw diagrams or picture of how they visualised concepts, it was possible that even the presence of an interviewer might hinder the expression of the very thoughts that we wished to research.

A questionnaire survey method was therefore selected, for completion by the subjects in their own time. To encourage expression, the questions were phrased to reinforce the visual response expected, with appropriately sized blank boxes for responses rather than lined areas, and by use of words such as "draw", "show", "picture" and "visualise".

3.2   Results

Of the 20 professionals who originally agreed to participate, one (female) failed to return the questionnaire, citing as the reason "lack of time". The sample then consisted of six academics, two medical practitioners, three lawyers, a bank official, four managers in industry and two in commerce, and a head teacher. Over half the sample (10) had received no training at all in time management.

The diary views used by the sample group were One-day (2), Two-day (1), Four-day (4), Week (9), Fortnight (4) and More-than-Two-week (2). Kelley & Chapanis had also reported the most popular format as week-to-view, although two respondents reported using the "pin it on the wall" method that was rejected as a valid diary method by Kelley & Chapanis. However, this is a form of multi-dimensional diagramming that permits very easy re-arrangement.

Diary Methods. The respondents all reported a similar method of accepting a new appointment into their diary system. This consisted simply of locating a free space on the diagram and making an entry. Only one respondent marked entries to show the degree of difficulty that would be experienced in altering or cancelling the appointment, and four volunteered that they used memory for this. A further one made a note of the outcome of the entry in the diagram, and five scored through the entry when completed.

These points illustrate the limit of a diary diagram as a time planning tool. The user is fully aware of the context of the entry, which is functioning as a prompt for images retained in memory. This ability may be fairly short term as the significance of the diagram after the passage of months or years was not examined in this study. In taking decisions on the addition of further entries, essential information on importance, flexibility, duration and priority is being made use of, although this information is not recorded in the diagram. This information must be being recalled from memory by association once the diagram has triggered the recollection. None of the respondents mentioned any entering of details covering a block of time, or of the length of time likely to be taken up by individual entries.

These responses would tend to indicate that the planning of time usage was in all cases being carried out mentally, rather than diagrammatically. The statements about making new entries tacitly assume that an assessment of the time commitment related to the individual entries is being carried out, and that the simple diary entry is just a physical pointer to this more detailed mental record.

Time Visualisation. The respondents were asked about their visual images of time when they thought about future events, rather than when writing them down. Only four respondents answered this section of the questionnaire entirely in words. A few included doodles or cartoons showing an attempt to answer graphically even though they did not have an image that they thought was relevant to the questions being asked.

Three Week View. The respondents were asked to show how they saw the next three weeks without detailing specific appointments. The majority of diagrams (9) were continuously linear in form (Fig 1).

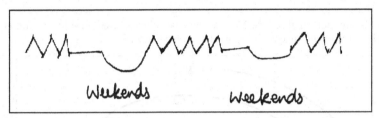

Fig. 1 Visualisation of three weeks. A continuous linear view with weekdays represented by an aggressive saw-tooth wave and weekends by a smooth curve

The representation was often divided up by complex embellishments representing weekends, days off, and different types of working day (Fig 2). The symbolism used in many of these diagrams is suggestive that weekends were viewed differently, and in a more relaxed manner, from the weekdays. This is most clearly represented in the "ramparts" image (Fig 3). The differing view of weekends can also be providing a marker by which the position on the continuous line is established.

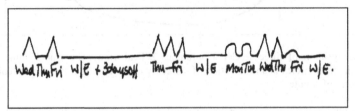

Fig. 2 Visualisation of three weeks. A continuous linear view with detailed symbols for activity on weekdays. Weekends and days off are represented by a simple unembellished line, suggesting that a separate diagram is being used for social activities

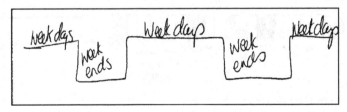

Fig. 3 Visualisation of three weeks - A continuos linear view with a clear distinction of weekend. There is an interesting symbolism in the weekdays being "up" and the weekends "down"

Two individuals very similarly saw their view as tapering off into infinity (Fig 4), and one as a long helix (Fig 5).

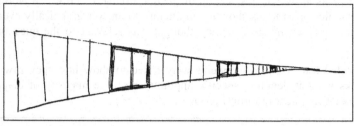

Fig. 4 Visualisation of three weeks - A continuous linear view tapering to infinity. The highlighting of weekends may be being used as a time marker

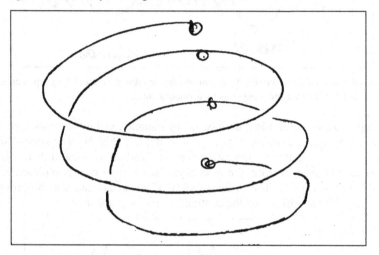

Fig. 5 Visualisation of three weeks. Continuous helical view. The embellishment of the diagram with small circles is an indication of a weekly occurrence that the respondent may be using as a form of counter

Most (5) showed some form of solid blocking in the diagram (Fig 6) to indicate the periods of time already committed.

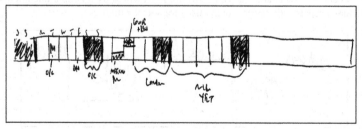

Fig. 6 Three-week view - solid blocking to show time periods already committed. The solid blocking of weekends shows that these are clearly considered as being not available

These images are not in any way similar to the diagrammatic form of printed diaries or commonly available computer software

The second most commonly held view (2) was of weeks as discrete slats (Fig 7), which is similar to the diagrammatic representation used in wall charts and calendars. Those who answered textually described a simple chronological list (Fig 8).

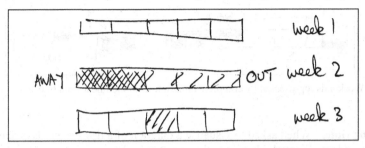

Fig. 7 Three-week view - Use of discrete slats to diagram separate weeks. This is very similar to a wall planner view

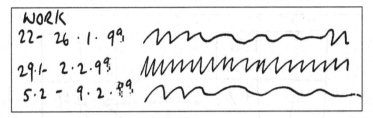

Fig. 8 Three-week view - Use of a simple chronological list

It was also common for the respondents to show in their diagrams a different view of weekends. Usually this feature was shown as a highlight (Fig 9), but some represented weekends by complete exclusion from the view (Fig 10).

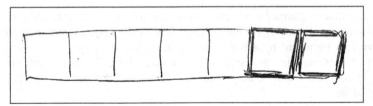

Fig. 9 Week ends represented by a highlight

The different treatment of the weekend in many diagrams by omission or representation as a plain line suggests that these respondents keep separate mental images of work time and leisure time. Among the textual views were comments such as "*weekends considered as recreational/personal work*", "*weekend planning is left to my wife*" and "*weekends are only considered on Fridays!*".

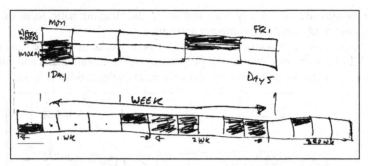

Fig. 10 Weekends represented by complete omission from the linear time line

Counting time.   When asked to visualise a diagram of a day eleven days ahead, only one respondent counted ahead to the weekend, than a week, and counted the remainder. Another respondent envisaged a diagram very similar to a conventional calendar in layout, and moved a week downward, and then four days across (Fig 11).

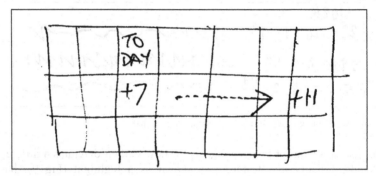

Fig. 11 Counting 11 days ahead by the one down, four across method

Four informants claimed that they would need to use a calendar, while six simply counted ahead continuously for eleven days. The others used a linear diagram model, with five moving ahead by a hop of seven days, followed by another of four days (Fig 12). The remainder managed the seven-day hop, but then counted off the remaining four days individually (Fig 13). Three persons did not respond to this question at all

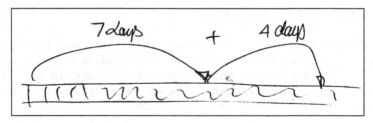

Fig. 12 Counting 11 days ahead by the two-hop method

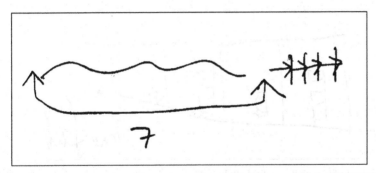

Fig. 13 Counting 11 days ahead by a hop of a week plus counting of remaining days individually

Periods of Time. In considering a task that would require a number of periods of time to be spent on it during the next few weeks, and the periods of time that would need to be devoted to completing it, the most common diagram (8) was of specific blocks of the linear diagram being associated with the task (Fig 14). The other responses all differed.

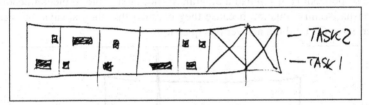

Fig. 14 Linear model with rows of blocks indicating time allocated to individual tasks

Schedules. When asked to explain how their concept of time would show them being behind or ahead of schedule in completing this task, only one respondent had clear visual images (Figs 15, 16) of the situation. The textual answers varied from "*don't visualise this concept*", to complex explanations of rescheduling other work.

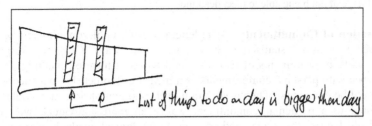

Fig. 15 Image of being behind schedule

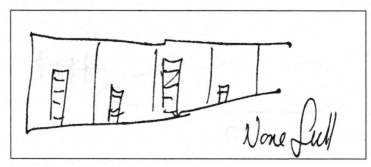

Fig. 16 Image of being ahead of schedule

Reminders.   The diary was given by most (14) respondents as the method they used to remind themselves about approaching deadlines. Two relied on memory, one on countdown lists and one on pieces of paper stuck to the window. When a deadline had been missed, several (12) claimed that they would do nothing about it, six would reschedule the task, and one claimed that he could not miss deadlines and stay in the job.

One respondent (Fig 17) had a diagram of not being able to meet a deadline. Six did not visualise this situation because they claimed that they always met deadlines. However, most (11) did not answer this question.

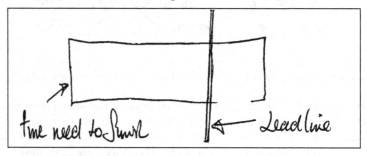

Fig. 17 Image of not being able to meet deadline

Visualisation of Commitments.   The length of time until commitments would be completed was also visualised in highly individual ways. Only two respondents related this to measurements of real time, and two to a form of counting. One used interference with personal commitments as a measure, another the sequence in which they should be completed. The visualisation of the amount of time that would need to be devoted to a particular commitment was even more individual, with no two respondents having the same method. The more logical methods were based on priority of commitment; degree of commitment; and experience.

When adding a new commitment, most (13) of the respondents visualised the other commitments in their diagram being adjusted to accommodate the newcomer by adding it in and rearranging the priorities. One commented that the amount of effort available to each commitment would decrease; one that he would still finish the existing task first; and another that tasks that could wait would be postponed.

Of the twelve respondents who showed in their diagram the amount of time that would need to be devoted to particular tasks to achieve completion, twelve used the size of space allocated and three remembered the time they had allocated. Dependencies between commitments were shown by visual links (4), lists (2), and the sequence in the diagram. Three respondents indicated that they did not do this but would like to do so in future.

The respondents had very individual perceptions of commitments. The only shared view was that two respondents viewed them as chains of appointments. The other views were: a specific tasks; one off event; a deadline; point of work; long appointment; vehicles in motion travelling in parallel at different speeds; just work; learning/experience outcomes; large blocks of time needed away from "*normal*" appointments; "*long string things*"; "*list of things I have promised*"; blocks of time; and parallel lines.

No other ways of planning or viewing commitments were used by eight of the respondents. The others also used notes and lists (4), wall chart (3), and regular review (1).

4   Discussion

This research has differed from other enquiries through investigating mental diagrams of time rather than planning and diary usage. It has indicated that people conduct their time planning, and management of commitments, on a very intuitive basis. This process is revealed as egocentric, with time planning by others either completely excluded, or maintained within a distinctly separate model. While similar models are in use, personal variations appear in most areas.

Within the limitations imposed by this research being undertaken with a fairly homogenous sample, for another purpose, and without investigation of factors such as employment content and changes over time, some preliminary indications can be derived.

Much information about commitments is not being recorded in diary diagrams as individuals rely heavily on memory, triggered by viewing blank (but actually already occupied) space in their diary diagram. The vast selection of traditional paper and electronic diary formats available produce a diagram of a specific length of time, which is repeated for consecutive periods. While periods vary from day-to-view to a year or more, they all have this identical feature of repeating the same diagram time and time again. The insight to a different form of mental diagram that individuals use for visualising time calls into question the whole diagrammatic construct of diary usage, and whether traditional diagrams are as effective as they might be in providing a workable framework for time planning.

This has major implications; firstly for the form and extent of the information that will be required for computer based time planning [7], and secondly in the design of the operation of the next generations of wearable computing appliances, where the familiar screen and keyboard are replaced with more intuitive methods of human-computer interaction

4.1 Explanation

An explanation for the difference between the two forms of diagram can be found in the viewpoint from which the model is observed. The traditional diary diagram can be seen as a detached external view, similar to that of an architect's drawing of a building. It is rigid and formalised, with information divided into clear and rigidly defined sections. The diagrams expressed by many of our respondents are internal and dynamic, like a video image from inside a living organ. They are moving, fluid, flexible and continuous.

There are no planning tools in an appropriate format to support those who have a continuous style of personal time line diagram. This lack of appropriate format may go some way to explaining the consistent dissatisfaction with diary formats that is regularly reported by a proportion of respondents in studies. These users visualise future time in a way that is not supported by any of the repetitive diary diagram formats. Hence, all that they record in their "unsuitable" diary are some triggers for the continuous real image that they cannot physically record, but keep in their preferred visualisation format in their own memory.

In considering the apparently anomalous situation where some users do not record details of whole blocks of important regular cyclical activities (such as their teaching timetable), we may be seeing the result of individuals trying to combine both systems. The mental diagram pointed to by their diary entry may be in turn cross referenced back to the printed image of their timetable or regular schedule, or this information may be being held mentally in the form of a further visual diagram.

4.2 Questions Raised by the Research

The insight that this study has provided into operational mental diagrams poses more questions that it answers. The pictorial responses showing the visualisation of future time have produced some startling different diagrammatic images. In addition to use of traditional diary diagrams, there are at least three variations on the linear diagram in use.

Further research will need to address the known limitations of this work, with more developed techniques to confirm the phenomena that has been shown here, and to investigate the existence of others. The extent to which these models are in general use in preference to the traditional diary representations should be investigated with a view to developing more suitable tools. Only with results that can be subjected to rigorous statistical analysis can these findings then be imputed to the whole population.

However, the linear images of time that were displayed in respondents' answers are clearly very different from the form of representation provided by diaries and time planning aids. This research has shown that current diary and time planning systems do not represent the mental diagrams of their users.

References

1. Holder, R., *Not enough hours in the day: whose fault is it anyway?*, in *Works Management*. 1996. p. 48-51.
2. Ehrlich, S.F., *Strategies for Encouraging Successful Adoption of Office Communication Systems*. ACM Transactions on Office Information Systems, 1987. **4**(4 Oct 87): p. 340-357.
3. Payne, S.J., *Understanding calendar use*. Human Computer Interaction, 1993. **8**: p. 83-100.
4. Kelley, J.F. and A. Chapanis, *How Professional Persons keep their calendars: Implications for Computerization*. Journal of Occupational Psychology, 1982. **55**(4): p. 241-256.
5. Kincaid, C.M., P.B. Kayne, and A.R. Kaye, *Electronic Calendars in the Office: An Assessment of User Needs and Current Technology*. ACM Transactions on Office Information Systems, 1984. **3**(1.Jan 85): p. 89-102.
6. Haag, Z., R. Foley, and J. Newman. *A Deontic Formalism for Co-ordinating Software Development in Virtual Software Corporations*. in *7th IEEE Workshop on Enabling Technologies: Infrastructure for Collaborative Enterprises (WETICE98)*. 1998. Stanford CA: IEEE Computer Society.
7. Clink, S., *Commitment Tracking: Relating Commitments to Calendar*, in *Department of Computing*. 1996, Caledonian University: Glasgow. p. 123.

Lines, Blobs, Crosses and Arrows:
Diagrammatic Communication with Schematic Figures

Barbara Tversky[1], Jeff Zacks[2], Paul Lee[1], and Julie Heiser[1]

[1]Stanford University
[2]Washington University at St. Louis

Abstract. In producing diagrams for a variety of contexts, people use a small set of schematic figures to convey certain context specific concepts, where the forms themselves suggest meanings. These same schematic figures are interpreted appropriately in context. Three examples will support these conclusions: lines, crosses, and blobs in sketch maps; bars and lines in graphs; and arrows in diagrams of complex systems.

1   Some Ways that Graphics Communicate

Graphics of various kinds have been used all over the world to communicate, preceding written language.   Trail markers on trees, tallies on bones, calendars on stellae, love letters on birch bark, maps on stone, and paintings in caves are some of the many remnants of graphic communications. Many graphics appear to convey meaning less arbitrarily than symbols, using a number of spatial and pictorial devices. Maps are a prime example, where graphic space is used to represent real space. Graphic space can also be used metaphorically to represent abstract spaces.   Even young children readily use space to express orderings of quantity and preference (Tversky, Kugelmass, and Winter, 1991). Space can be used to convey meanings at different levels of precision. The weakest level, the categorical level, uses space to separate entities into groups, such as lists of players on two baseball teams or, in writing, the letters belonging to different words.   Greater precision is needed for ordinal uses of space, as in listing restaurants in order of preference or children in order of birth. Often, the distances between elements as well as their order is indicated, as in the events of history or the skill of athletes where the differences between events or athletes are meaningful in addition to the ordering between them.

Space is not the only graphic device that readily conveys meaning. The elements in space do so as well. Many elements bear resemblance to the things they convey. Both ancient and modern examples abound. Ideographic languages conveyed things, beings, and even actions through schematic representations of them, just as airport signs and computer icons do today. Ideograms and icons also represent through figures of depiction, concrete sketches of concrete things that are parts of or associated with what is to be conveyed, as a scepter to indicate a king or scissors to indicate delete.

M. Anderson, P. Cheng, and V. Haarslev (Eds.): Diagrams 2000, LNAI 1889, pp. 221-230, 2000.

2   Meaningful Graphic Forms

Our recent work on diagrams suggests another kind of element that readily conveys meaning in graphics, more abstract than sketches of things and beings, yet more concrete than arbitrary symbols like letters.   In his self-proclaimed, but also generally recognized, "authoritative guide to international graphic symbols," Dreyfuss (1984) organized graphic symbols by content, such as traffic, geography, music, and engineering.   But Dreyfuss also organized symbols by graphic form, notably circle, ellipse, square, blob, line, arrow, and cross, all in all, only 14 of them, some with slight variants.   These graphic forms appear in a number of different contexts, with meanings varying appropriately.   Circles, for example, represent gauges, plates, warnings, and nodes, among other things.   Lines stand for barriers, piers, railroads, streets, limits, boundaries, divisions, and more.   We will call this class of graphic forms that readily convey more abstract meanings "meaningful graphic forms."

Why only a dozen or so forms, and why these forms?   One characteric of these forms is their relative simplicity.   They are abstractions, schematizations, without individuating features.   They have a useful level of ambiguity.   As such, they can stand for a wide variety of more specialized, more individuated forms.   A circle can stand for closed spaces of varying shapes, two- or three-dimensional.   When the individuating features are removed from a closed form, something like a circle is left.   A line can stand for a one-dimensional path or a planar barrier, of varying contours.   When the individuating features are removed from a path, something like a line remains.   A cross can represent the intersection of two lines.   These abstract forms can take on more particular meanings in specific contexts.   Using them seems to indicate that either the individuating feature omitted are not relevant or that the context can supply them.   In many cases, the forms themselves are embellished with more individuating features, especially when similar forms appear in the same context.

Another perspective is to regard graph readers as implicit mathematicians in interpreting depictions.   In other words, they interpret the primitive shapes in terms of their mathematical properties.   A circle is (a) the simplest, and (b) the most efficient form (shortest path) that encloses an area of a given size.   Thus interpreting a circle in a diagram invites the inference that nothing more is to be specified than that it depicts a closed area.   A blob departs from simplicity and efficiency in an unsystematic fashion. Thus, it invites the additional inference that the area depicted is not a circle or other systematic shape.   Similarly, a straight line is the simplest and most efficient form connecting two points.   Thus using the thinnest reasonable line invites the inference that an edge is indicated rather than an area, and making it straight invites the inference that nothing more is to be specified than that the ends are connected/related.   A squiggle departs from simplicity and efficiency in an unsystematic fashion.   Thus it invites the additional inference that the area depicted is not a straight line or other systematic shape.

These schematic forms, then, seem to depict abstractions, as if denoting concepts such as closed form or path.   Yet they are not arbitrary symbols like the word concepts they loosely correspond to.   Rather their very forms suggest those more general concepts.   A circle is a closed form, and something like a circle would be obtained from averaging shapes of many closed forms.   Similarly, a line is extended in one-dimension or on a plane, and something like a line would be obtained from averaging many one-dimensional or planar extensions.

3   Sketching Route Maps: Lines, Curves, Crosses and Blobs

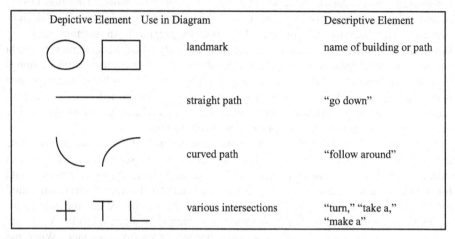

Depictive Element Use in Diagram		Descriptive Element
landmark		name of building or path
straight path		"go down"
curved path		"follow around"
various intersections		"turn," "take a," "make a"

Fig. 1: Core Elements of Route Depictions and Descriptions

The sketch maps that people provide when asked to give directions to some destination look quite different from regional maps.   Like route descriptions, route maps provide only the information needed to get from point A to point B, eliminating information extraneous to that goal. Tversky and Lee (1998, 1999) stopped students outside a dormitory and asked if they knew how to get to a nearby fast-food restaurant. If they responded affirmatively, they were asked to sketch a map or write directions to the place. The maps and directions they produced were analyzed according to a scheme developed by Denis (1997) for segmenting route directions. For both route maps and route directions, a small number of elements were used repeatedly by most participants. Moreover, these elements mapped onto one another. See Figure 1 for the map elements and the corresponding discourse elements.

The primary elements of route maps seem to be landmarks, paths, and intersections. Although the landmarks varied in size and shape, they tended to be represented by blobs, circle-like closed contours of indistinguishable shapes. Presumably, such blobs are intended to convey that there is a striker, but that the exact form of the structure is not important.   Although the streets varied in curvature, they were represented by either a straight line or a curved one. Again, the exact curvature does not seem to be an issue. Interestingly, the verbal descriptions made a two-way distinction as well. The straight lines corresponded to "go down," whereas the curved lines corresponded to "follow around" in the descriptions. A similar phenomenon occurred for intersections. They tended to be portray as crosses or partial crosses (T's, L's) depending on the number of streets in the intersection. However, the angle of the intersection was not reliably represented.

The depictions, then, schematized information about shapes of structures, curvatures of paths, and angles of intersections. Although the depictions had the potential to represent the spatial elements in an analog fashion, they did not. In fact, the depictions made very few, if any, critical spatial distinctions that were not made in the descriptions, a symbolic rather than spatial medium. On the whole, the depictions and the descriptions represented the same spatial elements and made the same distinctions among them. This suggested to us that it might be feasible to translate automatically between route depictions and descriptions. As a start in that direction, we gave participants either depictive or descriptive tool kits, containing the basic elements of the route directions. We also gave them a large set of routes, spanning large and small distances, complex and simple paths, and told them to use the tool kit to construction directions for them, supplementing the tool kits wherever necessary. In fact, for the vast majority of cases, the tool kits were sufficient, suggesting that the semantic elements frequently used in route directions and route maps are the essential elements for representing known routes.

4   Graphing Data: Bars and Lines

Line and bar graphs are the most frequent visualizations of data in popular as well as technical publications (Zacks, Levy, Tversky, and Schiano, in press). In many cases, they are used as if equivalent, though the purists insist on lines for only continuous variables and bars for discrete variables. By contrast, people—college students, that is—use bars and lines consistently but according to a different principle. Bars are closed forms and can be viewed as containers; they enclose one kind of thing, separating that kind of thing from other kinds of things, which may be in another bar. Lines, on the other hand, can be viewed as paths or connectors. Since bars contain and separate, it seems natural for them to convey discrete relationships. And since lines form paths and connect separate entities, it seems natural for them to convey trends.

To establish whether people associate bars with discrete comparisons and lines with trends, we ran two kinds of experiments: interpretation and production. In the interpretation experiments, students were shown line or bar graphs of height as a function of either a discrete variable, men or women, or as a function of a continuous variable, age 10 or 12 years (Zacks and Tversky, 1999). Examples of the graphs appear in Figure 2.

In fact, both the form of the depiction and the nature of the independent variable affected people's interpretations of line and bar graphs, but, surprisingly, the effect of form of depiction was greater. As determined by blind coders, bar graphs elicited proportionately more discrete comparisons. Some discrete comparisons were: "male's height is higher than that of female's" and "twelve year olds are taller than ten year olds." Similarly, line graphs elicited more descriptions of trends, such as: "height increases from women to men," "height increases with age," and even, "the more male a person is, the taller he/she is".

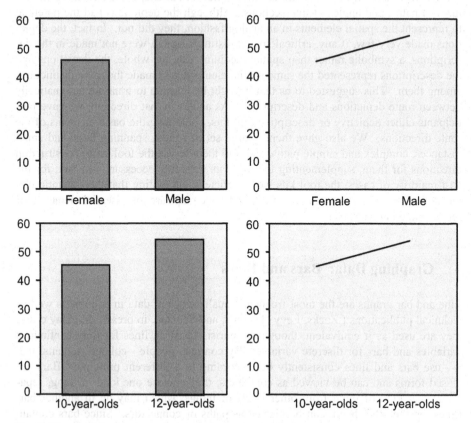

Fig. 2: Y-axes show *height* in inches. Bar and line graphs are presented to students for interpretations (Zacks and Tversky, 1999)

As before, descriptions of the relationships played a larger role in the graphic form selected than the nature of the underlying variable. When presented with a discrete comparison, such as "height for males (12 year olds) is greater than height for females (10 year olds)," students tended to construct bar graphs. However, when presented with a description of a trend such as "height increases from females (10 year olds) to males (12 year olds)," students tended to construct line graphs.

Lines and closed figures, namely bars, have readily available interpretations in the context of graphs, as trends or as discrete comparisons. These interpretations stand in sharp contrast to the interpretations lines and closed figures would be given in the context of other graphic forms, such as maps. In maps, lines are interpreted as and produced for roads or boundaries and closed figures are interpreted as or produced for structures.

5   Diagramming Complex Systems: Arrows

The early uses of arrows in diagrams remain obscure, but they did appear in diagrams to indicate direction of movement by the 18th century (e. g., Gombrich, 1990). There seem to be at least two physical analogs for arrows that indicate directionality. One is the arrow shot from a bow. The second is the arrow-like junctures that occur as liquid flows downhill (Tversky, in press).   Inferring direction from arrows seems like a small leap. Similarly, it seems a small leap to infer temporal direction from spatial direction. After all, much of the way we talk about time comes from the way we talk about space (e. g., Clark, 1973; Lakoff and Johnson, 1980).   Yet a more abstract but related use of arrows is in the sense of implication as in logic and in the sense of causality as in diagrams.   Indeed, temporal necessity is often regarded as a prerequisite for causal necessity.   Philosophy aside, Michotte (1963) has elegantly demonstrated that people readily make inferences from appropriate temporal relations to causal ones.   One dot moves next to another; if the second dot moves quickly in the same direction, the first dot is seen as "launching" or causing the movement of the second dot, much like billiard balls.   Arrows, then, seem well designed to indicate direction in space, time, and causality.

Uses of arrows have not been restricted to direction in space and time. In his extensive survey of diagrams, Horn (1998) counted 250 meanings for arrows, including and metaphoric uses, such as increases and decreases. Mapping increases to upward arrows and decreases to downward arrows is cognitively compelling.   Increasing quantities make higher piles, piles that go upwards.

Diagrams of complex systems, such as those in Figures 3, 4, and 5, are common in textbooks and in instructions for their operation.

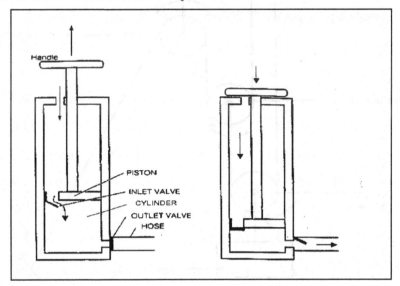

Fig. 3: Diagram of bicycle pump with arrows

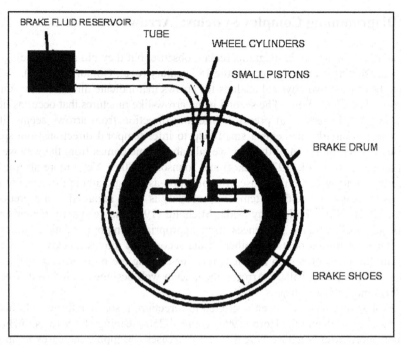

Fig. 4: Diagram of a car brake with arrows

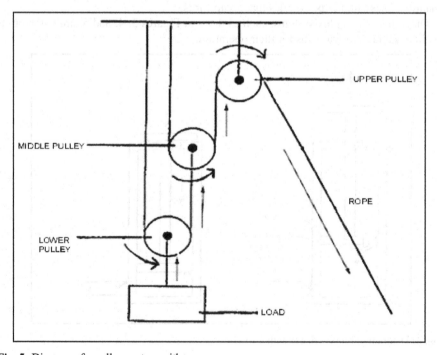

Fig. 5: Diagram of a pulley system with arrows

Note that each of these diagrams contain arrows. The arrows function to show the route and sequence of events in the operation of the system. Without the arrows, the diagrams primarily illustrate the structure of the systems; that is, what the parts are and how they are spatially interrelated. With the arrows, and some mechanical knowledge, the temporal sequence of events in the operation is more apparent. Together with the structure, the temporal sequence of events is a strong clue to the functioning of the system. Put differently, from the temporal sequence, combined with general knowledge, the causal chain of events in the operation of the system can be inferred.

How are arrows used in the interpretation of diagrams? The previous analysis suggests that the presence of arrows should encourage causal, functional interpretations of the diagrams whereas the absence of arrows should encourage structural descriptions of the diagrams. To ascertain the effects of arrows in diagrams, Heiser and Tversky (in preparation) presented one of the three diagrams either with or without arrows to undergraduates. We asked them to examine the diagram and write a description of it. The descriptions were coded without knowledge of diagram condition. Students who observed diagrams with arrows included nearly twice as much functional information as students who saw diagrams without arrows. Conversely, students who saw diagrams without arrows included more than twice the structural information as students who saw diagrams with arrows. For example, one participant who saw a diagram of the bicycle pump without arrows wrote a primarily structural description: "I see a picture of some type of machine or tool that has many different parts which are called handle, piston, inlet valve, outlet valve, and hose. Also the diagram shows a similar tool or machine but the parts are not labeled and are in different positions than the machine on the left. Contrast this with another participant's description of a pulley system, depicted with arrows: "By pulling the rope, which is part of the upper pulley, the clockwise motion of the upper pulley causes the middle pulley to move counterclockwise. The lower pulley also moves counterclockwise and lifts the load. All the pulleys are connected by the same rope." Or this description of the bicycle pump, by a participant who saw a diagram with arrows: "Pushing down on the handle pushes the piston down on the inlet valve which compresses the air in the pump, causing it to rush through the hose."

Complementary findings were obtained when students were given descriptions of systems and asked to produce diagrams. The descriptions of the bicycle pump, car brake, or pulley system were either structural, that is, they described the parts and their spatial interconnections, or functional, that is, they described the causal sequence of events in the operation of the system. Students who read functional descriptions were more likely to include arrows in their diagrams than students who read structural descriptions.

6   Meaningful Graphic Forms Again

Arrows, then, join the class of diagrammatic forms that readily convey a restricted set of meanings in context. Those meanings seem to derive in part from the graphic form and in part from the context. The forms of enclosed figures like blobs, circles and bars, of lines, of crosses, and of arrows suggest certain physical properties that have cognitively compelling conceptual interpretations. Enclosed figures suggest the possibility of containing certain elements, separating those elements from others. Correspondingly, we have found that people interpret and produce bar graphs for discrete comparisons between variables. Closed figures also suggest two- or three-dimensional objects whose actual shapes are irrelevant, thus schematized. This was revealed in their use to represent landmark structures in sketch maps. Lines suggest connectors, as seen in their interpretation and production for trends in data as well as their use to represent roads and paths in sketch maps. Crosses suggest points where paths intersect, also revealed in sketch maps. Finally, arrows suggest asymmetry, direction, in space, in time, in motion, in causality. Consonant with this, arrows in diagrams encouraged causal, functional interpretations of the systems depicted. Conversely, diagrams of causal, functional descriptions of systems were more likely to contain arrows than diagrams of structural descriptions of the same systems. The graphic forms suggest a class of possible meanings; more precise meanings are developed in specific contexts.

7 In Sum

Diagrams are often composed of schematic figures that serve as graphical primitives. A figure communicates meaning beyond that given by its' location in the diagram and beyond the local conventions established by the diagram. That is, schematic figures carry semantic weight. In sketch maps, blobs, straight lines, curved lines, and crosses are used systematically to convey information about geographical features. In graphs, bars indicate discrete comparisons while lines indicate trends. In mechanical diagrams, arrows signify order of functional operation. In each case, the meaning of the diagram as a whole is conditioned on the individual elements' ability to convey meaning on their own.

Diagrams seem especially suited to conveying a broad array of concepts and conceptual relations. They use characteristics of elements and the spatial arrays among them to convey meanings, concrete and abstract. Abstract meanings are metaphorically based on the concrete ones. Just as spatial language has been adopted to express abstractions, so space and the elements in it readily express abstractions. One reason that diagrams are useful is that they provide cognitively appealing ways of mapping elements and relations that are not inherently visual or spatial. Yet another reason that diagrams are useful is that they capitalize on the efficient methods people have for processing spatial and visual information.

References

1. Clark, H. H. (1973). Space, time, semantics, and the child. In T. E. Moore (Ed.), Cognitive development and the acquisition of language. Pp. 27-63. New York: Academic Press.

2. Denis, M. (1997). The description of routes: A cognitive approach to the production of spatial discourse. Cahiers de Psychologie Cognitive, 16, 409-458.
3. Dreyfuss, H. (1984). Symbol sourcebook: An authoritative guide to international graphic symbols. N. Y.: John Wiley & Sons.
4. Gombrich, E. (1990). Pictorial instructions. In H. Barlow, C. Blakemore, and M. Weston-Smith (Editors), Images and understanding.. Pp. 26-45. Cambridge: Cambridge University Press.
5. Horn, R. E. (1998). Visual language. Bainbridge Island, WA: MacroVu, Inc.
6. Lakoff, G. and Johnson, M. (1980). Metaphors we live by. Chicago: University of Chicago Press.
7. Michotte, A. E. (1963). The perception of causality. N.Y.: Basic Books.
8. Parkes, M. B. (1993). Pause and effect. Punctuation in the west. Berkeley: University of California Press.
9. Tversky, B. (In press). Spatial schemas in depictions. In M. Gattis (Editor), Spatial schemas in abstract thought. Cambridge: MIT Press.
10. Tversky, B., Kugelmass, S. and Winter, A. (1991) Cross-cultural and developmental trends in graphic productions. Cognitive Psychology, 23, 515-557.
11. Tversky, B. and Lee, P. U. (1998). How space structures language. In C. Freksa, C. Habel, and K. F. Wender (Editors), Spatial cognition: An interdisciplinary approach to representation and processing of spatial knowledge. Pp.157-175. Berlin: Springer-Verlag.
12. Tversky, B. and Lee, P. U. (1999). Pictorial and verbal tools for conveying routes. In Freksa, C., and Mark, D. M. (Editors). Spatial information theory: Cognitive and computational foundations of geographic information science. Pp. 51-64. Berlin: Springer.
13. Zacks, J., Levy, E., Tversky, B., and Schiano, D. (In press). Graphs in use. In Anderson, M., Meyer, B. & Olivier, P. (Editors), Diagrammatic reasoning and representation. Berlin: Springer.
14. Zacks, J. and Tversky, B. (1999). Bars and lines: A study of graphic communication. Memory and Cognition, 27, 1073-1079.

Animated Diagrams:
An Investigation into the Cognitive Effects of Using Animation to Illustrate Dynamic Processes

Sara Jones and Mike Scaife

Cognitive and Computing Sciences, University of Sussex, Brighton, UK
saraj@cogs.susx.ac.uk
mikesc@cogs.susx.ac.uk

Abstract. With increased use of multimedia and computers in education, the use of animation to illustrate dynamics is becoming more commonplace. Previous research suggests that diagrams may reduce cognitive processing as all information is perceptually available, making it more explicit and therefore requiring less inferencing (e.g. Simon and Larkin 1987). Animation, therefore, may be expected to enhance learning, especially when illustrating dynamic processes, as motion is depicted more visually explicitly, thus reducing cognitive processing. However, although animation may increase explicit perceptually available information, it may not automatically improve understanding. Visual explicitness itself does not necessarily guarantee accurate perception of specific information, nor does perception of information guarantee comprehension. Initial studies suggest that certain characteristics of diagrammatic animation have significant effects on cognitive interaction with material and therefore on comprehension. Current computer technology not only enables improved graphical animated illustration, but also provides the facility to physically interact with information on the screen. This in itself may influence the kind of learning that takes place. This paper presents research investigating how different ways of both representing and interacting with animated diagrams influence the kinds of cognitive interactions that may take place.

1. Introduction

Multimedia enables novel ways of representing information through different media formats and advances in graphical technology provide the facility for dynamic processes to be represented in animated as well as static form. A pervading assumption seems to be that the more explicit depiction of motion offered through animation should be beneficial for learning. However, good detailed accounts of the value of animation for learning are lacking.

M. Anderson, P. Cheng, and V. Haarslev (Eds.): Diagrams 2000, LNAI 1889, pp. 231-244, 2000.
© Springer-Verlag Berlin Heidelberg 2000

Diagrammatic animation can be seen as a particular form of external representation, which supports external cognition (Zhang and Norman 1994). Diagrams can convey "whole processes and structures, often at great levels of complexity" (Winn 1987, p.153) and research into the use of diagrams generally has shown cognitive benefits over other formats such as sentential descriptions. One reason for this is that many important aspects for comprehension are perceptually available in a diagram, which reduces cognitive processing, precluding the need to 'work out' such aspects as relational components (Larkin and Simon 1987). Hegarty and Kozhevenikov (1999) have also found that learners used external diagrammatic symbols to 'offload' information, as a strategy to reduce memory load. Diagrams should, therefore, facilitate understanding, as information is more explicit and requires less inferencing. If visual explicitness is important in reducing cognitive processing and in facilitating understanding of information, then the visual explicitness of motion inherent in an animated diagram should facilitate understanding of dynamic processes.

However, research into the effectiveness of computer animation for enhancing learning has been varied, with inconsistent results often due to methodological problems and variance across studies (Park 1998). Examples include the use of competing theories (dual coding, single coding, mental models), the large number of variables involved (text and animation vary according to lots of criteria - colour/black and white, degree of realism, type and complexity, textual reading level), individual differences (adults versus children, spatial ability, prior knowledge, intellectual ability, cognitive style), varied use of testing measures (recall, recognition, MCQ, problem solving tasks), research environment (situated versus experimental) (Large 1996). Research into the use of animation for learning is, therefore, inconclusive. Its use in depiction of dynamic processes thus encourages diverse aspects for investigation. It has also been noted that different representational formats of the same external structure "can activate completely different processes" (Zhang and Norman 1994 p.118). Different processes, in turn, may result in different conceptual understanding. Thus, the representational format may be critical in the way comprehension is realised.

One way to investigate this is to compare learning from both animated and static diagrams. However, research into this area shows no consistent evidence of improved understanding, suggesting that provision of a moving image does not necessarily facilitate overall conceptual understanding. Several researchers have cited the possible complexities of processing dynamic illustration and multimedia (e.g. Kaiser et. al. 1992, Hoogeveen 1997, Large et. al. 1994, Stenning 1998). This research suggests that animation may be problematic for comprehension for several reasons: use of animation on diagrams results in an unmanageable increase of perceptually available information, especially where multidimensional dynamics are involved (e.g. Kaiser et al 1992); use of multimedia representations can result in cognitive /memory overload due to too much information (Kaiser et. al. 1992), due to problems in having to integrate multiple representations not always simultaneously available on the screen (Rogers and Scaife 1997, Hoogeveen 1997). Interestingly, Kaiser et al. (1992) found that where only one dimension of the dynamic information was present then animation facilitated accurate observation. This suggests that 'parsing' of dynamic events may be important in facilitating correct dynamical judgement.

Despite the discovery of potential disadvantages of animation, Mayer (1991, 1992) has demonstrated the importance of simultaneous presentation of visual and verbal material in animated presentations, and Park (1998) maintains that animation provides several important instructional roles. For example, to attract and direct attention, to represent domain knowledge involving movement and in explaining complex knowledge phenomena, e.g. structural and functional relationships among system components. However, it may be true that animation by virtue of its motion, attracts attention, but does it attract attention merely to the overall moving image, thus, detracting from the individual components necessary to understand the system? Or does it draw attention to the actual dynamics taking place, especially if it is a multidimensional or consists of multifaceted dynamical system? Animation, by enabling visually explicit representation of temporal aspects, may 'explain' complex phenomena, but are these explanations explicit enough for the learner to obtain adequate comprehension?

Hegarty et al (1999) have investigated understanding of motion by looking at mental animation, i.e. the ability to infer motion from static diagrams, in terms of mental model construction of movement. This has been shown to be related to individual spatial ability and the type of system or diagram used in terms of its complexity (one moving component or several components moving simultaneously). Learners have to mentally simulate movement of a system, thus requiring them to 'work out' the dynamics. By contrast, providing explicit visual representation of motion (through animated diagrams) precludes the need to 'work out' dynamical aspects. Does provision of this information facilitate learning about dynamic systems?

Multimedia also provides the opportunity for innovative ways of interacting with information, not available with traditional media formats. A pervasive assumption is that the more complex and varied the technology the better the interaction and the better the learning. Large et. al. (1994) states that, "It is frequently claimed by... producers (and emphasised by reviewers) that multimedia capabilities enhance educational value" (p.528). However, there is no body of empirical evidence that this assumption is generally valid.

1.1   Theoretical Background

When working with diagrams, learners need to focus on and collect various pieces of information, which need to be integrated both with previous (internal) knowledge and with the current (external) information they are using. The focus on a general level, in understanding the effects of different diagrammatic representations on comprehension, then, is the relationship between internal and external representations and the mapping of meaning across the two. The 'Cognitive Interactivity' framework developed by Scaife and Rogers (1996) to investigate the cognitive value of graphical representations emphasises the interplay between internal and external representations on cognition through (i) the way that new information is integrated with existing knowledge and re-represented and (ii) the identification of the cognitive benefits and costs of particular forms of representations identifying the properties of external representations in terms of their computational offloading.

On a general level, computational offloading refers to the ways in which different forms constrain the amount of cognitive effort required to solve informationally equivalent problems. For example; graphical constraining occurs when different graphical forms of representations of the same information will limit the kind of inferences that are likely to be made, thus guiding thinking towards the concept to be acquired; Re-representation occurs when different external representations, that have the same abstract structure, make problem-solving easier or more difficult; Temporal/spatial constraining occurs when representations can make relevant aspects of processes and events more salient when distributed over time. In this framework, diagrams need to be investigated by examining particular aspects of learning such as, the kinds of inferences made, what information is lost, misinterpreted or correctly construed, how understanding is integrated with previous knowledge, and the effect of using different representations of the same structure.

On a more specific level, computational offloading may be a result of design issues or issues that relate specifically to the (design) properties of the diagrams. According to Green analysis of the semantic properties of diagrams will reveal particular cognitive benefits 'of particular diagrammatic representations for particular contexts. These properties are termed 'cognitive dimensions of notation', a cognitive dimension being "a characteristic of the way that information is structured and represented". Although Green (1989) cites several cognitive dimensions of notation, some emerge as an appropriate starting point for investigating the cognitive benefits of using animation. For example, hidden dependencies refer to information links between entities that are not visible when the entities themselves are viewed. Role expressiveness refers to the visibility and parsability of a meaningful structure. This may interpreted as the visibility of particular aspects of the representation and as the parsability of the components within the information. In terms of an animated diagram this might refer to which aspects are most visible, and how easy or difficult it is for the learner to parse all the pieces of information imparted into salient groupings and into a salient order. Perceptual cueing refers to 'redundant' coding of important attributes of the information, for example, bolding text.

1.2   Animation of Dynamic Systems: A Brief Account of Important Aspects

The overriding characteristic that distinguishes animated from static representations is the depiction of movement. The need to illustrate motion arises when representing a dynamic process or system. Thus, the use of diagrams portraying dynamics are particularly relevant for investigating the cognitive effects animation. However, just as with diagrams generally there are different ways of using animation to depict various kinds of information, which may effect the perception and readability of the diagram.

Design Dimensions: The (complexity) amount of information that animation can display is variable, according to several factors including the 'dimensions' of the representation. Thus movement can be multidimensional or unidimensional depending on the number of moving parts. With multidimensional representations this could be number of moving parts or simultaneous multiple movement to and from different places, or even movement occurring on multiple planes.

Representation: There are several different ways of depicting motion according to the system as well as designer choice. For example, flow may be depicted using arrows showing direction, dots as fluid, solid use of colour filling the track of the flow. Motion in a system is not just confined to representation of something moving from one place to another, but also moving components in the system allowing or perpetuating this motion, for example, opening and closing of valves in a pump system or moving of cogs in a mechanical system.

Temporal aspects: Not only does animation mean that motion is depicted visually more explicitly, but also that the temporal aspects of the process become overt. The depiction of speed and temporal relations of movement, may have prime importance in understanding an overall concept.

Specific hypotheses;
Semantic properties or the form of animation will effect the learning or types of inferences that can be made about dynamic systems.
These in turn will effect the amount of computational offloading.
Is computational offloading always beneficial?

The following studies used the domain of cardiac circulation to investigate the value of animation when learning a dynamic process. Although it may be possible to get real pictures of the heart beating or images of the inside of the heart, the most usual representations for learning this process are schematic. Schematic representations also have been found to be more conducive to learning than pictorial representations (Hegarty and Kozhevnikov 1999). The studies reported here used schematic diagrams of the heart with students aged thirteen and fourteen, learning the principles of cardiac circulation

2   Study 1: Comparison of Animated and Static Formats of Diagrammatic Representation

The aim of this study was to investigate the effects of media type, paper-based diagram versus computer-based animation, and of design of the learning task. Diagrammatic representation of blood flow through the heart was used in CD ROM and paper format, to compare animated with static representation of the concept. As CD ROM presentation allows more freedom in exploring information, whereas formal teaching situations use specific task questions, two different task presentations were used. (i) an open task where pupils were required to find out: how the blood flows in through and out of the heart; about blood that is rich in oxygen, and blood that has less oxygen and about the different chambers of the heart, and (ii) a structured task where pupils were given a worksheet with specific aspects for them to investigate.

2.1 Method

One hundred and twelve pupils aged thirteen and fourteen, from two Sussex community colleges participated in the study. Fifty three were male (mean age = 13 years 10 months; sd=3.2 months), and 59 were female (mean age = 13 years 11 months; sd=3.6 months). The independent variables were media format (CD ROM or static) and task type (open or structured). The dependent variable was the test score. A pilot study was undertaken to ensure that information was appropriate and to formulate task instructions and the test. In the main study half of the participants worked with information using the computer (animated) the other half worked with the paper (static) format. Half of each of these groups worked from the 'open' task, the other half used the structured worksheet.

Pupils were allocated into pairs by the teacher on the basis of working well together. This method was adopted as data was wanted, not only on the quantitative analysis of test scores, but also on the qualitative analysis of the kinds of information noticed and the ways in which pupils gleaned this information. Working in pairs allows discourse between children, facilitating the collection of verbal protocols. All participants worked for fifteen minutes on their allocated task, followed immediately by the test.. No pupils had completed formal teaching on cardiac circulation.

Animated condition: The 'heart' section of 'What's the Secret?' CD ROM from 3M Learning Software was used. The programme covered varying aspects about the heart, including animated video presentation of blood flow. Information needed for overall understanding was found in separate parts of the programme. Pupils were guided to the relevant areas prior to beginning their own investigations, and pupils could ask for guidance during the task if they were unable to find the information they needed. Participants worked in pairs for fifteen minutes, prior to completing individual assessments.

Static condition: Three sequential static colour printed diagrams were taken from the heart section of 'What's the Secret' CD ROM. This design ensured coverage of information was apparently equivalent to the CD ROM information needed to complete the test. Half of the pupils were given the 'open' task, the other half used the structured worksheet. Participants worked with the information in pairs for fifteen minutes, prior to completing individual assessments.

Post test: This consisted of a black and white version of the diagram taken from the CD ROM. Pupils had to complete the diagram using arrows to show where the blood enters, flows through, and leaves the heart, using conventional colouration, a blue pen for deoxygenated blood and a red pen for oxygenated blood. Pupils were also required to put the following eight labels - 'to body', 'from body', 'to lungs', 'from lungs', 'right atrium', 'left atrium', 'right ventricle', 'left ventricle' - in the appropriate places (see fig1).

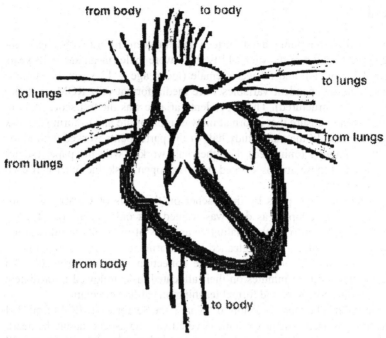

Fig. 1 Example of test diagram

2.2 Results

Scoring Method. Four different aspects of the diagram, relating appropriately to the task and the structured work sheet, were chosen for analysis. One point was allocated for each of the following; (i) oxygenated and deoxygenated blood represented in separate sides of the heart; (ii) depicting different types of blood in the correct sides; (iii) correct labelling using all eight labels; (iv) correct depiction of direction of blood flow using arrows, or from labelling if arrows were not used. Subjects could therefore receive a score from 0 to 4. An interobserver reliability score of 95% was achieved.

Analysis. The data were analysed to determine whether media presentation or task type affected learning about blood flow through the heart, and to determine the kinds of inferences made from the diagrams. The data are presented in three sections: analyses of the effects of both variables on the overall test score; analyses of the effects of variables on selected areas of the test from which scores were obtained; and analyses of errors made.

1. Analysis of both variables on the overall test score. A two way independent analysis of variance (ANOVA) was performed on the test scores. No significant main effect of media format was found. There was a significant main effect of task type (F=5.08; df =1,1; p=0.02), Scheffe (p= 0.0215); those using a structured work-

sheet performed significantly better than those using an open task, regardless of media presentation. No interaction was found between the two variables.

2. Analysis of the effects of variables on selected areas of the test A series of two-tailed chi-square analyses were performed separately on the four selected parts of the test to determine whether media presentation or task type affected performance in particular areas of the test. No significant differences were shown in any section: Blood in separate sides, chi-square 0.017, d.f = 1, not significant; Blood in correct sides, chi-square = 0.013, d.f = 1, not significant; Labels, chi-square = 0.505, d.f = 1, not significant; Directionality, chi-square = 0.108, d.f = 1, not significant.

3. Analysis of errors made. Despite the absence of overall significant differences, an inspection of individual scores showed that pupils made more errors with labelling and direction of blood flow. Analysis on these aspects revealed differences in errors of interpretation of blood flow. The two most noteworthy errors were; showing blood to flow into the heart deoxygenated but to flow out oxygenated or vice versa; and perceiving the blood to enter the heart through the ventricles and leave via the atria. A chi-square was performed to determine whether pupil interpretations were affected by the two independent variables, and was found to be significant for 'in deoxygenated/out oxygenated' (chi-square = 6.00; d.f = 1; Fisher exact p=0.0222), and approaching significance for 'blood entry through ventricles' (chi-square = 4.03; d.f = 1; Fisher exact p=0.0608).

2.3 Discussion

The results showed no significant difference in overall test performance between computer and paper diagrams. However, significant differences of task type were evident. This suggests primarily the importance of organising information for learning. Not only may it be important to guide learners towards specific pieces of information, but also to order the aspects they attend to, in such a way that in enhances integration of information. Focusing and sequencing relieves learners from deciding which aspects are important and in which order to 'read' information, which may reduce confusion and enable focus of attention on relevant aspects.

Errors made in the test were primarily concerned with blood flow and labelling. This suggests that these concepts were less clear from the diagrams or were more conceptually complex to grasp. This may be due to several factors: 1) The static diagram did not explicitly show direction of blood flow, as it lacked appropriate motion cues. Direction of flow had to be inferred from the labels (fig.1): 2) On the animation there is copious movement of different parts and different types of blood making it difficult to see where each element was going and what each component was doing. Pupils made comments such as 'stop moving' and used the diagram in stationary format to work out the blood flow. This supports evidence that multidimensional problems may not be aided by animation (Kaiser et. al. 1992). It also suggests that the design of the diagram in terms of its cognitive properties may effect understanding, and that attention to the cognitive dimensions is important in facilitating learning. For example, in terms of role-expressiveness (Green 1989), the visibility and parsability of salient pieces of information from an animated diagram may be insufficiently available to the learner.

Analysis also showed that errors made differed according to presentational format. This verifies that different representations not only influence whether problem solving is easier or more difficult, but also generate different understanding, confirming that different representational formats of a common structure "can activate completely different processes" (Zhang and Norman 1994, p.118). This also suggests that certain aspects of a concept may benefit from being represented either in static or in animated format.

Overall, improved understanding of the circulation in the heart was not apparent from the use of animation. Although use of animation may be assumed to increase perceptually available information thus reducing cognitive processing, cognitive effort and 'working out' of dynamics may be reduced to such a degree that learning is ineffectual. Although animation may be more explicit "our perceptual appreciations do not spontaneously form the basis of our conceptual understanding of dynamics" (Kaiser et. al. 1992, p.686). Anderson (1995) proposes that conceptual knowledge is based on the meaning of a representation. Errors in pupil learning suggest that comprehension of the function of the heart was lacking. This information was not clear from either presentation format. If meaning is important in conceptual understanding then awareness of the function of the heart, or its components, may constrain inferences made from the diagram. For example, if the function of the ventricles to 'pump' blood out of the heart is made explicit, then this constrains the way that the blood is likely to flow. The function of a represented structure may critically influence diagram interpretation.

This could be seen as analogous to Green's (1989) concept of hidden dependencies. In the context of this research understanding the way the system works may be dependent on understanding the functions of particular components in their relation to the functioning of the system as a whole and understanding links between the heart and other processes within the lungs and the body. These links and relationships may be 'hidden' in this information. Thus, the functions of individual components of a dynamic system may be considered as hidden dependencies, and a process within a process may be a hidden dependency for understanding the whole concept of cardiac circulation.

3   Study 2: Investigation of Knowledge Construction from Animated Diagrams

Study 1 demonstrated that different representational formats of the same process result in different understanding, that understanding multidimensional dynamics is not necessarily facilitated by animation alone, comprehension being dependent on other factors. Study 2 examined more specifically using qualitative data, the differences in ways that pupils worked with animation and static diagrams of blood flow, which parts they found hard or easy to understand, and investigate cited claims about animation.

Their task was to find out about blood flow, how the heart pumped blood and how the valves worked.

3.1 Method

Twenty two pupils aged 13 and 14 years (mean age = 14 years 1 month) participated in the study. Thirteen were male and nine female. This study primarily used verbal protocols, therefore pupils were again allocated in pairs by the teacher. They also worked either with a computer presented animated diagram or with a static paper diagram. All participants worked with the researcher for approximately half an hour on their allocated information, but those using the static diagram were also given an opportunity to briefly work with the animated version, to investigate any further information. All sessions with pupils were video-recorded. Two conditions were used:

(i) animated condition: The same CD ROM programme was used as in study 1, but pupils worked only with the animated diagram. Text information was obscured and introduced only if pupils found it too difficult to retrieve information from the diagram alone. Pupils interacted with the programme only by clicking to view the animation again.

(ii) static paper condition: This consisted of a single diagram from study 1, initially without accompanying text. Pupils were provided with pens to make notes on the information if they chose. One of each pair was given a few minutes to study the diagram alone. Time allowed was dependent on the individual pupil, but generally did not exceed five minutes. Pupil 1 was then required to explain to pupil 2 what the diagram was showing. They then worked with the information together participating in discussions between themselves. The researcher intervened only when pupils were unable to glean any more information, by asking open questions or more specific questions depending on pupil progress. The data was collected in this way to try to find out more specifically the kinds of information that was clear or unclear, and the ways in which information was pieced together.

3.2 Results

Animation appeared to generate artificially high confidence levels, increasing complexity and preventing learners from paying appropriate attention to the information. For example, all pupils working with animation expressed confidence of their understanding of the diagram, but when asked specific questions e.g. about blood pathways, they were unable to give clear or complete answers. This suggests that animation provides inappropriate computational offloading precluding the possibility of effective learning, by giving the learner false certainty of gained knowledge. Only when learners had to explain the process or were asked specific questions, were they aware of their lack of understanding.

Two pairs made inaccurate interpretations of flow, perceiving blood flowing in one colour but flowing out the other. This was interpreted as perceiving oxygen exchange taking place inside the heart. Furthermore, these pupils stated that the function of the heart was to 'clean the blood' i.e. to oxygenate the blood. This equates with errors found in study 1, also suggesting that meaning and interpretation of a representation may be related, and that understanding is dependent on aspects from the diagram that were 'hidden' i.e. functional relationships and the processes within the system.

Some pupils expressed difficulty with focusing, suggesting that directing attention to pertinent pieces of information is important. With each viewing of the animation pupils were able to explain blood flow in more detail, and combine this information with contraction and valve action. Thus, they appear to take in small amounts of information at one time, gradually combining their knowledge with previous pieces of information, which highlights the importance of parsing or sequencing knowledge acquisition. However, distinct patterns of attention became apparent when pupils were asked questions about specific aspects. For example, all pupils primarily focused on blood flow. After one more viewing four of the pairs noticed the heart contracting. No pupils spontaneously noticed valve action and were specifically directed to this component by the researcher. In terms of the concept of role-expressiveness this suggests that certain areas may be identified as salient components to parse, and may have a particular coherent order for parsing.

Despite some disadvantages, animation may provide benefits over use of static diagrams alone. When attention was focused and pupils were directed, animation imparted more information about the dynamics of the system than pupils could obtain from a static diagram, for example, they could construct clearer models about how the valves worked and how the heart 'pumped'. Analysis confirms that animation, can be too complex - being too fast, lacking in learner control, and impeding attention focus. This suggests that animation alone may be insufficient to aid comprehensive learning of a dynamic process.

4   Discussion

Overall these studies suggest that the use of animated diagrams to facilitate understanding of a dynamic process is not merely a matter of providing visually explicit information. Visual explicitness may result in increased availability of information about dynamics. In terms of computational offloading visual explicitness should reduce cognitive processing through increased availability of information. However, visual explicitness can result in an overconfidence in the knowledge accrued from the diagram inhibiting efforts at comprehensive learning. Visual explicitness also seems to contribute to the complexity of the representation due to the increased amount of information. Complexity of the representation may be due to several inherent characteristics of animation: multidimensional dynamics, speed of motion, multifaceted moving components, transience of information, difficulty in focusing attention, salience. Stenning (1998) suggests that such complexity may be a result of the following characteristics of animation; i) evanescence, the problem of memory load is enhanced by the 'transient form' of animation, which means that all sequences of movement need to be held in memory to integrate with new pieces of information to understand the process;

ii) expressiveness (as defined by Stenning 1998) ; Animation demands a more expressive representation, as all aspects of a process must be shown simultaneously, making focus on one aspect difficult; iii) control, the evanescent form means that learners are unable reaccess pieces of information as they are no longer perceptually available. Learners, therefore, have no control over what information is viewed, or for how long, as information is continuously passing by. Thus, these aspects may be compounded by lack of learner control over the information presented preventing reaccess to information required, and the tendency of visual explicitness to generates artificially high confidence levels giving the learner false certainty of gained knowledge.

In terms of the cognitive dimensions approach, it seems that certain the properties of animated diagrams may be identified and noted for further investigation. For example, there seems to be evidence for hidden dependencies - information links between entities that are not visible when the entities themselves are viewed. In this context understanding the system is dependent on understanding relational functions and processes within the primary process. These aspects appear to be hidden. There is evidence that investigating the role-expressiveness of animated diagrams would be particularly pertinent in view of the importance of reducing complexity through parsing information. This research points to salient components to parse and coherent order of parsing.

Despite cited disadvantages, animation imparted more information about the dynamics of the system than pupils could obtain from a static diagram, for example, they could construct clearer models about how the valves worked, how the heart 'pumped' and the temporal relations between the two. There was evidence of patterns in attention focus, i.e. there were similar patterns across pupils in the order that information was selected to be parsed. Not only does this suggest the kinds of aspects that are more visually discernible, but also demonstrates the intuitive need to take in small amounts of information at one time, gradually combining their knowledge with previous pieces of information. This also suggests the importance of parsing or sequencing information for knowledge acquisition.

There are dichotomous findings here; results suggest that animation can provide more information about dynamics than a single static representation, but that it is too complex in a non-interactive, illustrative capacity, resulting in cognitive overload and ineffectual learning. However, animation depicts more information about motion changes, temporal sequences and relational effects of motion or temporal aspects on the dynamic system than a single static representation. Therefore, there is a conflict between providing more 'explicit' information of dynamics in combination with increased cognitive/ memory load and complexity, versus more implicit information about dynamics combined with reduced complexity, and as a consequence, with reduced dynamic information as well. Therefore, provision of an animation in and of itself may be insufficient to generate learning of a dynamic process or system, and more detailed investigation into the specific properties and 'cognitive dimensions' of animated diagrams is needed.

5   Conclusion

So far this research has exposed the complexities of presenting dynamic systems as animated diagrams for learning, but has also shown the enhanced level of information imparted over static representations of the same domains. It has also begun to try to identify particular properties of animated diagrams (e.g. visual explicitness, multidimensional movement, transient information, high quantities of information, lack of salience) that could be analysed in terms of identifying important cognitive dimensions for comprehensive understanding (e.g. role expressiveness, hidden dependencies, perceptual cueing) and investigating different design issues in relation to these dimensions.

Acknowledgements

This research is supported by an ESRC (Economic and Social Research Council) grant.

References

1. Anderson, J. (1995) *Cognitive Psychology and its Implications*. 4th edition. Freeman, New York.
2. Green, T. (1989) Cognitive Dimensions of Notation. In (eds) Sutcliffe and Macaulay *People and Computers V* . C.U.P.
3. Hegarty, M. and Kozhevnikov, M. (1999) Types of visual-spatial representations and mathematical problem solving. *Journal of Educational Psychology* vol. 91, No. 4, p.684-689.
4. Hegarty, M. and Kozhevnikov, M. (1999) Spatial Abilities, Working Memory, and Mechanical Reasoning. *University of California, Santa Barbara, http://www.arch.usyd.edu.au/kcdc/books/VR99/.*
5. Hoogeveen (1997) Towards a theory of effectiveness of multimedia systems. *International Journal of Human-Computer Interaction.* 9 (2),151-168.
6. Jones and Scaife (1999) Diagram Representation: A Comparison of animated and static formats. *Proceedings of AAEC Ed- Media 99* p.622 – 627.
7. Kaiser, M., Proffitt, D., Whelan, S. and Hecht, H. (1992) Influence of animation on dynamical judgements. *Journal of Experimental Psychology*: Human Perception and Performance. 18, 669-690.
8. Large, A., Beheshti, J., Breuleux, A., Renaud, A. (1994) Multimedia and Comprehension: A Cognitive Study. *Journal of the American Society for Information Science*, 45 (7), 515-528.
9. Large, A (1996) Computer Animation in an Instructional Environment. *Library and Information Science Research,* 18, 3, 23.

10. Larkin, J. & Simon, H. (1987) Why a diagram is (sometimes) worth ten thousand words. *Cognitive Science* 11, 69-100.
11. Mayer, R. and Anderson, R. (1991) Animations need narrations: An Experimental Test of a Dual-Coding Hypothesis. *Journal of Educational Psychology.* 83, 4, 484-490.
12. Mayer, R. and Anderson, R. (1992) The instructive animation: helping students build connections between words and pictures in multimedia learning. *Journal of Educational Psychology,* 84, 444-452.
13. Park, O. (1998) Visual Displays and Contextual Presentations in Computer Based Instruction. *Educational Technology Research &Development.*
14. Rogers, Y. and Scaife, M. (1997) How Can Interactive Multimedia Facilitate Learning? *Proceedings of First International Workshop on Intelligence and Multimodality in Multimedia Interfaces* (in press).
15. Scaife, M. and Rogers, Y. (1996) External Cognition: how do graphical representations work. *International Journal of Human-Computer Studies* 45, 185-213.
16. Stenning, K. (1998) Distinguishing Semantic from Processing Explanations of Usability of Representations: Applying Expressiveness Analysis to Animation. *Proceedings of Intelligence and Multimodality in Multimedia Interfaces Workshop.*
17. Winn, W. (1987) Charts, graphs, and diagrams in educational materials. In D. M. Willows & H. A. Houghton (eds)*The Psychology of Illustration*. Volume 1 Basic Research. New York, NY; Springer-Verlag.
18. Zhang, J. and Norman, D. (1994) Representations in distributed cognitive tasks. *Cognitive Science* 18, 87-122.

A Comparison of Graphics and Speech in a Task-Oriented Interaction

Patrick G. T. Healey[1], Rosemarie McCabe[2], and Yasuhiro Katagiri[3]

[1] Department of Computer Science, Queen Mary and Westfield College,
University of London, London E14NS, UK.
ph@dcs.qmw.ac.uk

[2] Academic Unit for Social and Community Psychiatry, East Ham Memorial Hospital
London, UK.
rosemariemccabe@csi.com

[3] Media Integration and Communication Laboratories, ATR International
Kyoto, Japan.
katagiri@mic.atr.co.jp

Abstract. The use of graphical media in synchronous communication has received relatively little attention. This paper reports the results of an experimental study of graphical communication that systematically compares interaction in a task-oriented dialogue with and without a shared virtual whiteboard. Observations of the interaction show that a wide variety of communicative functions can potentially be served by graphical interaction. Analyses of both overall performance and communicative process demonstrate that graphical communication can provide a clear transactional advantage in communication. The results also show that participants develop their use of graphics, producing progressively more abstract graphical representations as their experience increases.

1 Graphics in Interaction

Drawings and diagrams play a significant role in many routine communicative interactions. Perhaps the most familiar use of graphical representations in communicative interactions is in the sketch maps drawn to help people navigate their way to unfamiliar locations. Drawings are frequently deployed during interaction to elaborate verbal explanations and arguments and also used in more specific communicative contexts such as the negotiations between client and architect. The importance of graphical communication as an element of specific, task-related, interactions has been formally documented in a number of studies; for example, Tang's (   ) analysis of the integration of graphics, gesture and speech in simulated design meetings and Medway's (   ) analysis of architectural design meetings. Technological support for this kind of ad hoc graphical interaction is provided for by shared virtual whiteboards which have become a standard component of many multi-media conference systems and are often used as an integrated tool within multi-media applications designed for more specific task domains (e.g., Watson and Sasse,   ). Despite this, there is a paucity

M. Anderson, P. Cheng, and V. Haarslev (Eds.): Diagrams 2000, LNAI 1889, pp. 245–   , 2000.
© Springer-Verlag Berlin Heidelberg 2000

of research on the use of graphics in interactive communication. To date, the majority of research has focussed on the interpretation of fixed representational schemes or notations; ranging from the relative effectiveness of descriptive uses of graphical representations, such as bar charts and maps in conveying information, (e.g., Buckingham Shum, et. al.   , Levy, et. al.   ) to the properties of representations, such as Euler circles and Venn diagrams, that can function as inferential calculii (e.g., Stenning and Oberlander,   ). By contrast, our interest here is in the analysis of the properties of the synchronous, dynamic, use of graphical representations as a distinct communicative medium. The programmatic reason for focusing on the use of graphics in comunication is that, in contrast to the contexts mentioned above in which the graphical representations are employed under some fixed interpretation, the use of graphics in interaction often involves a process of (re)interpretation. Different graphical devices may take on a variety of roles and interpretations as an interaction proceeds (e.g., Neilson and Lee   ).

An initial, practical, question for the analysis of communicative media is to assess what they contribute to communication, in particular, whether they can be shown to enhance or improve the effectiveness of interaction. Although graphics are commonly used in a variety of exchanges only text-based and video-based media have received significant attention in this respect. Studies have found that, in general, while different communicative media can be shown to affect various aspects of the structure and conduct of communication they have equivocal, and not infrequently deleterious, effects on its success. For example, in studies of synchronous text-based communication in small groups Straus and colleagues (e.g., Straus,   ,   ; Straus and McGrath,   ) found that, for a range of tasks, text-based communication shows no clear task outcome differences from face-to-face communication. Text-mediated groups take approximately twice as long to achieve the same levels of task performance as face to face groups but use half as many words. The absence of a clear advantage in task outcome is perhaps unsurprising since, intuitively, the overhead created by typing outweighs the possible benefits that might accrue from, for example, the relative permanence of the textual record compared to speech or the ability to make simultaneous contributions. Text-mediated communication may promote a more equal balance of contributions from participants and an increase in the proportion of task-related communication however these findings may be a simple consequence of the greater difficulty of communicating in the text-based condition rather than a property specific to the medium itself (cf. Straus,   ). In addition to the extra overhead involved in text-mediated communication there is a concomitant loss of visual cues such as gaze, gesture, and other non-verbal signals that are often cited as important to the maintenance of a smooth flow of interaction.

Surprisingly, the findings for video communication links have also proved relatively negative. Anderson et. al. (   ) review a number of experiments that compare face-to-face and video-mediated communication in an information giving task and a negotiation based task. For the information giving task, comparison of face-to-face with audio only communication shows that face-to-face communication requires approximately 28% fewer turns and 20% fewer words to achieve the same level of task performance (Boyle, Anderson and Newlands,   ). For the travel task a similar advantage of 20% fewer words for face-to-face over audio only is also reported for the same level of task performance (Anderson et. al.   ). Despite this reliable advantage for face-to-face communication in the information giving task, Doherty-Sneddon et. al. (   ) report that even high quality video-mediated communication requires more turns (11%) and more words (10%) than audio only for the same level of task performance. Anderson et. al.   also report that audio only compared with audio + video do not differ in task performance or dialogue length. In summary, the advantage that can be demonstrated for face-to-face communication is not reproduced in situations in which visual contact is sustained by video links (although see Daly-Jones et. al.   ).

In contrast to text and video, whiteboard communication has received very little attention. Although some informal observations on the use of whiteboards have been reported (e.g., Traum and Dillenbourg,   ), as far as we are aware, the current study is the first attempt to provide an experimental, comparative, analysis of the characteristics of a shared whiteboard as a distinct communicative medium. Shared virtual whiteboards provide a novel means of exploiting a relatively natural and familiar form of communication. They also provide for an interesting contrast with video links since, intuitively, whiteboards might be expected to support a different communicative function. Drawing on a distinction introduced by Brown and Levinson (   ), video links primarily provide support for the interactional coherence of communication through the transmission of signals such as gaze, gesture and orientation. By contrast, whiteboards might be expected to contribute primarily to transactional coherence, providing participants with a supplementary means of conveying task related information, but contribute only indirectly to interactional coherence. Conventional cues to the pace, structure and focus of interaction are less clearly supported by whiteboard technologies, although equally, some adaptation to new interactional cues, such as the speed or location of drawing, is also possible. Like text, graphical communication has the potential to provide the advantages of allowing simultaneous contributions by different participants and also providing a semi-persistent trace of the state of the interaction. However, a shared whiteboard, used in conjunction with speech, does not impose the same bottlenecks on information flow as text-mediated communication.

2 The Block Task Experiment

The present study was designed to provide a systematic, experimental, exploration of the use of a graphics in task-oriented interaction. The initial general prediction was that, in contrast to other media, a shared whiteboard should significantly improve task performance by supporting the transactional coherence of task-oriented communication.

2.1 Methods

This study compares communication in two different media conditions; aural (audio only) and graphical (audio plus whiteboard). A simple construction task was chosen in which individuals are paired into dyads. One member of the dyad, the Giver, is provided with a picture of a model that consists of a configuration of blocks (see figure) while their partner, the Follower, is provided with a collection of blocks. The basic task is for the Giver to collaborate with the Follower to build, within a given time limit, the model depicted. This task provides simple measures of task performance and communicative process, comparable to previous studies. Pilot studies had shown that it was engaging and sufficiently demanding for participants. Naturally, choice of task is also constrained by considerations of ecological validity, however, the exploratory nature of the study and the goal of providing a systematic comparison were the primary motivating factors in the design.

A significant concern in studies involving the use of novel media relates to how individuals may adapt their performance over time. This concern was partially addressed in two aspects of the design. Firstly, participants were screened to ensure they were familiar with the use of paint programs that have a GUI style of interaction similar to that of the shared whiteboard. Secondly, experience was directly manipulated as a factor in the experimental design to make the study more sensitive to any possible trends or changes in the use of the whiteboard medium with experience.

2.2 Materials

The materials for the task were drawn from a commercially available childrens construction set and consisted of a set of 9 coloured magnetic blocks of different shapes and sizes. Eight different possible configurations were chosen as models for the participants to build, each involving a subset of only 7 blocks. Pictures of each model were produced for the Givers to use, an example is given in figure .

All conversational interaction in both the aural and graphical conditions was recorded onto standard audio cassettes for subsequent transcription. In both conditions the Follower was also videotaped to provide a detailed record of their progress in constructing the model. A still from the video in the graphical communication (whiteboard) condition is illustrated in figure .

The software used in the whiteboard condition was from the Conference component of Netscape Communicator 4.05 running on two Macintosh powerbooks connected via a local area 10MHz ethernet. Only the whiteboard component of communicator was used, a sample display can be seen in figure   below. Because of uneven network traffic there was a variable delay between the update of the sending screen, which was immediate, and the receiving screen. This delay in response time averaged at about 1 second during the period of data gathering. Subsequent experience with this software has suggested that this is not unrepresentative of normal response time on a local area network. The whiteboard interactions were recorded on VHS video.

Fig. 1. Graphical Communication (Whiteboard) Condition

2.3 Procedure

24 participants, 16 females and and 8 males aged 18-37years (average age; 19.7yrs), were recruited from amongst undergraduate and post-graduate psychology students at the University of Nottingham. At the beginning of the experimental session participants were given twenty minutes of instruction. The block task was explained to them and, in the appropriate condition, the whiteboard software was demonstrated. Participants were advised that all interactions would be videotaped and audiotaped for subsequent analysis and that they were free to withdraw if this presented them with any problem.

In both conditions, participants were seated at two desks, approximately 1 meter apart, with a screen between them which prevented visual contact. In the aural condition the only additional equipment on the desks were the microphones, in the graphical condition screens were also placed on each desk together with a mouse. The setup for the graphical condition is illustrated in figure . On each trial participants performed the task twice, once in the role of Giver and once in the role of Follower. A two minute limit was set for completion of each model after which the pair switched roles, repeating the task with a new model. Consequently, each trial took up to a total of 4 minutes and the whole procedure was repeated for four trials with participants paired into new dyads on each trial. Participants were randomly assigned to conditions and dyads, subject to the constraint that no two individuals were paired together twice. Materials were assigned to dyads such that no individual saw a model, or its picture, twice and every individual was involved in the construction of every model as either a Giver or a Follower. The experimenter was present in the room during the task on all trials. Overall the experiment involved a mixed factor design with a between subjects factor of communicative modality; audio only vs. audio + whiteboard and a within subjects factor of task experience with four levels corresponding to each of the four trials.

3 Results

3.1 Task Performance

From the videotapes, each model that a pair built was scored between 0 and 7 for a) the number of blocks correctly identified as parts of the model within two minutes (IDs) and, b) amongst those blocks correctly identified, the number of pieces assembled in the correct orientation to one another (Accuracy). An analysis of variance, with a single between subjects factor of communicative media, was performed on the number of pieces correctly identified. The two scores for each pair were averaged for analysis with pair as a between subjects factor. This indicated no reliable difference between the audio and the audio + whiteboard conditions ($F_{(1,46)} = 1.85, p = 0.18$) in the accuracy of their IDs. However, the parallel analysis for accuracy of placement of blocks demonstrated a reliable advantage for dyads in the graphical condition of 13% ($F_{(1,46)} = 4.18., p = 0.04$).

Because of equipment failure two dialogues failed to record. The remaining 96 dialogues, each dialogue corresponding to the construction of one model, were transcribed for analysis. For each dialogue, the number of turns and number of words used until either completion of the model or the two minute limit expired was counted up. To test for a dialogue length advantage in the audio + whiteboard condition a single factor analysis of variance was performed. Scores for both dialogues for each dyad were again pooled to avoid violating the assumption of independence. For both number of words and number of turns the expected advantage for the whiteboard condition was clearly demonstrated (words: $F_{(1,44)} = 34.9, p = 0.00$; turns: $F_{(1,44)} = 32.09, p = 0.00$). Overall, dyads in the graphical condition used approximately 21% fewer turns and 18% fewer words than pairs in the audio only condition.

3.2 Use of the Graphical Channel

The analysis above demonstrates that the whiteboard substantially improved the efficiency of exchanges, nonetheless, limitations on its use were observed. Six models in the whiteboard condition were constructed without any use of the whiteboard at all. This was associated with a different model in each case but involved only two participants who were sceptical that drawing was a useful means of solving the problem. More interestingly, it was observed that although, in principle, both individuals could interact freely via the whiteboard, the structure of the task promoted a fairly rigid division of labour. Individuals would use the whiteboard to generate task-related representations when acting as Giver but would use it only passively to guide their construction when acting as follower. The Followers occasionally made explicit requests for a picture to help them follow the instructions but very rarely initiated any drawing activity themselves. Direct graphical interaction via the whiteboard was only observed in the construction of three models, out of a total of 48, and was initiated by the same individual, in their role as follower, in each case.

Fig. 2. Example Picture of a Model

Inspection of the whiteboard videos indicated a surprisingly wide variety of ways in which drawings were being used to support communication. Sometimes drawings were used only to assist in clarifying the properties of particular blocks, more commonly the entire configuration of the model would be represented. Most drawings employed a two-dimensional side view but occasionally attempts were made to represent the three-dimensional structure of a model, possibly influenced by the three-quarter view used in the pictures. Colour was employed in a number of functionally distinct ways including; emphasis, grouping, deixis, topic management and figurative identification of blocks. For example, heavy lines would be drawn on one face of the outline of a block to emphasise the point at which another piece should connect to it. Critical elements of the diagram could also be identified either by drawing a small patch of colour on them or by completely filling a particular area in a particular colour. Amongst these functions, the use of colour as an extra, figurative, cue to the identification of blocks, (i.e., a blue block would be drawn using a blue colour from the whiteboard palette) was the most common. On the first trial all drawings were constructed so that each block was colour coded. Interestingly, this use of colour declined as experience with the task increased. In early trials givers produced diagrams in which some attempt was made to reproduce the colour of each piece of the model as illustrated in figure . By contrast, in later trials individuals progressively deviated from this pattern tending toward the production of drawings in which models were depicted in a single colour. This trend is illustrated in figure   in which the proportion of pieces drawn in a non-matching colour is plotted against trials.

To assess the pattern of change in the use of colour, the videos of each whiteboard interaction were used to score each model for the number of blocks which were drawn in a non-iconic colour. Scores for both models for each pair were averaged and entered in an analysis of variance with trial number as a

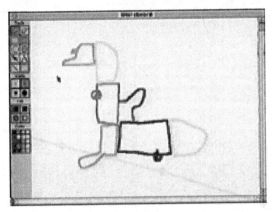

Fig. 3. Drawing of a Model Involving Figurative use of Colour

between subjects factor. This confirmed the presence of a reliable increase in the number of miscoloured blocks across trials; omnibus $F_{(3,20)} = 2.822, p = 0.06$, linear trend: $t_{(20)} = 2.80, p(\text{one-tailed}) = 0.00.$.

The observed change in the use of colour across trials can be interpreted in the context of the findings on task performance which indicate that the principal informational value of the whiteboard is in providing support for communication of the configurational characteristics of the models not the identification of individual blocks. As a result, the progressive reduction in the use of colour to assist in identifying blocks may be attributable to dyads increasing sensitivity to those properties of the whiteboard which provide support for their experience with the task develops. The selection of colours demanded relatively high levels of accuracy with the mouse and caused a small delay; factors which could discourage their use where no advantage was gained. This is not intended to imply that this was necessarily a conscious trade-off made by participants. Nobody commented explicitly on the distinction between identifying blocks and determining their configuration, however, in a few cases a Giver would explicitly remark that they were using the wrong colour for part of a model.

3.3 Communicative Coherence

In view of the changes in whiteboard use observed with experience only dialogues from the last two trials were analysed for comparison. It was judged that these would be more representative of the patterns of interaction occurring in each condition while still providing a sufficient number of data points to be tested statistically. During transcription all the dialogues were coded according to standard transcription conventions by the second author including overlapping speech, lengthened vowel sounds, short pauses, and rapid turn changes without overlap (latching). This encoding was done blind to the possible hypotheses concerning interactional coherence. In order to index the relative smoothness of the

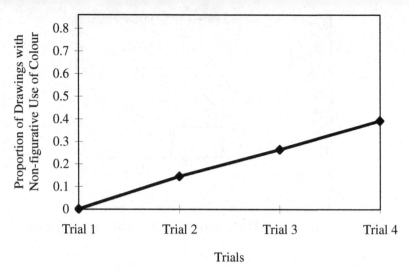

Fig. 4. Changes in the use of Colour with Experience

interactions in the each condition we the frequency of occurrence of overlapping speech and latching, or rapid 'run-throughs' during change of speaker. All instances of overlapping speech in the trial four dialogues were identified and classified into either a) backchannels; utterances such as "yeah" "aha" "mmm" which are understood to signal continued attention and provisional understanding and b) interruptive overlaps . It is important to note that the category of interruptive overlaps is broad and includes dialogue phenomena such as collaborative completions which, although interruptive, do not necessarily indicate a problem with the coherence of the interaction (see also the discussion in Anderson et. al.,   ). The difficulty of reliably discriminating and coding such phenomena precludes a more sophisticated treatment here. For current purposes, the categories of latching and backchannels were used as indices of relatively high interactional coherence and the category of interruptive overlaps, with the qualifications about its interpretation noted, was treated as an index of relatively low interactional coherence.

One pair from the whiteboard condition that failed to use the whiteboard were dropped from the analysis. In order to correct for distortions due to the reliably different dialogue length in the two conditions, the scores for backchannels, interruptions and latching for each remaining pair were converted into ratios of relatively low coherence (interruptions) to relatively high coherence: (backchannels and latching) . To test the general prediction that the whiteboard should have no effect on dialogue coherence, the resulting ratios were entered into an analysis of variance with media condition (whiteboard vs. audio) as a single factor. This was marginal ($F_{(1,22)} = 3.19, p = 0.08$) with a mean of 0.85 in the whiteboard condition suggesting lower interactional coherence than in the audio only condition which had a mean ratio of 0.52.

4 Discussion

The basic finding of the current study is that the data collected here demonstrate a clear task performance advantage for graphical communication on the shared whiteboard. While it did not significantly improve dyads accuracy in identifying components of the models, the availability of a graphical communication channel did enhance their ability to reproduce the correct overall configuration. This improvement in task performance was matched by an advantage in terms of dialogue length with, on average, 21% fewer turns and 18% fewer words used in the whiteboard condition. Together, these findings provide clear support for the general prediction that the whiteboard should enhance the transactional coherence of the interaction. This finding makes intuitive sense since relative spatial orientation is particularly difficult to communicate linguistically but relatively easy to represent graphically (cf. Oviatt,   ). Intuitions are not, of course, always a reliable guide to the effectiveness or value of a communicative medium; as illustrated by the difficulty in demonstrating a comparable advantage for video links. The present results suggest that the design and development of graphical media would perhaps repay greater attention within the area of computer-mediated communication. Video links are more expensive in terms of system demands and network bandwidth than shared whiteboards but may be less effective as a means of supporting communication in some circumstances.

An obvious question that arises in the interpretation of the present findings is the extent to which the advantage conferred by the graphical medium depends on characteristics that are more or less unique to the block task. This is an issue that can only really be resolved by further work however the observations on the use of the graphical channel suggest that, in principle, there are a range of ways in which graphics could potentially be exploited to support interaction. At least initially, participants used graphics in the service of a wide variety of communicative functions. For example, as noted above, colour was deployed amongst other things; as figurative or iconic cue to reference, to provide emphasis and to manage topic and focus. In the block task these possibilities appear to have been subordinated to the transactional use of drawings to communicate configurational information. Additionally, the potential contribution of the whiteboard to other aspects of interaction was mitigated by the delay between the update of the two whiteboard screens. This reduced the value of any potential cues to, for example, the cadence of the interaction or the management of turn-taking and made the coordination of drawings with referring expressions difficult. This may partly account for the finding that the whiteboard had at best a neutral, and more likely negative, effect on interactional coherence while nonetheless providing a significant transactional advantage.

Perhaps the most interesting aspect of the current data was the finding that the graphical representations produced on the whiteboard became less figurative and more abstract as task experience increased. In addition to the loss of colour, it was noted that the shape of pieces also tended to become conventionalised. In a few cases the shape of a piece was even reduced to a schematic line drawing that picked out only the principle axis of the block but gave no cues to its volume. This raises the possibility that a process of contraction and conventionalisation of representations was occurring similar to that reported for referring expressions in language (e.g., Clark and Wilkes-Gibbs, 1996). It also indicates that the conventions for the use and representation of space, and the interpretation of the drawings and their components were being modified during the course of the interactions. We are currently exploring these and other issues in further work.

Acknowledgements

The authors wish to thank the staff of ATR Media Integration and Communications Laboratories for providing an environment supportive to basic, interest driven, research. Thanks also to the Centre for Research in Development, Instruction and Training, Department of Psychology, University of Nottingham, especially Dr. Claire O'Malley and Pat Brundell, for invaluable support and advice during data-collection for this research.

References

[1997] Anderson, A.H., O'Malley, C., Doherty-Sneddon, G., Langton, S., Newlands, A., Mullin, J. Fleming, A.M. Van der Velden, J. (1997) The Impact of VMC on Collaborative Problem Solving: An Analysis of Task Performance, Communicative Process, and User Satisfaction. pp.133–155 In K. Finn, Sellen, A. J. and Wilbur S.B. (eds.) *Video-Mediated Communication*. Lawrence Earlbaum Associates. London.

[1994] Boyle, E.A., Anderson, A.H. and Newlands, A. (1994) The effects of eye contact on dialogue and performance in a cooperative problem solving task. *Language and Speech*, **37**, (1), pp.1–20.

[1988] Brown, P. and Levinson, S.C. (1988) *Politeness*. Cambridge: Cambridge University Press.

[1997] Buckingham Shum, S.J., MacLean, A., Bellotti, V.M.E. and Hammond, N.V. (1997) Graphical Argumentation and Design Cognition. *Human-Computer Interaction*, **12**, (3), pp.267–300.

[1996] Clark, H.H. and Wilkes-Gibbs, D. (1996) Refering as a collaborative process. *Cognition*, **22**, pp.1–39.

[1988] Daly-Jones, O., Monk, A. F., and Watts, L. A. (1998). Some advantages of video conferencing over high quality audio conferencing: fluency and awareness of attentional focus, *International Journal of Human-Computer Studies*. **49**, 21–58.

[1997] Doherty-Sneddon, G., Anderson, A., O'Malley, C., Langton, S. Garrod, S. and Bruce, V. (1997) Face-to-face and video-mediated communication: A comparison of dialogue structure and performance. *Journal of Experimental Psychology: Applied*. **3**, (2), pp105–125.

[1996]Doherty-Sneddon, G. and Bruce, V. (1996) Comparison of face-to-face and video-mediated interaction. *Interacting with Computers*. **8**, (2), pp.177–192.

[1996]Levy, E., Zacks, J., Tversky, B. and Schiano, D. (1996) Gratuitious graphics? Putting preferences into perspective. *CHI'96*, April 13-18, Vancouver, BC Canada. pp.51–50.

[1996]Medway, P. (1996) Writing, Speaking, Drawing: The Distribution of meaning in Architects Communication. pp.25–42 in In M. Sharples and T. van der Geest (eds), *The New Writing Environment*. London: Springer Verlag. O'Malley, C., Langton, S. Anderson, A.

[1994]Neilson, I. and Lee, J. Conversations with graphics: implications for the design if natural language/graphics interfaces. *International Journal of Human-Computer Studies*, **40**, pp.509–541.

[1997]Oviatt, S. (1997) Multimodal interactive maps: Designing for human performance. *Human-Computer Interaction*. **12**, (1 and 2), pp.93–129.

[1997]Roth, S.F., Chuah, M.C., Kerpedjiev, S.,Kolojejchick, J.A.and Lucas, P. (1997) Toward an information visualization workspace: Combining multiple means of expression. *Human-Computer Interaction*. **12**, (1) and 2, pp.131–185.

[1995]Stenning, K and Oberlander, J. (1995) A cognitive theory of graphical and linguistic reasoning: Logic and implementation. *Cognitive Science*, **19**, (1). pp.97–140.

[1997]Straus, S.G. (1997) Technology, group process, and group outcomes: Testing the connections in Computer Mediated and Face-to-Face Groups. *Human-Computer Interaction*, **12**, pp.227–266.

[1996]Straus, S.G. (1996) Getting a clue: The effects of media and information distribution on participation and performance in computer-mediated and face-to-face discussion groups. *Small Group Research*, **27**, pp.115–142.

[1994]Straus, S.G. and McGrath, J.E. (1994) does the medium matter? The interaction of task type and technology on group performance and member reactions. *Journal of Applied Psychology*, **79**, (1), pp. 97–97.

[1991]Tang, J. C. (1991) Findings from Observational Studies of Collaborative Work. *International Journal of Man-Machine Studies*. **34**(2), pp.143–160.

[1996]Traum, D.R. and Dillenbourg, P. (1996) Miscommunication in multi-modal collaboration. *Workshop on Detecting, Repairing and Preventing Human-Machine Miscommunication*, AAAI'96, August 4th, Portland, Oregon.

[1995]Tversky, B. (1995) Cognitive Origins of Graphic Conventions. In F.T. Marchese (ed.) *Understanding Images*. New York: Springer-Verlag. pp.29–53.

[1996]Watson, A. and Sasse, A. (1996) Assessing the usability and effectiveness of a remote language teaching system. *Proceedings of ED-MEDIA'96*, Boston, July 1996. AACE (Association for the Advancement of Computing in Education). pp. 685–690 mark-clayton@tamu.edu

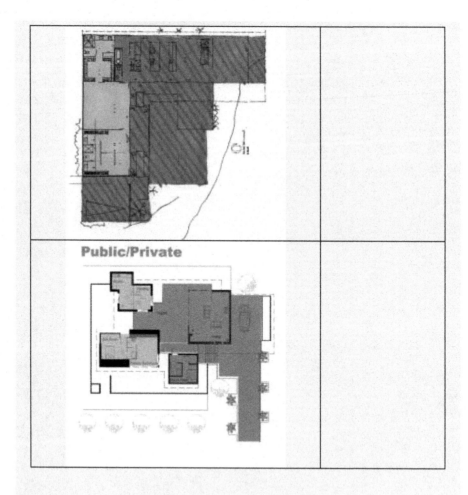

Public/Private

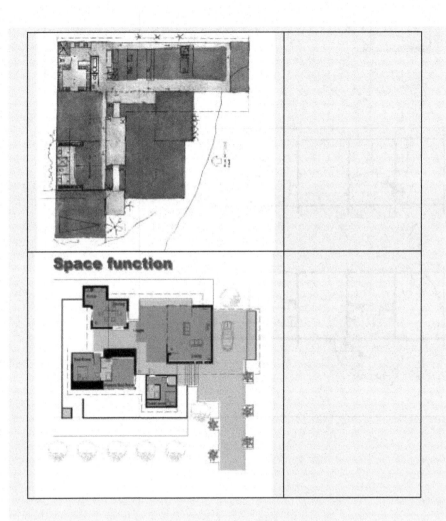

Space function

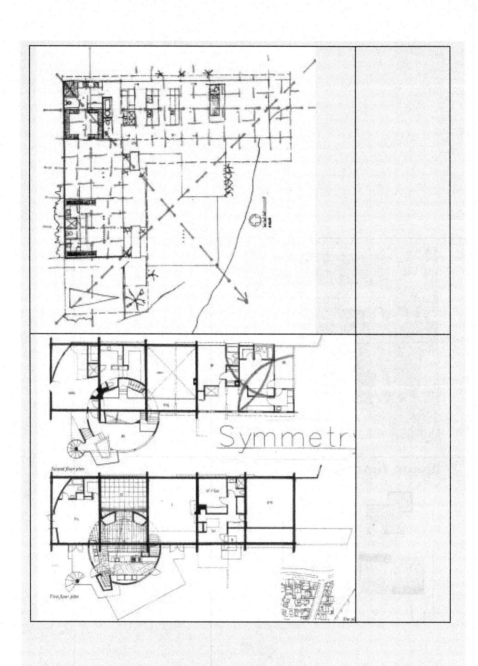

Symmetry

Second floor plan

First floor plan

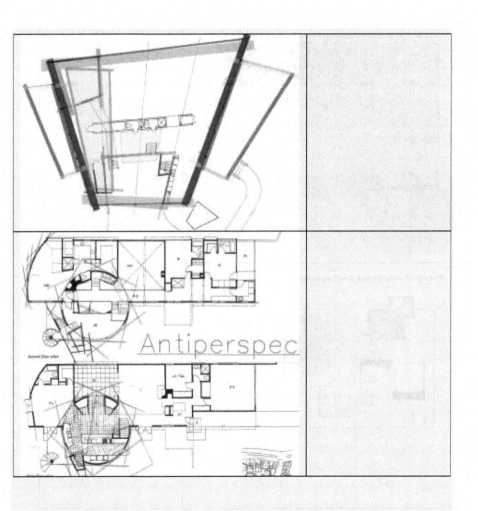

Antiperspec

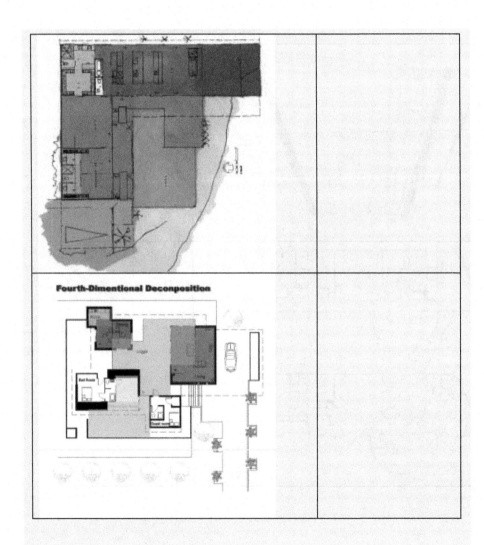

Fourth-Dimentional Deconposition

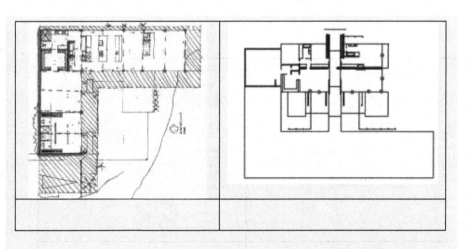

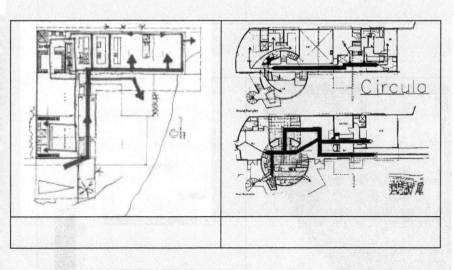

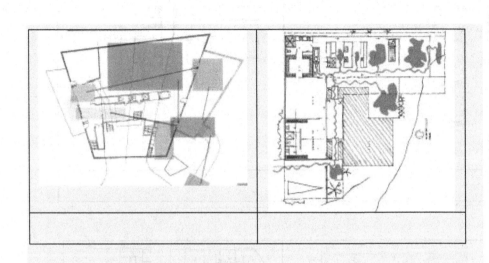

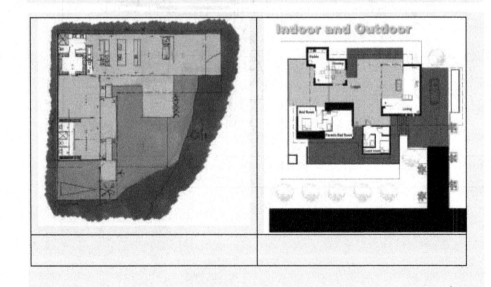

Indoor and Outdoor

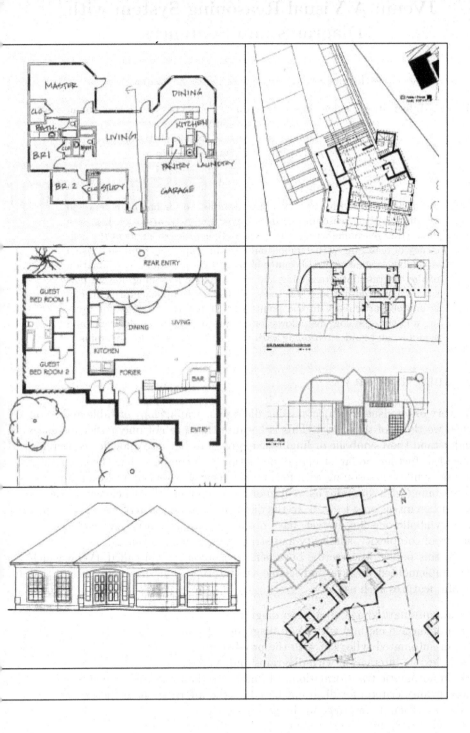

JVenn: A Visual Reasoning System with Diagrams and Sentences

Hajime Sawamura and Kensuke Kiyozuka

Dept. of Information Engineering and Graduate School of Science and Engineering,
Niigata University,
8050, Ninocho, Ikarashi, Niigata, 950-2181 Japan
{sawamura}@cs.ie.niigata-u.ac.jp

Abstract. Deduction by a computer studied so far has been centered around symbolic reasoning with formulas. Recently, attention has been directed to reasoning with diagrams as well, in order to augment the deficiency of reasoning with symbols only. In this paper, we propose a visual reasoning system called JVenn which attains a unique amalgamation of the diagrammatic reasoning and the symbolic reasoning, having perspicuity of diagrams and strictness of symbols complementarily. JVenn is unique particularly in the points that it has the strategy for proving a chain of syllogisms, allows for an interplay between diagrams and symbols, and guides reasoning with the beauty measure for diagrams.

1 Introduction

We very often visualize objects, using diagrams, graph, chart or table and so on when we think of them [,]. This is because they are not only easier for us to understand than symbolic or linguistic expressions but also allow for conceiving new ideas further. So far, studies of deduction by a computer have been centered around symbolic reasoning with formulas. Recently, attention has been directed to reasoning with diagrams as well, in order to augment the deficiency of reasoning with symbols only [, ,]. In the qualitative comparison of visual reasoning with symbolic one, we, in fact, have obtained some advantageous evidences on the proof complexity, similarity and strategy in visual reasoning [,].

In this paper, we describe our visual reasoning system called JVenn which was implemented in Java and usable via Internet. This has many useful and helpful features such as

- an automatic drawing of Venn diagrams
- a deductive method with Venn diagrams [ ]
- an automated syllogism with the proof strategy [ ]
- a flexible interaction with diagrams and symbols
- an automatic transformation of formulas to diagrams and vice versa
- a beauty criteria for diagrams based on the information aesthetics [ ]
- a user-friendly interface for human reasoners
- a high portability and usability, etc.

M. Anderson, P. Cheng, and V. Haarslev (Eds.): Diagrams 2000, LNAI 1889, pp. 271–   , 2000.

The paper is organized as follows. In Section 2, we briefly describe the logical system for visual reasoning by Hammer which underlies JVenn, and the Aristotelian syllogism [,] which is a primary object dealt with in JVenn. In Section 3, we describe the unique features of our reasoning system JVenn, the outline of the implementation method and the system configuration. In Section 4, we procedurally describe how to use JVenn, using two typical syllogistic reasoning. The final section includes comparisons with other similar systems and future work.

2 Preliminaries

As a logical basis of our visual reasoning system JVenn, we first introduce Hammer's diagrammatic system for Venn diagrams and sentences, closely following his paper [ ]. Next, we briefly describe Aristotelian syllogism [ ], a logic of categorical propositions (terms).

2.1 Outline of Hammer's System

Well-Formed Diagrams Both symbols and diagrams can be used in the diagrammatic system by Hammer [ ]. The following symbols that are used in first order logic are used in the diagrammatic system as well: (1) logical connectives: $\forall, \exists, \rightarrow, \wedge, \vee, \neg, \lambda$, (2) constant symbols: $a, b, c, a_1, b_1, c_1, \cdots$, (3) variable symbols: $x, y, z, x_1, y_1, z_1, \cdots$, (4) predicate symbols: $A, B, C, D, E, N, M, P, Q, R, S, \cdots$. The diagrammatic elements to draw Venn diagrams: "Rectangle", "Closed

Fig. 1. Diagrammatic symbols

curve", "Shading", "Line", "$\otimes$" and symbols are dipicted in Figure . The combination of line and $\otimes$ is called "$\otimes$-sequence". The combination of line and constant symbols is called "Constant-sequence". $\otimes$-sequence and constant-sequence are generically called "sequence". A rectangle denotes all elements of a set and a closed curve denotes a subset. A shaded region denote an empty set. The region containing a sequence represents that the region is not empty, in other words, sort of disjunction. The region enclosed by a rectangle or one closed curve is called "basic region", and a region which has no other region as a proper part is called "minimal region". In order to name closed curves, we tag on them such labels as $\lambda x A$, $\lambda x A$ and so on, of the form of λ-terms.

Definition 1. *The well-formed diagram (wfd) is defined in the following way: (1) Any rectangle is a well-formed diagram, (2) suppose D is a wfd and C is* *a closed curve with one label not in D. The diagram added C to D is a wfd, provided that the region in C and the regions enclosed by the closed curves in D intersect exactly once, (3) suppose D is a wfd and b is any constant symbol. D added a b-sequence or ⊗-sequence is a wfd, provided that all sequences are in the rectangle, each ⊗ or b of the sequence must be laid in a minimal region, and (4) suppose D is a wfd. The diagram added shade to the wfd D is a wfd, provided that any shade must fill some minimal region.*

The Following defines a relation between labels and regions of a diagram [ ].

Definition 2. *Suppose D is a wfd and x, y, z are variables. $\cong$ between labels and regions is a relation satisfying the following two conditions: (1) Suppose a closed curve on D is tagged $\lambda x \varphi$ and r is a region enclosed by the closed curve. If y is a free variable in $\varphi(y)$, $r \cong \lambda y \varphi(y)$, and (2) if $r \cong \lambda x \varphi(x)$, $s \cong \lambda y \psi(y)$ and z is a free variable in $\varphi(z)$ and $\psi(y)$, the relation $\cong$ is defined as follows.*

- $\bar{r} \cong \lambda z \neg \varphi(z)$
- $r \cap s \cong \lambda z (\varphi(z) \wedge \psi(z))$
- $r \cup s \cong \lambda z (\varphi(z) \vee \psi(z))$
- $r - s \cong \lambda z (\varphi(z) \wedge \neg \psi(z))$
- $\bar{r} \cup s \cong \lambda z (\varphi(z) \rightarrow \psi(z))$

Diagrammatic Rules The rules of the diagrammatic system consist of three types: (1) Drawing rules of Venn diagrams, (2) conversion rules between diagrams and formulas, (3) rules of first order logic. The following is an outline of these rules, based on [ ]. New diagrams are depicted by the repeated applications of them.

Erasure of a part of a Venn diagram Suppose D is a wfd. If a part of sequence lies on a shaded region of D, erase such a part of sequence. If the sequence is divided after this erasure, reconnect the remaining sequences.

Extend a sequence Add an extra sequence to some sequence of D.

Erasure of diagrammatic elements Apply one of the following three operations to a wfd D: (1) Erase a closed curve on the wfd D, (2) erase any shaded region on the wfd D, and (3) erase any entire sequence, provided that together with these operations, erase the shade without enclosure, leave only one ⊗ if more than two ⊗'s occur in a minimal region, or reconnect the remaining sequences if the sequence is divided.

Introduce a new basic region Make a new rectangle containing closed curves without any shade and sequence.

Conflicting information If a wfd D has a shaded region containing a whole sequence, any wfd may be drawn.

Unification of two wfds A wfd D is obtainable from two wfd D_1 and D_2 by this rule, provided wfds are satisfying the following conditions: (1) The label set of D is the same as those of D_1 and D_2, (2) if the region r on D_1 or D_2 is shaded, the counterpart region of D is shaded. Conversely, if the region r

on D is shaded, the counterpart region of D_1 or D_2 is shaded, and (3) if the region r on D_1 or D_2 contains a sequence, the counterpart region of D contains the sequence. Conversely, if the region r on D contains a sequence, the counterpart region of D_1 or D_2 contains the sequence.

The following seven rules allows to convert formulas to diagrams and diagrams to formulas. Suppose D is a wfd.

$\forall$-**Apply** Suppose we have a diagram D and a universal formula $\forall x \varphi$. If there is a region r of D such $r \cong \lambda x \varphi$, shade any region $\bar{r}$ or add a sequence to r, obtaining a new diagram.

$\forall$-**Observe** A universal formula $\forall x \varphi$ is obtainable from D if there is a region r of D such $r \cong \lambda x \varphi$ and $\bar{r}$ is shaded.

$\exists$-**Apply** Suppose we have a wfd D and an existential formula $\exists x \varphi$. If there is a region r, add a $\otimes$-sequence to the region, obtaining a new diagram.

$\exists$-**Observe** An existential formula $\exists x \varphi$ is obtainable from D if there is a region r of D such $r \cong \lambda x \varphi$ and either the region r contains a $\otimes$-sequence or $\bar{r}$ is shaded.

Constant-Apply Suppose we have a diagram D and a formula $\varphi(b)$. If there is a region r, add a b-sequence to the region r, obtaining a new diagram.

Constant-Observe A formula $\varphi(b)$ is obtainable from D if there is a region r of D such $r \cong \lambda x \varphi$ and the region r contain a b-sequence.

Inconsistent-Information Any formula is obtainable if any shaded region contains a whole sequence.

We can apply rules and axioms of first-order logic to formulas. These rules make it easier to reflect the contents of formulas to diagrams by transforming formulas to more simple forms of formulas (if possible). This system is known to have soundness and completeness [,].

2.2 Aristotelian Syllogism

The syllogism is made up of judgments (categorical propositions) which consist of the following four types: Universal affirmative judgment (A) of the form, All s is p, Universal negative judgment (E) of the form, All s is not p, Existential affirmative judgment (I) of the form, Some s is p, and Existential negative judgment (O) of the form, Some s is not p [,]. These judgments are expressed in terms of modern logic as, $\forall x(s(x) \rightarrow p(x))$, $\forall x(s(x) \rightarrow \neg p(x))$, $\exists x(s(x) \wedge p(x))$, $\exists x(s(x) \wedge \neg p(x))$ respectively. Here is a typical example of the syllogism: From $\forall x(M(x) \rightarrow P(x))$ and $\exists x(S(x) \wedge M(x))$, we have $\exists x(S(x) \wedge P(x))$.

Figure   shows a diagrammatic proof for the syllogism above where only Venn diagrams are used in the diagrammatic reasoning system. We call such a system System1. Figure   shows a diagrammatic proof for the same where Both Venn diagrams and sentence are used. We call such a system System2.

In Figure , we first have Venn diagrams drawn corresponding to premises. Then, two premises are unified by the rule of "Unification of wfds". Since $\otimes$ is on a shaded region, $\otimes$ can be erased on the diagram by the rule, "Erasure of a part

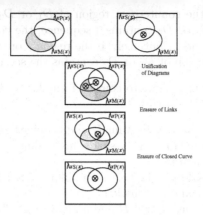

Fig. 2. Proof in System1

of a Venn diagram" . Next, the closed curve labeled with $\lambda x M(x)$ are erased by rule of "Erasure of diagrammatic elements", since we want a closed curve labeled $\lambda x S(x)$ and $\lambda x P(x)$. The final diagram is a conclusion. On the other hand, in Figure , we first draw a new basic diagram by the rule "Introduce a new basic region". Then, we apply the formula $\forall x(M(x) \rightarrow P(x))$ to the diagram, using $\forall$-Apply. In the sequel, we apply the formula $\exists x(S(x) \wedge M(x))$ to the resulting diagram, using $\exists$-Apply. The remaining part becomes the same as in System1. These proofs are essentially same except the way to initially draw diagrams from formulas.

3 Implementation of a Visual Reasoning System JVenn

In this section, we will describe only the most interesting functional features of JVenn, choosing from among all the aspects of JVenn. The system configuration is shown in Figure .

It mainly consists of two parts: The interface part of the main window and the Venn diagrams drawing window, through which we can communicate with JVenn, and the processing part consisting of the two classes: the Venn diagram manager and the Venn diagram.

3.1 The Interface Part – Window Construction –

Figure is the main window of JVenn where all the icons of Venn diagrams we drawn are appeared keeping the proof structure. We can drag and drop an icon on other icons on this window, in order to unify two diagrams. This is an very easy way to apply the rule "Unification of two wfds", looking at the overall proof structure. Obviously we easily discard unnecessary icons of a Venn diagram into the trash on the main window. By double-clicking an icon of a Venn diagram, the hidden actual diagram will be made present.

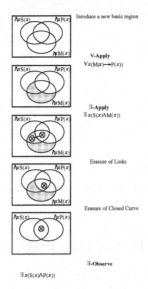

Introduce a new basic region

V-Apply
$\forall x(M(x) \rightarrow P(x))$

∃-Apply
$\exists x(S(x) \land M(x))$

Erasure of Links

Erasure of Closed Curve

∃-Observe

$\exists x(S(x) \land P(x))$

Fig. 3. Proof in System2

Figure   is a window where a Venn diagram is drawn and displayed. The front window shows the GUI components for the inference rules, and the back window shows the GUI components for the other diagrams drawing tools. The role of each button is as follows:

(1) window control menubar
(2) button to add closed curves
(3) text box to input labels of closed curves
(4) button to change the buttons display
(5) button to add shades
(6) button to change the buttons display
(7) button to transform formulas to Venn diagrams
(8) button to transform Venn diagrams to formulas
(9) button to check the information conflict
(10) button to delete shades
(11) button to delete closed curves
(12) button to delete elements of a sequence
(13) button to add elements of a sequence
(14) drawn Venn diagrams
(15) display box for beauty measure, formulas, etc.

(1) controls various windows such as displaying editing buttons, hiding windows, deleting diagrams and so on. (7) to (13) are for applying inference rules. (2) to (6) are for editing Venn diagrams. Figure   is a snapshot of JVenn running on the Web browser of Macintosh.

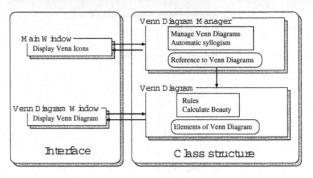

Fig. 4. The system configuration of JVenn

3.2 The Management Part of Venn Diagrams

This controls all the Venn diagrams on JVenn, having a reference to each Venn diagram. The main interface of this part is the main window described above where we can add or delete Venn diagrams and furthermore invoke an automated syllogism with the proof strategy given in [ ].

3.3 Handling Venn Diagrams

3.3.1 Internal representation of Venn diagrams
Venn diagrams consist of various constituents such as closed curves, shades and sequences. Obviously, it is closed curves that play a central role in representing Venn diagrams. Therefore, they can be considered the most basic data. Then a minimal region can be represented by the closed curves enclosing it, and a region can be represented by a set of minimal regions. A shade is represented by a region, a ⊗ is represented by a minimal region and a sequence is represented by a list of the minimal regions to which each ⊗ belongs.

3.3.2 Transformation of diagrams to formulas and vice versa
This can be done from the relation between labels and regions (see Definition 2) and the conversion rules between diagrams and formulas (see the subsection 2.1.2). The outline of the method is as follows:

1. Seek the minimal regions $r = r_1, \cdots, r_n$ constructing a region r.
2. Seek the relation of each minimal region to its label. This is feasible since each minimal region knows the closed curves enclosing it.
3. Obtain a label of the region r by the relation of ∪ regions and ∨ labels in (Definition). We use some propositional simplification rules if necessary.

Let us consider the example in Figure and see how this method goes, in case of ∃-Observe. At the step , there are two minimal regions, at ( ) there are two

Fig. 5. Main window

labels respectively, $\lambda x(A(x) \wedge \neg B(x))$ and $\lambda x(A(x) \wedge B(x))$, and finally at ( )
we have $\lambda x(A(x) \wedge \neg B(x)) \vee (A(x) \wedge B(x))$. After some simplifications, we have
$\exists x A(x)$.

Similarly for the transformation of formulas to diagrams. It is only necessary
to find the diagram with the same label as the given formula, and seek all the
minimal regions of the closed curve with the label.

3.3.3 Measuring beauty of Venn diagrams based on the information aesthetics

It is natural that we feel inclined to say how beautiful a picture is when we
paint or look at. This would be no exception in visual reasoning as well. Here we
attempt to give a beauty criterion to diagrams and reasoning with them. This is
not only a beauty measure but also expected to determine a direction of reasoning, in other words, a new way to guide proofs. Here, we give a calculus of beauty
that Venn diagrams might have, consulting the information aesthetics [ ], where
the beauty measure is defined to be $M_{\ddot{A}} = O/C$(called Burkoff's quotient),
where C stands for the complexity and O the order that a diagram might have,
similarly to the definition of information in information theory. For our purpose,
however, we do not require the additivity on the beauty measure, and use the
refined version of the following [ ].

$$M_{\ddot{A}} = K \sum_{i=1}^{m} M_{\ddot{A}_i} = \frac{1}{m} \sum_{i=1}^{m} \frac{\sum_{j=1}^{n_i} E_{ij}}{n_i C_i}$$

where m is the number of aesthetic objects (the viewpoints of the beauty measure), $K = \frac{1}{m}$ is a constant, n_i is the number of attributes that determine the

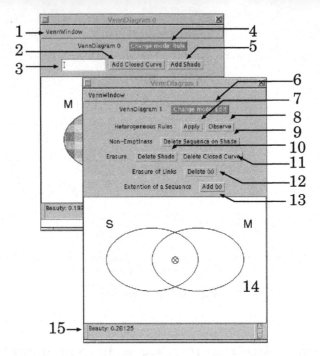

Fig. 6. Venn diagrams window

order by a specific viewpoint i, and E_{ij} and C_i are the order and the complexity from the viewpoint i.

We take into account the following viewpoints for the case of the beauty of Venn diagrams: (1) Kinds of elements in Venn diagrams, and (2) Configurations of elements in Venn diagrams. From these two viewpoints ($m = 2$), we define each E_{ij} in the above equation as follows.

(i) From the viewpoint of the diagrammatic elements:

$C_1 = $ The number of elements.
$E_{11} = $ The order of shades, $\frac{2^{cc}-S}{2^{cc}}$, where S is the number of shades and cc is the number of closed curves.
$E_{12} = $ The order of $\otimes$, $\frac{2^{cc}-t}{2^{cc}}$, where t is the number of $\otimes$.
$E_{13} = $ The order of lines, $\frac{t-l+1}{t+1}$, where l is the number of lines.

(ii) From the viewpoint of the configuration of diagrammatic elements:

$C_2 = $ The number of closed curves.
$E_{21} = $ 1 if a diagram is symmetric with respect to some line, 0 otherwise.
$E_{22} = $ The symmetricity with respect to some rotation, $\frac{360-r}{360}$, where r is an angle of rotation.
$E_{23} = $ 1 if shades are adjacent or do not exist, 0 otherwise,

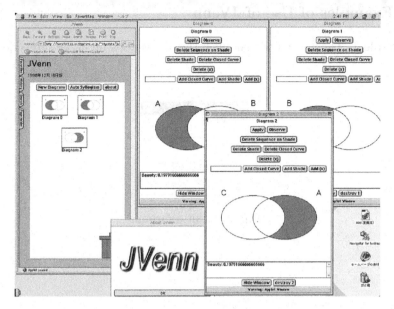

Fig. 7. A snapshot of JVenn

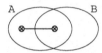

Fig. 8. From diagrams to formulas

where we define E_{ij} so as to be $0 \leq E_{ij} \leq 1$. Then our diagrammatic rules are classified into three groups: (i) beauty increasing rules: Erasure of a part of a Venn diagram and Erasure of diagrammatic elements, (ii) beauty decreasing rules: Extend a sequence, Unification of two wfds, $\forall$-Apply and $\exists$-Apply, (iii) rules that do not commit to beauty: the other rules. In a word, the more complicated diagrams are, the smaller the value of the beauty measure is. This might seem to be strange since the beauty generally has nothing to do with the complication of diagrams. However, this turns out to interact well with the flow of typical proofs like in Figure 2 and 3. In order to get to an conclusion of the syllogistic form from some assumptions, we had better pay attention to the direction in which the beauty value increases at some proof steps. This can be viewed as sort of a proof guide. In the next section, we exemplify how the beauty is changing in the process of syllogistic reasoning, and discuss its role in visual reasoning.

4 Examples of Use of JVenn

Example 1. The first proof example is the one we took up in the subsection   .

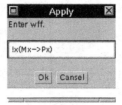

Fig. 9. Input a formula

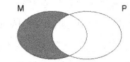

Fig. 10. $\forall x(M(x) \rightarrow P(x))$

Fig. 11. $\exists x(S(x) \wedge M(x))$

Invoking JVenn first, the main window like Figure   will appear. Clicking the button "New Diagram" on it, the window like Figure   appears with no Venn diagrams on it.

Doing this operation twice, we have two windows with no Venn diagrams on them displayed. Clicking the button "Apply" on it and displaying the window like Figure   , we input two premises through the two windows respectively, and transform them into the diagrams. They are seen in Figure   and Figure   .

In order to unify these two diagrams, drag one icon onto the other on the main window, obtaining Figure   . Since there is a part of the ⊗-sequence on a shade, delete the part of a ⊗-sequence by clicking the button 9 in Figure   ,

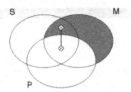

Fig. 12. The unified diagram

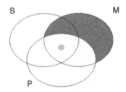

Fig. 13. Erase the ⊗ on the shade

obtaining Figure   . We do not have any concern about the predicate M, and hence delete the closed curve tagged M by clicking the button 11 in Figure   , obtaining Figure   . Finally we can obtain a formula in the message box by clicking the button "Observe" in Figure   . In this case, it is the only formula $\exists x(S(x) \land P(x))$ JVenn could read from the Venn diagram.

Figure   shows a transition of beauty in this diagrammatic proof. As can be seen, the beauty is lowest just after the unification of the two diagrams, and the conclusion resumes the higher or equal marks of beauty than or to those of the premises (this reminds us of the truth-preserving property of inference rules). If we erased the other closed curve at the last step, we would have a conclusion with somewhat lower beauty value, that did not lead to our desired one. Therefore we would say that such a beauty measure could play a role of a proof guide proper to visual reasoning, which we think should deserve further studies in the future.

Example 2. We take up the following chain of syllogistic reasoning from three premises to a conclusion.

Premise 1	$\forall x(B(x) \rightarrow C(x))$
Premise 2	$\exists x(A(x) \land B(x))$
Conclusion 1	$\exists x(A(x) \land C(x))$
Premise 3	$\forall x(C(x) \rightarrow D(x))$
Conclusion 2	$\exists x(A(x) \land D(x))$

In JVenn, we first draw three Venn diagrams for the three premises. Then by clicking the button "AutoSyllogism" on the main window, we have the window (see Figure   ), where we input the two predicate names corresponding to the desired conclusion. By this operation, there will be produced intermediate conclusions and a final conclusion we wanted (see Figure   ).

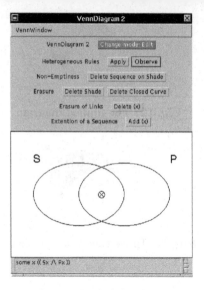

Fig. 14. $\exists x(S(x) \land P(x))$

5 Concluding Remarks

In this paper, we proposed a visual reasoning system JVenn which attained a unique amalgamation of the diagrammatic reasoning and the symbolic reasoning, having perspicuity of diagrams and strictness of symbols complementarily. This aspect would be particularly important for human reasoner.

Two logic-based visual reasoning systems have been known so far, as far as we know. Venn 2.1 [ ] is such a visual system that students can learn the meaning and validity of syllogisms expressed in natural language sentences. It allows us only to draw Venn diagrams, but incorporates no deductive apparatus like the Hammer's system. HyperProof 1.0 [ ] is the first hybrid system that allows to deal with both diagrams and symbols, aiming at a logic tutoring system where we can learn concepts like truth and derivation in a simple block world. Compared with these systems with different design philosophy, JVenn is unique particularly in the realization of

- the transformation between diagrams and symbols
- the automated syllogism under the proof strategy [ ], and
- the beauty measure for diagrams and its use in guiding proofs.

In the future, it is planned to combine JVenn with other powerful symbolic reasoning systems [ ]. We think such a combination is the most promising direction to an ideal system for computer-supported reasoning.

JVenn, although still experimental, can be downloaded from the following address: http://www.cs.ie.niigata-u.ac.jp/Research/jvenn/

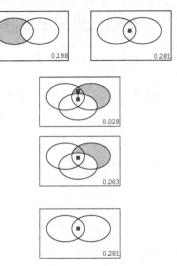

Fig. 15. Change of beauty

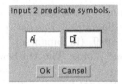

Fig. 16. Input predicate names

Acknowledgements

The authors would like to thank the anonymous reviewers for their critical but helpful comments on an earlier version of this paper.

References

1. G. Allwein and J. Barwise, editors. *Logical Reasoning with Diagrams.* Oxford University Press, 1996. ,
2. J. Barwise and J. Etchemendy. *HyperProof.* CSLI Publications, 1994.
3. M. Bense. *Einführung in die informationstheoretische Ästhetik.* Rowohlt Taschenbuch Verlag, 1969. (Japanese translation by K. Kusabuka, Keisoshobo, 1997). ,
4. E. Hammer. Reasoning with sentences and diagrams. *Notre Dame Journal of Formal Logic,* 35(1):73–87, 1994. , , ,
5. E. Izuhara, T. Yoshida, and H. Atsumi. *Systems of Diagrams - Diagrammatic Thinking and its Representation.* Nikkagiren, 1986. (in Japanese).

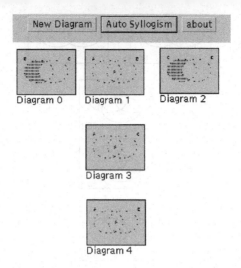

Fig. 17. Automated syllogism with the proof strategy

6. K. Kiyozaka and H. Sawamura. Considerations on the complexity of visual reasoning with sentences and diagrams. *Journal of Japanese Society for Artificial Intelligence*, 14(4):646–656, 1999. (in Japanese). ,

7. R. B. Nelsen. *Proofs Without Words*. The Mathematical Association of America., 1993.

8. T. Nishida and Y. Anzai. Special issue on diagrammatic inference. *Journal of Japanese Society for Artificial Intelligence*, 9(1):182–215, 1994.

9. J. Nolt and D. Rohatyn. *Theory and Problems of Logic, Schaum's Outline Series*. McGraw-Hill, Inc., 1988. ,

10. H. Sawamura and K. Kiyozuka. *Considerations on the Complexity and Beauty of Visual Reasoning with Diagrams and Sentences*. 2000. (unpublished manuscript).

    ,
11. S.-J. Shin. Situation-theoretic account of valid reasoning with venn diagrams. In [1], pages 81–108.

12. T. Taniguchi. *Philosophical Logic*. Gakubunsya, 1979. (in Japanese). ,

13. C. Tinelli. Combination methods in automated reasoning. http://acsl.cs.uiuc.edu/combination/.

14. R. Wesley and J. Moor. *Venn - A Philosophy Tutor, Courseware Development Group*. Dartmouth College, Kinko's, 1986.

A Proposal for Automating Diagrammatic Reasoning in Continuous Domains

Daniel Winterstein[1], Alan Bundy[1], and Mateja Jamnik[2]

[1] Division of Informatics, University of Edinburgh
80 South Bridge, Edinburgh, EH1 1HN, UK
danielw@dai.ed.ac.uk
A.Bundy@ed.ac.uk
[2] School of Computer Science, University of Birmingham,
Birmingham, B15 2TT, UK
M.Jamnik@cs.bham.ac.uk

Abstract. This paper presents one approach to the formalisation of diagrammatic proofs as an alternative to algebraic logic. An idea of 'generic diagrams' is developed whereby one diagram (or rather, one sequence of diagrams) can be used to prove many instances of a theorem. This allows the extension of Jamnik's ideas in the DIAMOND system to continuous domains. The domain is restricted to non-recursive proofs in real analysis whose statement and proof have a strong geometric component. The aim is to develop a system of diagrams and redraw rules to allow a mechanised construction of sequences of diagrams constituting a proof. This approach involves creating a diagrammatic theory. The method is justified formally by (a) a diagrammatic axiomatisation, and (b) an appeal to analysis, viewing the diagram as an object in $\mathbb{R}^2$. The idea is to then establish an isomorphism between diagrams acted on by redraw rules and instances of a theorem acted on by rewrite rules. We aim to implement these ideas in an interactive prover entitled RAP (the Real Analysis Prover).

1 Introduction

There are some conjectures which people can prove by the use of geometric operations on diagrams, so called diagrammatic proofs. Insight is often more clearly perceived in these diagrammatic proofs than in the algebraic proofs.

It is not surprising that geometry (and geometric reasoning) was the original form of mathematics. For example, Pythagoras' Theorem was proved circa 500BC. The Pythagoreans' proof was lost, and as with anything relating to Pythagoras, it is impossible to know just what was done, when and by whom. However his proof would certainly have been geometric [ ]. The elegant proof in Figure is due to an unknown Chinese mathematician writing ~200 BC [ ]. By comparison, algebra is a recent invention, usually attributed to al-Khwarizmi in 830AD [ ]. The modern algebraic formalism is barely a hundred years old, the

---

[1] One could argue it began with Diophantus c250AD, but this does not affect our argument.

M. Anderson, P. Cheng, and V. Haarslev (Eds.): Diagrams 2000, LNAI 1889, pp. 286–   , 2000.
© Springer-Verlag Berlin Heidelberg 2000

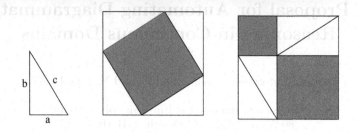

Fig. 1. The geometric proof of Pythagoras' Theorem is a classic example of diagrammatic reasoning. How can such a proof be formalised?

result of the axiomatisation project of Hilbert, Frege, Russell et al. A side-effect of this great project is that diagrams have fallen out of favour as acceptable methods of proof. Only algebra is regarded as formal. The current monopoly of algebraic formal mathematics is summed up by Tennant [ ]:

> "[the diagram] is only an heuristic... it has no proper place in the proof as such... For the proof is a syntactic object consisting only of sentences arranged in a finite & inspectable array."

This insistence on algebraic formalism is a curious position, as Barker-Plummer observes: "most mathematicians deny that diagrams have any formal status, but on the other hand, diagrams are ubiquitous in mathematics texts" [ ]. In the light of this, it is interesting to note that in al-Khwarizmi's work, the use of algebra is justified with geometric proofs [ ].

2 Axiom: *n*. Received or Accepted Principle; Self Evident Truth

- Collins Gem English Dictionary

In spite of current doctrine, the concepts of axioms and proof are not inherently restricted to sentential reasoning. The only necessary criterion for rigorous proof is that all inferences are valid. That is, any conclusions reached are genuinely implicit in the hypothesis.

However, the problem remains that formally, drawings *cannot* prove anything about algebra, and vice versa. $\mathbb{R}^2$ is an abstract object and not identical to an (infinite) sheet of paper, which is subject to physical laws and limitations, i.e. a line does not define a set in $\mathbb{R} \times \mathbb{R}$, nor does $\{(x,y)|y = ax + b\}$ define a line. People can make the connection and reason with visual representations, but syntactic manipulations only act on algebra. Since we wish to make inferences about (algebraic) objects beyond the diagram, an interpretation of the diagram is

---

[2] $\mathbb{R}$ denotes all real numbers, and hence $\mathbb{R}^2$ denotes the real plane.

needed. This can be justified by "homomorphic" mappings of diagrams to target domains, as described by Barwise & Etchemendy ("homomorphic mapping" is not a technical term, but it denotes any mapping where significant parts of diagram structure are preserved) [ ].

Our basic framework will be to establish two isomorphisms:

$$\text{diagrams} \longleftrightarrow \text{objects in } \mathbb{R}^2 \longleftrightarrow \text{mathematical target domain}$$

Only the second mapping can be proved to be an isomorphism. In spite of this we will talk interchangeably of concrete diagrams drawn on paper, blackboards, computer screens, etc. and diagrams 'drawn' on a bounded subset of $\mathbb{R}^2$. The universal use in mathematics of $\mathbb{R}^2$ to represent flat surfaces, and the obvious mapping between them, justify this laxness. Where we wish to distinguish, we will use 'pure diagrams' to refer to drawn diagrams, and 'Real diagrams' for the equivalent objects in $\mathbb{R}^2$.

3   An Example Proof

Figure   gives an example proof demonstrating the use of diagrams with redraw rules in place of algebra with rewrite rules. The example is a typical analysis theorem (it assumes some familiarity with the subject).

Theorem 1 (Metric space continuity implies topological continuity).
For X, Y metric spaces, $(\forall x \in X, \varepsilon > 0 \ \exists \ \delta > 0$ such that $|x - y| < \delta \Rightarrow |f(x) - f(y)| < \varepsilon)$ implies $(\forall S \subset Y, S$ open in $Y \Rightarrow f^{-1}(S)$ open in $X)$.

Our proposed approach is to give diagrammatic definitions for the mathematical concepts in the form of "redraw rules" – rewrite rules with diagrammatic pre- and post-conditions. An individual diagram can be suggestive and convey much to the viewer, but it is only the behaviour of diagrams that can produce rigorous proofs. We ignore the degenerate case $f^{-1}(S) = \emptyset$ – where the theorem is true by definition – by excluding it from the antecedent. The proof is not a single diagram but an ordered sequence of diagrams, in the same way that an algebraic proof is not a single statement but an ordered list of statements. The comments explain the steps taken, but are not necessary to the proof process. As we will show, the reasoning involved can be given entirely in diagrammatic inference rules. This approach requires defining diagrams, redraw rules and interpretations. The behaviour of the diagrams under the redraw rules can then be shown to be isomorphic (via the interpretation) to the behaviour of objects in the target domain.

---

[3] A function that is both one-to-one and onto is called an isomorphism.
[4] This theorem roughly says that if a function maps nearby points to nearby points, then its inverse will preserve open sets.

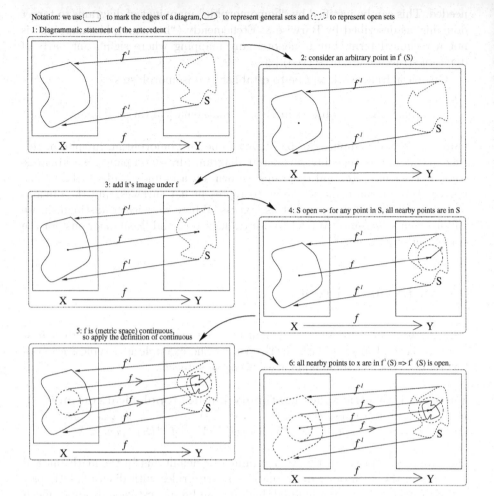

Fig. 2. An example diagrammatic proof: metric space continuity implies topological continuity

4 Related Work

Diagrammatic reasoning is a relatively new area of research and there is little directly related work. Barwise and Etechemendy's HYPERPROOF system, which currently sets the standard for the educational applications of theorem provers [ ], uses diagrams to great effect. It combines a first order logic prover with a visual representation for reasoning in a blocks world domain [ ]. However, although HYPERPROOF mixes diagrammatic logic with sentential (predicate) logic, it does not have any diagrammatic inference rules. To date, most systems have concentrated on using diagrams to guide an essentially algebraic proof [ ]. For example, systems such as "&"/GROVER use diagrams as heuristics [ ].

The exception to this is the DIAMOND system, which proves theorems in natural number theory. It uses the constructive ω-rule to generalise from diagrammatic proofs of individual cases by showing that the given proof defines a uniform proof procedure for any instance [ ]. The constructive ω rule works via meta-induction, and is therefore restricted to countable domains. However, we claim in this paper that the idea of using a uniform proof procedure to show that all cases can be proved (and are therefore true) can be adapted to continuous domains. We also draw on ideas from Shin's work for Venn Diagrams, where results from real analysis are used to show soundness of diagrammatic inference rules [　].

5 Some Formal Notation for Scribbling

Here we present our ideas for the formalisation of a diagrammatic system that reasons in continuous domains. Perhaps the main purpose of such formalisations is to avoid questions of human interpretation and intuition. This is analogous to the development of formal methods for sentential logic, allowing for truly rigorous proofs. Let $\mathcal{L}$ be a language. We now define a *drawing* and a *drawing function*.

Definition 1. *A* drawing D *is a set* $\{(X_1, t_1, i_1), (X_2, t_2, i_2), ...\}$ *where* $X_j \subset \mathbb{R}^2$, $t_j \in \mathcal{L}$, $i_j \in \mathbb{N}$.

Fig. 3. An example drawing: $D = \{(\{(x, y) \mid x^2 + y^2 = 2\} \cup \{(x, y) \mid y = x, x \in [0, \sqrt{2}]\}, closed\_ball, 1), (\{(1, 1)\}, point, 2)\}$ Implicit in this conversion is that the scale of the diagram does not matter

Definition 2. *A* graphic object *or* drawing function *is a partial function* $d_n :$ $(D, P) \rightarrow D'$ *where* D, D' *are drawings,* P *are parameters in* $\mathbb{N}$, $\mathbb{R}$ *or* $\mathcal{L}$, *and* $D' = D \cup \{(X, n, i)\}$ *such that* $i = 1 + max\{j : (Y, m, j) \in D\}$. i *is called the* instance number. *An instantiated or drawn object is a particular value for a drawing function.*

Often a construct depends upon a previously drawn part of the diagram. For example, in the definition of a continuous function:

$$\forall x, \varepsilon > 0 \exists \, \delta(x, \varepsilon) \text{ such that } |x - y| < \delta \Rightarrow |f(x) - f(y)| < \varepsilon$$

δ is dependent on ε and x. In our framework, these dependencies are handled implicitly: objects are assumed to be dependent on everything that was drawn before them, and independent of anything drawn afterwards. This information is 'stored' in the instance numbers of each object (e.g., see Figure).

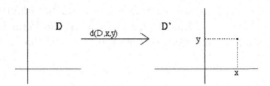

Fig. 4. One possible point drawing function $d_1(D, x, y) \to D \cup \{\{(x, y)\}, point, 1\}$

We now define a label and a diagram.

Definition 3. *A label is a partial function* $l : \{drawings\} \to \mathcal{L} \times \mathbb{R}$.

Let $d_1(p_1) \circ d_2(p_2)$ denote $d_2(d_1(\emptyset, p_1), p_2)$. Given a set of graphic objects and labels $P = d_1, d_2, ..l_1, l_2, ...$, we now define a *diagram type*.

Definition 4. D_P *is a diagram of type* P *if* $D_P = \{d = d_1(p_1) \circ d_2(p_2) \circ ..., l_1, ..., l_m)$ *where the* l_i *are label functions, such that* $l_i(d) = (n, x), l_j(d) = (n, y) \Rightarrow x = y$ *(we will refer to this as the labelling condition).*

Let $\mathbb{D} = \mathbb{D}(P)$ denote the set of all diagrams of type P. Or, to put it another way, D_P is a drawing of various objects in specific positions, marked according to object type and instance. Labels are used to show equal values for lengths or other such properties. The extra structure in the form of labels and diagram types on top of $\mathbb{R}^2$ prevents ambiguity such as in the diagram in Figure .

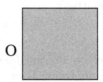

Fig. 5. Ambiguity in diagrams: is the 'O' a label for the area of the rectangle, a label for the length of a side, or just a passing circle?

Definition 5. *A diagrammatic theory is a tuple*

$$< \mathcal{L}, \{objects\}, \{labels\}, \sim, \{R_1, R_2, ..,\} >$$

where the $R_i : \mathbb{D} \to \mathbb{D}$ are called redraw rules.

We say a diagram D is within a theory, if the theory contains redraw rules $R_1, R_2, ...$ such that $D = R_1 \circ ... \circ R_N(\emptyset)$.

5.1 Generic Diagrams

One big stumbling block in diagrammatic reasoning is the problem of universally quantified variables. Diagrams are inherently existentially quantified by the fact that they are drawn, and therefore specific objects. We cannot draw an abstract object. For any concept we wish to express there will usually be a continuum of different instances, and we can only ever draw a finite selection of these. For example, a theorem may mention a line of any length, but a drawn line must have a fixed length.

There are two solutions to this, and we will use a combination of both in this project. One is to let the interpretation do some of the work. The other is to define equivalence classes of diagrams, and work with proof processes which can be shown to be valid for all equivalent diagrams. In this way, each diagram is allowed to stand for a class of related diagrams.

Consider the two triangle diagrams in Figure . Are they equivalent? This

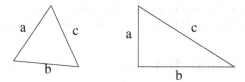

Fig. 6. Are these diagrams equivalent?

depends on what we are trying to show. The proof that the internal angles of a triangle add up to 180° can be drawn with either. On the other hand, Pythagoras' Theorem is only true for the right angled triangle. It is envisioned that the automated theorem prover we propose here would have access to several related diagrammatic theories. It would choose an appropriate one to work in when given a conjecture.

A useful category of equivalence relations is one that includes equivalence relations induced by groups of geometric transformations. Given a group G of transformations of $\mathbb{R}^2$ we can define an equivalence relation $\sim$ over the diagrams.

Definition 6. *If $D = \{(X_1, t_1, 1), ...\}$, $D' = \{(X'_1, t'_1, 1), ...\}$ then $D \sim D'$ iff $\forall (X_j, t_j, i_j) \in D$, $t_j = t'_j$ and $\exists g \in G, g(X_j) = X'_j, g^{-1}(X'_j) = X_j$ or $\exists$ redraw rule R, diagrams C, C' such that $C \sim C', R(C) = D, R(C') = D'$.*

For example, in Figure , the diagrams would be equivalent if G contained rotations and stretches. We call G the group of unimportant deformations, and say a diagram D is a *generic diagram* for the equivalence class of D.

Definition allows any equivalence relation, but in practice only a few are of interest. In an informal survey, students were presented with a collection of diagrams and asked to judge which ones should be considered equivalent. There was almost unanimous agreement as to which transformations should and should not be allowed in G. Translation, rotation, reflection are always considered valid. Transformations that do not preserve topological properties (i.e. inside/outside) are never valid. Those which do not preserve shapes, or affect one area of the diagram differently to another are accepted on diagrams containing 'amorphous blobs', but not on those composed of 'rigid' straight line shapes. These informal results can be summarised as: G should preserve the apparent properties (topology, shapes, etc.) of the diagram. It seems that diagrams are assumed to give all the relevant information: i.e. if a clearly recognisable shape is drawn, then this is an important feature. Otherwise, shape does not matter. Such cognitive issues do not affect the validity of a formalisation, but do affect it's usefulness.

We have used the redraw rules in defining $\sim$ in such a way as to ensure that $\forall D, D' \in \mathbb{D}$, R a redraw rule, $D \sim D' \Rightarrow R(D) \sim R(D')$. It is this property, which we call the *generic diagram property*, that allows us to generalise from a proof of a theorem for one instance, to a proof for all equivalent instances. A sequence of redraw rules must apply equally to all members of an equivalence class, thus guaranteeing equivalent proofs. If we define the relation $\sim$ differently – as we will need to for some areas – the generic diagram property must be proved. This should be possible using standard maths results and techniques.

6   Are Truth Values Relevant?

Sentential (predicate) logics do have many advantages, not least of which is the existence of well developed methods for checking their validity. Statements are associated with truth values and inference rules are valid if and only if they are sound. That is, P can be deduced from Q only if, in all models where Q is true, P is also true. This property of being sound can be tested quite simply using truth tables.

It is not clear that the values 'true' and 'false' have any meaning when applied to diagrams. Treating diagrammatic statements as predicate statements with a few spatial relations gives rise to 'diagrams' such as the one in Figure   taken from [ ]. Here the diagram is viewed as an existentially quantified statement which can then be judged true or false. However this approach is quite unnatural, in that the objects considered are not diagrams themselves but sentential descriptions of diagrams. As such it is more an attempt to develop a predicate calculus with a visual interpretation. Our approach differs in that we do not

$B : box \Rightarrow L_1, L_2, L_3, L_4 : line$
such that L_1 connects_to $L_2 \wedge L_2$ connects_to $L_3 \wedge$
L_3 connects_to $L_4 \wedge L_4$ connects_to L_1
and set $B.lines = \{L_1, L_2, L_3, L_4\}$.

$A : and\_gate \Rightarrow B : box, L : label$ such that L inside B
with $L.text =$ "&" and set $A.frame = B.lines$.

$N : nand\_gate \Rightarrow A : and\_gate, P : point$ such that P on $A.frame$
and set $N.frame = A.frame, N.out = P$.

$S : sr\_latch \Rightarrow N_1, N_2 : nand\_gate, L_1, L_2, L_3, L_4 : line, L_5, L_6 : polyline$
such that L_1 touches $N_1.frame \wedge L_2$ touches $N_2.frame$
$\wedge L_3$ touches $N_1.out \wedge L_4$ touches $N_2.out$
$\wedge L_5$ touches $L_3 \wedge L_5$ touches $N_2.frame$
$\wedge L_6$ touches $L_4 \wedge L_6$ touches $N_1.frame$
and set $S.set = L_1, S.reset = L_2, S.out_1 = L_3, S.out_2 = L_4$.

Fig. 7. Ceci n'est pas une diagramme

try to interpret or parse diagrams. Instead we look at inference rules that act directly on diagrams.

A diagram cannot be true or false. How could we draw a false diagram? We can only talk of the truth and falsity of algebraic statements *associated* with the diagram by an 'interpretation' or 'logic mapping'. We therefore do not define soundness in diagrammatic reasoning at all. Hilbert would approve:

> "Mathematics is a game played according to certain simple rules with meaningless marks on paper."[ ]

Instead we define *validity of interpretation*.

Definition 7. *Consider interpretations of the form* $I : \mathbb{D} \to L$ *where* L *is some conventional logic for the target domain and* I *is an injective function. We say* I *is valid iff* $\forall D \in \mathbb{D}, D \xrightarrow{R} D' \Rightarrow I(D) \models I(D')$.

Proofs of validity will vary. For example, if we wish to prove a theorem about a property p (e.g., area), then it suffices to show that the relevant redraw rules preserve property p.

If we do not have 'false' diagrams then we cannot use proof by contradiction. Such proofs can often be reformulated as proofs of the contrapositive without explicitly using contradiction. It is hoped that using proof of the contrapositive will eliminate the need for proof by contradiction.

7 The Constructive $\aleph_1$ Rule

The *constructive ω-rule* allows us to deduce $\forall x P(x)$ by providing a uniform procedure to generate proofs for every x. In practice, this involves meta-induction:

---

[5] If we have a proof of $P \Rightarrow Q$ by $[P, \neg Q] \Rightarrow R$, $[P, \neg Q] \Rightarrow \neg R$. Then we can prove $\neg Q \Rightarrow (P \Rightarrow R)$ and $\neg Q \Rightarrow (P \Rightarrow \neg R)$. So $\neg Q \Rightarrow (\neg P \wedge R)$, $\neg Q \Rightarrow (\neg P \wedge \neg R)$ and by resolution (that is, a case split rather than contradiction) $\neg Q \Rightarrow \neg P$.

instead of induction on n, induction is carried out on the *proof of n* to show a valid proof exists for every case. This was introduced by Baker, and was used in Jamnik's DIAMOND system to generate general proofs from diagrammatic base case proofs [ ].

We wish to extend this idea to prove theorems with a continuum (i.e. an uncountable number) of cases. This gives the inference rule:

$$\frac{\{P(x)|x \in X\}}{\forall x P(x)}$$

where X may be of any cardinality. We are interested in the case where X has cardinality $\aleph_1$, which we call the *constructive $\aleph_1$ rule*. Whilst very similar in concept, it leads to completely different proofs. Instead of using meta-induction, the existence of a uniform proof procedure for all cases is proved by showing that a specific proof for x_0 defines a valid proof for an arbitrary case $x \in X$. In the formalism outlined above, this amounts to showing that a proof from one diagram defines a valid proof from any other diagram in it's equivalence class. Intuitively, it can be seen that the generic diagram property and validity of interpretation as defined above are exactly what is needed to do this. Future work on this project should include a rigorous treatment of this.

8 Another Example: Pythagoras' Theorem

The generality of the proof of Theorem   given in Figure   (see §) relied on multiple interpretations of diagrams. Here we demonstrate the use of equivalence classes for generalisation. Let G be the group of stretches along an axis, $90°$ rotations, and translations. Assume that we have definitions for the following redraw rules:

1. draw_right_angled_triangle(a,b) (draw the triangle $\{(0,0), (a,0), (0,b)\}$ and add length labels to the short sides),
2. translate_triangle,
3. rotate_triangle_90°,
4. copy_triangle (draw a copy of specified triangle object with identical labels),
5. label_square_area (add an area label to a square area, i.e. recognise an emergent property),
6. subtract_triangle.

We only define one interpretation – each triangle is interpreted as itself. By stretching, one triangle can become any right-angled triangle, and so all right angled triangles are in the same equivalence class. Thus the correct generalisation from the specific traingle used in the diagrams to the general theorem is set by the equivalence class. The use of labels keeps the copied triangles identical to the original (since if a stretch breaks the identical size of two triangles, the result will fail the labelling condition and therefore will not be a valid diagram). By applying these rules (*in order:* 1,4,3,2,4,3,3,2,4,3,3,3,2,5) we draw the diagram in Figure . We then use rule 2 to transform this diagram to the diagram in Figure .

Fig. 8. Pythagoras' Theorem I

Fig. 9. Pythagoras' Theorem II

Rule 6 strips away the extra triangles. Pythagoras' Theorem, as presented here, is an area theorem and so to show the validity of our interpretation, we need to show that the translate and rotate redraw rules preserve the area, and the copy and subtract rules cancel each other out. In the purely diagrammatic theory, this is true by definition. In the $\mathbb{R}$eal diagrams, it is trivial for translation, copy and subtract, but not for rotation. In general, the fact that rotation preserves area is a corollary of Pythagoras' Theorem. However, for 90° rotations, Pythagoras' Theorem is not necessary: we only require $\left|\left(\begin{smallmatrix} 0 & -1 \\ 1 & 0 \end{smallmatrix}\right)\right| = 1$. Thus the proof is valid for the instance shown. Also, the $\sim$ relation was induced from a group G as set out in §   . Hence the generic diagram property holds. Therefore the proof carries over from the proved instance (triangle $\{(0,0), (1,0), (0,2)\}$) to all equivalent instances. The equivalence class contains all right angled triangles, so we have proved the theorem.

9   Concept Overview for Real Analysis

Historically, real analysis was developed to justify the use of calculus. $\mathbb{R}^3$ was meant to be the real world, and the definitions were supposed to capture how the universe works. Arguably most of the work went into coming up with the right definitions.

As the universe is a notoriously geometric place, it is not surprising that many of the concepts are best understood geometrically. This is why we choose the theory of real analysis. It is an area whose algebra is often confusing to students who meet it for the first time. Therefore it gives a good demonstration

---

[6] The trivial cases, where the 'triangle' is a line or a point are in separate equivalence classes and so must be considered separately.

of these ideas, taking complex algebraic formulations and replacing them with the geometric concepts they represent.

Working in Euclidean plane geometry, as with Pythagoras' Theorem above, there is a canonical interpretation: that of each diagram as itself. Unfortunately there can be no such canonical interpretation for diagrams applying to $\mathbb{R}^n$, since $\mathbb{R}^2$ is not homeomorphic to $\mathbb{R}^n$ for all $n \neq 2$. Worse still, there are finite collections of open connected sets $S_1, .., S_N$ in $\mathbb{R}^n$, such that there do not exist $S'_1, ..., S'_N$ in $\mathbb{R}^2$ which are connected and have the same intersection relations as the S_i. These results mean that it is not possible to represent even finite collections of sets in $\mathbb{R}^n$ with sets in $\mathbb{R}^2$ without potentially losing some topological property. This is not a problem – it is actually quite convenient to ignore properties irrelevant to a theorem – but does introduce design choices and the need for several diagrammatic theories.

Analysis proofs often require reasoning about countably large sets of objects (e.g., infinite sequences or open covers). Whilst the framework we have outlined here does allow us to 'draw' countable sets of objects, they cannot physically be drawn. We therefore represent countable sets of objects by a single graphical object whose behaviour is correct with regard to the properties we are interested in (e.g., see Figure   ).

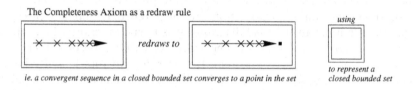

The Completeness Axiom as a redraw rule

redraws to   *using*

ie. a convergent sequence in a closed bounded set converges to a point in the set   *to represent a closed bounded set*

Fig. 10. Representing countable sets of objects: convergent sequences as illustrated by the (diagrammatic) completeness axiom

9.1 Example Proof for Theorem 1 Revisited

We can now begin to see how the example of a proof of Theorem   from §   can be formalised. Drawing objects are needed for general sets, open sets, open balls $B_\varepsilon(x) = \{y| \, |x - y| < \varepsilon\}$, points, and function application arrows, plus a redraw rule for each step (e.g., see Figure   ). Here we define the equivalence relation $\sim$ by $D \sim D'$ iff there is a bijection $f : D \to D'$ such that $f(X, t, i) = (X', t', i') \Rightarrow t = t', i = i'$ and f is an isomorphism with respect to the relations `inside` and `intersects`. Part of our future work is to prove that the redraw rules used obey the generic diagram property.

---

[7] Pythagoras' Theorem is also an analysis result, and part of this project's remit.
[8] Bizarrely, for all n there are continuous surjective maps of $\mathbb{R}^2$ *onto* $\mathbb{R}^n$, but these are not injective.

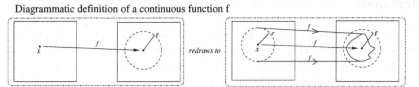

That is, given a point x and distance e, there is a distance r such that all points less than r from x are mapped by f to be closer than e from f(x)

Fig. 11. Redraw rules for proof of Theorem   from § : a diagrammatic definition for continuity

10   Limitations and Future Work

The next step in this project is to complete the formalisation of the ideas presented in this paper. We will then implement them in a prototype interactive theorem prover (called the Real Analysis Prover – RAP). Finally, we aim to implement an automated theorem prover for analysis theorems using diagrammatic reasoning. This will require developing heuristics appropriate to diagrammatic inference for guiding the proof search.

Working with actual diagrams, whilst possible, would be computationally very inefficient. The RAP system will therefore use a visual language of predicates with spatial relations.

Our current research and development of the formal structure is incomplete. One interesting question is whether the equivalence relation between diagrams can be relaxed to include so-called 'degenerate' or 'trivial' cases (such as empty sets or identical points). Currently these must be treated as separate cases. However, it is often possible to transform 'normal' diagrams into degenerate versions, but not vice versa. Ideally, the 'normal' proof would then carry over to degenerate cases.

So far, the proposed framework does not cover proofs of a recursive nature. In the future we hope to extend our framework to include them, perhaps by using a method used in DIAMOND, namely, generalisation via meta-induction.

Our aim in this project is to demonstrate the potential for applying diagrammatic reasoning in mathematical systems. The software we develop should also have a practical application in mathematics teaching, where we hope it will complement conventional methods.

The system outlined is only capable of incorporating algebraic manipulations in a crude way (by 'hiding' the algebra in redraw rules). Hybrid proofs, fluently combining diagrammatic and algebraic reasoning, are clearly a desirable goal. Such systems might finally get close to reproducing the reasoning methods of real-life mathematicians.

References

1. D. Barker-Plummer, S. C. Bailin, and S. M. Ehrlichman. Diagrams and mathematics, November 1995. Draft copy of an unpublished paper.

2. J. Barwise and J. Etchemendy. Heterogeneous logic. In J. Glasgow, N. H. Narayanan, and B. Chandrasekaran, editors, *Diagrammatic Reasoning: Cognitive and Computational Perspectives*, pages 211–234. AAAI Press/The MIT Press, 1995.

3. P. Hayes. Introduction to "Diagrammatic Reasoning: Theoretical Foundations". In J. Glasgow, N. H. Narayanan, and B. Chandrasekaran, editors, *Diagrammatic Reasoning: Cognitive and Computational Perspectives*. AAAI Press/The MIT Press, 1995.

4. M. Jamnik. *Automating Diagrammatic Proofs of Arithmetic Arguments*. PhD thesis, Division of Informatics, University of Edinburgh, 1999.

5. M. Jamnik, A. Bundy, and I. Green. On automating diagrammatic proofs of arithmetic arguments. *Journal of Logic, Language and Information*, 8(3):297–321, 1999.

6. B Meyer. Constraint diagram reasoning. In J Jaffar, editor, *Principles and Practice of Constraint Programming (CP99)*, number 1713 in Lecture Notes in Computer Science. Springer-Verlag, 1999.

7. R. B. Nelsen. *Proofs without Words: Exercises in Visual Thinking*. The Mathematical Association of America, 1993.

8. J. J. O'Connor and E. F. Robertson. Abu Ja'far Muhammad ibn Musa Al-Khwarizm, July 1999. From online software "The MacTutor History of Mathematics Archive". http://www-history.mcs.st-and.ac.uk/history/Mathematicians/Al-Khwarizmi.html.

9. J. J. O'Connor and E. F. Robertson. Pythagoras of Samos, July 1999. From online software "The MacTutor History of Mathematics Archive". http://www-history.mcs.st-and.ac.uk/history/Mathematicians/Pythagoras.html.

10. N. Rose. *Mathematical Maxims and Minims*. Rome Press Inc., Raleigh, NC, USA, 1988.

11. S. J. Shin. *The Logical Status of Diagrams*. Cambridge University Press, 1995.

Playing with Diagrams

Robert K. Lindsay

University of Michigan
205 Zina Pitcher Place
Ann Arbor, Michigan USA
lindsay@umich.edu

Abstract. This paper extends work that developed a programmed model of reasoning about geometric propositions. The system reasons by manipulating representations of diagrams and noticing newly emerged facts that are construed as inferences. The system has been explored as a means of verifying diagrammatic demonstrations of classical geometric propositions and for constructing diagrammatic demonstrations of conclusions supplied for the system. The process of discovering propositions to be demonstrated is a more difficult task. This paper argues that central to the discovery process is systematic manipulation of diagrams – playing – and observing consistent relations among features of the diagram as manipulations are made and observed. The play results in the creation of an "episode" of diagram behaviors which is examined for regularities from which a general proposition might be proposed. The paper illustrates this process and discusses the advantages and limitations of this system and of other computational models of diagrammatic reasoning.

1 Introduction

In work previously reported [1-5] I constructed a computer model that represents diagrams, both as pixel arrays and as propositional statements. For example a specific triangle would be represented as a specific set of marked pixels in an array, plus a set of data structures naming the triangle and recording facts about it (perhaps that it is equilateral and hence isosceles) and constraints upon it (perhaps that it must remain isosceles throughout this use of the diagram). These representations, internal to computer memory, are treated as the equivalent to the computer of an actual physical diagram plus associated knowledge to a human. Instructions to the model result in the construction of specific representations and in manipulations of them by moving components about, adding other elements, etc., subject to the joint constraints imposed by the structure of two-dimensional space (geometric constraints) and those imposed for the moment by instruction (these latter are called situation-constraints).

The model, which I call ARCHIMEDES, is supplemented with additional information as well. For example it has a limited knowledge of algebra, and of the taxonomic hierarchy of geometric figures (all squares are rectangles, and so forth). It can also be provided with some kinds of additional geometric knowledge such as the side-angle-side congruency theorem or the fact that the measures of the interior angles of a triangle always sum to 180 degrees. The model can be instructed to construct a diagram and manipulate it in a sequence of ways. It notices new facts about the

M. Anderson, P. Cheng, and V. Haarslev (Eds.): Diagrams 2000, LNAI 1889, pp. 300-313, 2000.

manipulated representation and records this newly observed information. Its attention can be called to other facts that it has not noticed, and which it can verify by observation. In this manner, the model can verify that a demonstration of an abstract principle, such as the Pythagorean Theorem, is correct. That is, the model can determine whether or not the manipulations reflect the truth of the proposition for the particular case at hand.

Verifying diagrammatic demonstrations in this manner uses diagrams in an essential way rather than merely as a heuristic device for guiding exploration of a problem tree of formal deductions, which is the more customary approach to geometry by computer modeling, beginning with Gelernter [6, 7] and followed by, among others, [8], [9], and [10, 11]. However, a demonstration is not a proof, first of all because a proof is by definition propositional, and second because it literally involves only one, or at best a finite number, of instances. Human observers, however, often construe such demonstrations as convincing arguments because they can see that they could be applied to any member of the class of objects depicted, for example to any right triangle in the case of the Pythagorean Theorem. The knowledge that enables that inference goes beyond the abilities of the model as thus far described; see however [12].

A more complex task is to ask that the model, rather than just verifying a demonstration, should itself construct a sequence of constructions and manipulations that could demonstrate a given proposition. I have explored extensions to my model that can handle problems of a limited class in this manner as well [13].

A yet more complex task is to discover conjectures that might then lead to the construction of demonstrations. This is an open-ended and extremely difficult extension that blends into the problem of modeling mathematical creativity itself. As such, there is little chance that progress will be made on this task in the general case until breakthroughs have been made on fundamental questions about the nature of intelligence.

However, one suggestion often made is that a person discovers interesting conjectures not just by proving propositions at random, but by studying the behavior of diagrams as they are actually drawn and manipulated, or as such operations are performed in the imagination. I call such activities playing with diagrams.

The current model is able to perform one sort of play: systematic variation of some parameters of a diagram, recording of the values of affected parameters, and the discovery of patterns in the resulting record.

This approach is similar to that explored in Eurisko [14] and BACON [15-17]. The former worked in the context of elementary number theory and the latter with sets of physical data. The present work is also inspired by work on qualitative reasoning about physical systems [18].

2   ARCHIMEDES at Play

2.1   Example 1

Consider the classic theorem of geometry, the Pythagorean Theorem, that the lengths of the sides of right triangles are related by the equation $c^2 = a^2 + b^2$ where c is the length of the hypotenuse and a and b are the lengths of the other sides. There are literally hundreds of diagrammatic demonstrations of this theorem that involve construction of the squares on the sides of a right triangle followed by various manipulations that demonstrate the theorem (e.g., see [19]). Each such demonstration is performed on a particular instance of a right triangle. To convince oneself of its generality, demonstrations might be performed on a variety of exemplars, provided they were chosen in such a way that no "accidental" property was essential to the demonstration. Thus a 45 degree right triangle might be a special case for which the theorem is true simply because 45 degrees is a special angle measure. There is no procedure for choosing exemplars that guarantees that all such special cases have been eliminated and the proposition is true in general for all right triangles (although see [20]). Propositional proofs avoid some of these problems by using size, position, and orientation independent characterizations; basically some of the problems are defined away.

Suppose one wished to gain a deeper understanding or a greater sense of assurance that the Pythagorean Theorem is true. One way is to note that it is not true of (some examples of) non-right triangles. Furthermore, one may notice that the relation between hypotenuse square and the sum of the other squares changes *monotonically* with the critical angle, and the relation holds exactly and only as the critical angle passes through 90 degrees.

This observation can be made diagrammatically by constructing a triangle and allowing one of its angles to increase from acute to obtuse while observing the sizes of the squares. My diagrammatic system performed the following experiment. Start with two line segments of arbitrary length and use them to construct a degenerate triangle, that is, one with one straight angle and two null angles. Then increase the measure of one of the null angles while holding the original sides fixed in length. Increase the angle until it is 180 and the other two angles are null. See Figure 1.

ARCHIMEDES performed this experiment and took "snapshots" at various stages including the critical values of 0, 90, and 180 for the altered angle. For each snapshot the diagram was examined and several parameters were recorded. Among these were size of the critical angle cba, critical angle class, and the areas of the three squares on the sides. Also the program looked for symmetries in the diagrams, but found none.

The sequences of parameter values recorded is called an episode. For this experiment a portion of it looked like this:

EPISODE FROM PT WATERSHED DEMO length 13:
conditions= {angle-cba varied}

(<cba class)= {null acute acute acute acute acute right obtuse obtuse obtuse
obtuse obtuse straight}

(<cba size)= {0   13   32   53   72   87   90 103 122 143 162 177 180}

(abde area)= {900 900 900 900 900 900 900 900 900 900 900 900 900}

(bcfg area)= {1600 1602 1597 1600 1588 1604 1600 1602 1597 1600 1588
1604 1600}

(achj area)= {100 162 457 1060 1768 2384 2500 3042 3757 4420 4768 4804
4900}

symmetries= {nil nil nil nil nil nil nil nil nil nil nil nil nil}

A series of noticing functions then examined this episode to look for key features. It found the following.

EPISODIC REGULARITIES NOTICED:

CONSTANTS:
(abde area):value=900 position=all
(bcfg area): value=1600 position=all
symmetries:value=nil position=all

INCREASING SIZES
(achj area): max=4900 position=13

MAXIMA FOR ANGLE SIZES:
(<cba size): max=180 position=12
          000 013 032 053 072 087 090 103 122 143 162 177 180)

MINIMA FOR ANGLE SIZES:
(<cba size): min=000 position=0
          000 013 032 053 072 087 090 103 122 143 162 177 180)

Thus the program has noticed that two of the squares have remained approximately constant in area while the third has increased monotonically. (All measures in the model are considered approximate because of the discrete array used. Equality is determined if differences fall within a tolerance that is determined by a parameter.)

A suite of additional noticing functions is also routinely applied to the regularities. In this case it notices that the critical angle is a right angle only once, and at that point it verifies the PT relation, whereas the relation is not true for any other values of the critical angle.

For a person, these observations comprise an understanding of the critical, or watershed, condition. That is, not only is the PT relation true of right triangles, it is not true of any others because of the monotonic increase of the area of the hypotenuse square. The observations might also suggest for one who is mathematically more sophisticated than the model that, since the relation proceeds smoothly from a less-than inequality through equality to a greater-than inequality, there might be a more general relation for which the PT is a special case.

The program does not in fact understand any of this deeper insight nor could it perform the generalization. What is can do is what it has done, explore the relation among triangle sides and angles by diagram manipulation.

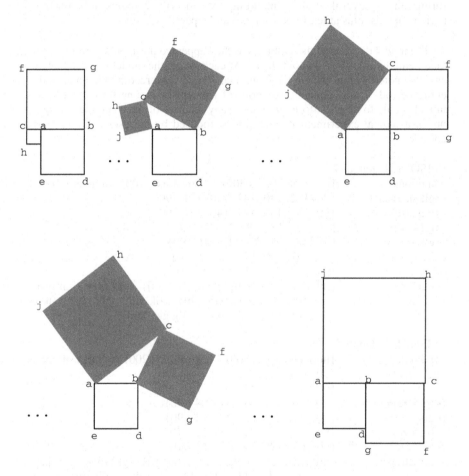

Fig. 1. Increasing the size of angle cba from zero to 180 degrees

2.2   Example 2

In a second example, the model was instructed to explore the behavior of angles inscribed within ellipses as the dimensions of the ellipse were systematically altered. The model constructed a sequence of ellipses with a fixed horizontal axis of length

300 units and vertical axes of 100, 200, 300, 400 and 500 (the third ellipse was thus a circle). See Figure 2.

For each ellipse a sequence of pairs of points on its perimeter was selected beginning with each end of the horizontal axis and proceeding counter-clockwise along the top and bottom in equal steps. For each pair of points in turn each point was connected to the ends of the horizontal axis. Thus for the first pair the connecting lines were of zero length. At each subsequent stage this results in the formation of several angles, including two inscribed angles subtended by the horizontal axis, plus at least two triangles and a parallelogram.

The model discovered these objects as they appeared during the construction, rather than being explicitly told about them. At each stage, the model recorded angle sizes and also noted symmetries in the diagrams. These were recorded in an episode. The entire period of play generated a series of five episodes, one for each ellipse. The model is able to record information about any features of the changing configuration, but in a typical experiment its attention is focused by the user to limit the size of episodes.

EPISODE 1 length 7
conditions=   {(horizontal-axis 300) (vertical-axis 100) (number of points 7)}
(<glf size)=   {90 132 140 142 140 131   90}
(<fkg size)=   {90 132 140 142 140 131   90}
(<fgl size)=   { 0   8   13   19   27   41   90}
(<gfk size)=   { 0   8   13   19   27   41   90}
(<lfg size)=   {90   40   27   19   13   7   0}
(<kgf size)=   {90   40   27   19   13   7   0}
symmetries=   {nil ((flg gkf)) ((flg gkf)) ((fgk gfj) (gkl fjh) (ghl fjh) (gkj
                      fjh) (flg ghf) (fjg ghf) (flk ghj) (flh ghj) (fjk ghj)
                      (fjh ghj)) ((flg gkf)) ((flg gkf)) nil}

EPISODE 2 length 7
conditions=   {horizontal-axis 300) (vertical-axis 200) (number of points 7)}
(<glf size)=   {90 104 111 113 111 104   90}
(<fkg size)=   {90 104 111 113 111 104   90}
(<fgl size)=   { 0   13   23   34   47   64   90}
(<gfk size)=   { 0   13   23   34   47   64   90}
(<lfg size)=   {90   63   46   34   23   12   0}
(<kgf size)=   {90   63   46   34   23   12   0}
symmetries=   {nil ((flg gkf)) ((flg gkf)) ((fgk gfj) (gkl fjh) (ghl fjh) (gkj
                      fjh) (flg ghf) (fjg ghf) (flk ghj) (flh ghj) (fjk ghj)
                      (fjh ghj)) ((flg gkf)) ((flg gkf)) nil}

EPISODE 3 length 7
conditions=   {(horizontal-axis 300) (vertical-axis 300) (number of points 7)}
(<glf size)=   {90   90   90   90   90   90   90}
(<fkg size)=   {90   90   90   90   90   90   90}
(<fgl size)=   { 0   16   31   45   60   75   90}
(<gfk size)=   { 0   16   31   45   60   75   90}

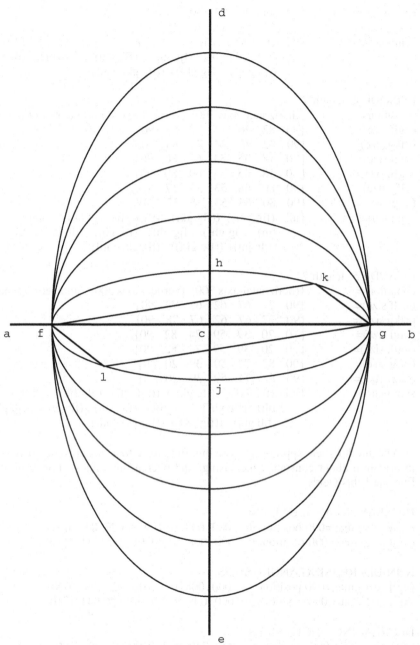

Fig. 2. A series of ellipses with inscribed angles. Points k and l move along an ellipse to generate an episode.

(<lfg size)=	{90 74 59 45 30 15 0}
(<kgf size)=	{90 74 59 45 30 15 0}
symmetries=	{nil ((flg gkf)) ((flg gkf)) ((fgk gfj) (gkl fjh) (ghl fjh) (gkj fjh) (flg ghf) (fjg ghf) (flk ghj) (flh ghj) (fjk ghj) (fjh ghj)) ((flg gkf)) ((flg gkf)) nil}

EPISODE 4 length 7

conditions=	{(horizontal-axis 300) (vertical-axis 400) (number of points 7)}
(<glf size)=	{90 82 76 74 77 83 90}
(<fkg size)=	{90 82 76 74 77 83 90}
(<fgl size)=	{ 0 18 35 53 69 80 90}
(<gfk size)=	{ 0 18 35 53 69 80 90}
(<lfg size)=	{90 80 68 53 35 17 0}
(<kgf size)=	{90 80 68 53 35 17 0}
symmetries=	{nil ((flg gkf)) ((flg gkf)) ((fgk gfj) (gkl fjh) (ghl fjh) (gkj fjh) (flg ghf) (fjg ghf) (flk ghj) (flh ghj) (fjk ghj) (fjh ghj)) ((flg gkf)) ((flg gkf)) nil}

EPISODE 5 length 7

conditions=	{(horizontal-axis 300) (vertical-axis 500) (number of points 7)}
(<glf size)=	{90 77 67 62 67 78 90}
(<fkg size)=	{90 77 67 62 67 78 90}
(<fgl size)=	{ 0 20 39 59 74 82 90}
(<gfk size)=	{ 0 20 39 59 74 82 90}
(<lfg size)=	{90 82 74 59 39 20 0}
(<kgf size)=	{90 82 74 59 39 20 0}
symmetries=	{nil ((flg gkf)) ((flg gkf)) ((fgk gfj) (gkl fjh) (ghl fjh) (gkj fjh) (flg ghf) (fjg ghf) (flk ghj) (flh ghj) (fjk ghj) (fjh ghj)) ((flg gkf)) ((flg gkf)) nil}

The model then applies its noticing functions to each episode, resulting in information about constants, increasing, and decreasing values. For example, for Episode 1 this yields

INCREASING ANGLE SIZES:
(<gfk size):max=090 position=6   000 008 013 019 027 041 090)
(<fgl size):max=090 position=6   000 008 013 019 027 041 090)

NON-DECREASING ANGLE SIZES:
(<gfk size):max=090 position=6   000 008 013 019 027 041 090)
(<fgl size):max=090 position=6   000 008 013 019 027 041 090)

DECREASING ANGLE SIZES:
(<lfg size):min=000 position=0   090 040 027 019 013 007 000)
(<kgf size):min=000 position=0   090 040 027 019 013 007 000)

NON-INCREASING ANGLE SIZES:
(<lfg size):min=000 position=0   090 040 027 019 013 007 000)
(<kgf size):min=000 position=0   090 040 027 019 013 007 000)

MAXIMA FOR ANGLE SIZES:
(<glf size):max=142 position=3   090 132 140 142 140 131 090)
(<fkg size):max=142 position=3   090 132 140 142 140 131 090)
(<fgl size):max=090 position=6   000 008 013 019 027 041 090)
(<gfk size):max=090 position=6   000 008 013 019 027 041 090)
(<lfg size):max=090 position=0   090 040 027 019 013 007 000)
(<kgf size):max=090 position=0   090 040 027 019 013 007 000)

MINIMA FOR ANGLE SIZES:
(<fgl size):min=000 position=0   000 008 013 019 027 041 090)
(<gfk size):min=000 position=0   000 008 013 019 027 041 090)
(<lfg size):min=000 position=6   090 040 027 019 013 007 000)
(<kgf size):min=000 position=6   090 040 027 019 013 007 000)

The remaining episodes yield similar results. For each episode it is noted among other things that angle <gfk increases from 0 to a maximum of ninety degrees, and that angle <fkg reaches an extreme value when the moving points lie on the vertical axis. Furthermore, the program notices that for the third case (the circle), angle <fkg is constant and furthermore is a right angle.

Finally, the model then examines the set of episodes itself in search of regularities across them, using similar methods. It notices that the extreme values reached by <fkg changes monotonically and achieves a critical value of right angle for episode 3, the circle. This is a potential watershed because horizontal and vertical axes are equal for this case and no other. That is, for the first two cases, <fkg increases to a maximum (which is smaller in the second case but always obtuse) and then decreases to its original value, while for cases four and five the angle decreases to a minimum (which is smaller in the fifth case and always acute) and then increases to its original value. Thus, the sequence of extreme values decreases monotonically as one proceeds through the cases, passing through 90 degrees for the circle.

The information in this series of episodes thus suggests that the circle is a watershed case for which the inscribed angles are right angles, which of course is a true proposition about circles. The program has not proven this nor even constructed a demonstration of it in the sense that term is used in this research. Rather it has found a conjecture.

Episodes may be examined for indication of other key features as well. Two examples are asymptotes and cyclic behavior. The latter is not of much value in geometric contexts, but could be very important in many other applications.

3   Discussion

Much work in Artificial Intelligence amounts to the development of programming languages in which one can describe a problem solving procedure for a class of tasks. A programmer can then pose tasks to the model which attempts to do them. Thus in a sense the creator of the model indirectly builds in the solution that the model later "finds" when asked to do so, much as an accounting program builds in the "bottom line" when the program is written and data are input.

Models differ in how well-hidden the building-in process is. At one extreme, this process is trivial: one tells the computer what the answer is and then asks the question. However, there are many ways to tell the computer the answer, and some of them are quite indirect and general, and hence more of the problem-solving function is performed by the model.

In adaptive systems models, such as the genetic algorithm, genetic programming and neural net models, the modeler provides a very general representation and a general adaptive (learning or evolving) process that, through interaction with an "environment" leads to an organization that deals with a class of tasks. The "environment" is the hidden door through which the modeler supplies more of the answer.

At the other modeling extreme are highly special-purpose programs that deal with a limited class of tasks. For those limited tasks, the model performance is often quite spectacular. It is readily possible to program a computer with a general method that will devise answers that its creator could not even if he attempted to follow the algorithm himself, since a person is by comparison slow and error prone on those tasks at which computers excel. Deep Blue, the chess machine, is a good example. Deep Blue can play better chess than any of its creators; nonetheless it is "just doing what it was told to do."

Cognitive architecture is a term that has come into use to describe a programming environment for the development of complex programs. The term is used in contrast to *programming language* to emphasize that it is more than an instruction set, that it also includes a host of conventions about information representation and access. Some architectures provide ways to represent logical expressions and supply a deduction/theorem-proving engine for manipulating them. Still others, such as SOAR [21] and the ACT series of models [22], provide mechanisms for stating general knowledge, and means for combining and exploring collections of knowledge. These are distinct architectural approaches and each entails and affords different kinds of behaviors.

In the realm of diagrammatic reasoning much of the interesting theoretical work can be seen as the specification of an architecture. To mention just two examples of many, Furnas's bitpict model [23] defines a pixel representation and a set of operations that can be performed on it. Various programs can be composed of these operations, for example, a program can be written that counts blobs by combining its primitive abilities into a sequence of operations that results in a picture that depicts (in Roman Numerals!) the number of objects. The work of Anderson and McCartney [24] can be seen in this same light.

What AI and related cognitive modelers are doing is, I think, best seen as explorations of cognitive architectures. That is, we are exploring very general and varied approaches, not with the hope of simulating human behavior totally or producing superior intelligence even in a limited domain, but rather as empirical explorations of the space of possible computational models.

ARCHIMEDES is a cognitive architecture in this sense, but it differs significantly from the others mentioned. The general architectures of which ARCHIMEDES is an example are sometimes called Tool Box models. In contrast to general and unified representations and algorithms – such as adaptive systems, logic-based systems, and production systems – Tool Box models see the mind as a collection of specialized tools and representations that are selected as need arises. This is an idea that is consonant with the view that evolution is a tinkerer, producing adaptations to specific situations and borrowing them for new uses. Thus my model proposes an architecture in which the computer builds and consults mental models of its object of investigations. The model borrows the perceptual and motor adaptations that are proven means for intercourse with the physical world, adds the abstraction made possible by language, and creates something new and powerful. The architecture explores the use of iconic representations that make use of the structure of space and allow the use of Gedanken experiments to explore deeper consequences of this structure, even though at the moment it leaves much of the work to me.

4   Conclusion

Geometric reasoning is historically among the earliest uses of diagrams for both practical and mathematical problem solving. It combines propositional and diagrammatic reasoning in an essential interplay.

Geometry is the study of the structure of space. Plane geometry is a part of this study, limited to the structure of two-dimensional space. Geometric diagrams, as the term is generally used, are two dimensional drawings that are abstractions intended to represent idealized objects with approximations constructed by drawing instruments, whether pencils or, today, computers. In geometric reasoning, diagrams are used as models of propositions about space, and propositions about space are intended to reflect the structure of diagrams, or more precisely, of the abstract situations the diagrams are intended to capture.

If we restrict the discussion to representations of truly two-dimensional objects, excluding perspective drawings and other depictions of three-dimensional objects, geometric reasoning is the potential fruit fly of modern diagrammatic reasoning research because it makes use of all of the properties of two dimensional space, and thus reveals the full power of the representation with a minimum use of arbitrary task-specific conventions for interpretation. Consider, in contrast, Euler/Venn diagrams. These use only some (topological) properties of two-dimensional space, specifically the notion of containment, which is mapped onto – represents – set inclusion (when one area is contained within another) and set membership (when a point is within a bounded area). The diagrams are often presented in a sequence, which uses the ordering of space to indicate the sequence of the argument. However the diagrams ignore metric properties of distance and direction, perpendicularity and parallelness, and so forth. This means that understanding these diagrams entails understanding this assignment of meaning to spatial structure plus knowing what features of structure are to be ignored. The situation is similar for scientific graphs, flow-charts, and most other diagrammatic representations, with each employing different mappings of meaning onto spatial structure and ignoring different aspects of structure. Geometric diagrams, in contrast, are intended to reflect all structural properties of two dimensional space, although this requires that the observer ignore line width and so forth and deal with the intended idealization. Since they are recorded in space (on a piece of paper for example) the mapping is iconic.

The intuitive structure of two-dimensional space is partially captured by Euclid's definitions, postulates, and axioms. Although it is now known that this is not the veridical description of actual space, for objects of human scale it is adequate both for reasoning and for successful intercourse with the environment, just as Newtonian mechanics is adequate for modeling physical processes that occur at a human scale though it is inaccurate at very large and very small scales.

There is no canonical representation of this Euclidean structure.   One representation is the Euclidean axiomatization, *augmented by extensive human knowledge*. A more complete representation is provided by Hilbert's axiomatization which makes explicit some of the knowledge Euclid assumed to be "common knowledge." It must also be recognized that the Hilbert axiomatization is not complete either, for it presumes un-specified abilities of an interpreter of the representation.

The structure of intuitive space is also captured in a different but equivalent form by analytic geometry. Geometry and algebra have been unified, beginning with Descartes' seminal insights. What this means for those of us who are attempting to model geometric reasoning is that we have a choice of underlying representation. In general, one chooses an analytic representation for the simple reason that conventional digital computers are a natural match for numerical representations. This does not mean, however, that diagrams represented numerically are not really diagrams. What makes them diagrams is not bits or voltages or axioms or CCD signals. What makes them diagrams is that they capture the structure of space. This is another way of saying that they enforce constraints on the behavior of the representations that reflect restrictions on the behavior of objects in space.

I suggest that the basic theory of diagrammatic reasoning should be based on understanding and employing the structure of space and interfacing this with symbolic reasoning ability.

Acknowledgment

This material is based on work supported by the United States National Science Foundation under Grant No. IRI-9526942.

References

1.   Lindsay, R.K.: Qualitative Geometric Reasoning. In: Proceedings of the Eleventh Annual Conference of the Cognitive Science Society (Ann Arbor, MI). Lawrence Erlbaum, Hillsdale, NJ (1989) 418-425

2. Lindsay, R.K.: Diagrammatic Reasoning by Simulation. In: Reasoning with Diagrammatic Representations.   Technical Report SS-92-02. American Association for Artificial Intelligence, Menlo Park, CA (1992) 130-135
3. Lindsay, R.K.: Understanding Diagrammatic Demonstrations. In: Proceedings of the Sixteenth Annual Conference of the Cognitive Science Society (Atlanta, GA). Lawrence Erlbaum, Hillsdale, NJ (1994) 572-576
4. Lindsay, R.K.: Using Diagrams to Understand Geometry. Computational Intelligence **14**(2) (1998) 222-256
5. Lindsay, R.K.: Knowing about Diagrams. In: Anderson, M., Olivier, P. (eds.): Reasoning with Diagrams. Springer-Verlag, Dordrecht (2000)
6. Gelernter, H.: Realization of a Geometry Theorem Proving Machine. In: International Conference on Information Processing. UNESCO House, Paris (1959) 273-282
7. Gelernter, H., Hansen, J.R., Loveland, D.W.: Empirical Explorations of the Geometry Theorem Proving Machine. In: Proceedings of the Western Joint Computer Conference. National Joint Computer Committee, New York (1960) 143-147
8. Nevins, A.J.: Plane Geometry Theorem Proving Using Forward Chaining. Artificial Intelligence **6** (1975) 1-23
9. Koedinger, K.R., Anderson, J.R.: Abstract Planning and Perceptual Chunks: Elements of Expertise in Geometry. Cognitive Science **14** (1990) 511-550
10. McDougal, T.F., Hammond, K.J.: A Recognition Model of Geometry Theorem Proving. In: Proceedings of the Fourteenth Annual Conference of the Cognitive Science Society (Bloomington, Indiana). Lawrence Erlbaum, Hillsdale, NJ (1992) 106-111
11. McDougal, T.F., Hammond, K.J.: Representing and Using Procedural Knowledge to Build Geometry Proofs. In: Proceedings of the Eleventh National Conference on Artificial Intelligence (Washington, DC). AAAI Press, Menlo Park, CA (1993)
12. Lindsay, R.K.: Generalizing from Diagrams. In: Cognitive and Computational Models of Spatial Reasoning. American Association for Artificial Intelligence, Menlo Park, CA (1996) 51-55
13. Lindsay, R.K.: Using Spatial Semantics to Discover and Verify Diagrammatic Demonstrations of Geometric Propositions. In: O'Nuallian, S. (ed.) Spatial Cognition. John Benjamins, Amsterdam (2000) 199-212
14. Lenat, D.: AM: An Artificial Intelligence Approach to Discovery in Mathematics as Heuristic Search. Doctoral dissertation, Stanford University (Computer Science Department Report CS-76-57) (1976)
15. Langley, P.: BACON-1: A General Discovery System. In: Proceedings of the Second National Conference of the Canadian Society for Computational Studies in Intelligence. (1978) 173-180
16. Langley, P., Bradshaw, G.L., Simon, H.A.: BACON-5: Discovery of Conservation Laws. In: Proceedings of the Seventh International Joint Conference on Artificial Intelligence (Vancouver, British Columbia). American Association for Artificial Intelligence, Menlo Park, CA (1981)
17. Langley, P., Bradshaw, G.L., Simon, H.A.: Rediscovering Chemistry with the BACON System. In: Michalski, R.S., Carbonell, J.G., Mitchell, T.M. (eds.): Machine Learning. An Artificial Intelligence Approach. Tioga Press, Palo Alto, CA (1983) 307-329

18. Bobrow, D.G., (ed.) Qualitative Reasoning about Physical Systems. Bradford Books, Cambridge, MA (1985)
19. Loomis, E.S.: The Pythagorean Proposition: Its Proofs Analyzed and Classified and Bibliography of Sources for Data of the Four Kinds of Proofs. 2nd edn. Edwards Brothers, Ann Arbor, MI (1940)
20. Miller, N.: Case Analysis in Euclidean Geometry. Department of Mathematics, Cornell University (2000)
21. Newell, A.: Unified Theories of Cognition. Harvard University Press, Cambridge, MA (1990)
22. Anderson, J.: The Architecture of Cognition. Harvard University Press, Cambridge, MA (1983)
23. Furnas, G.W.: Reasoning with Diagrams Only. In: Reasoning with Diagrammatic Representations. Technical Report SS-92-02. American Association for Artificial Intelligence, Menlo Park, CA (1992) 118-123
24. Anderson, M., McCartney, R.: Inter-diagrammatic Reasoning. In: Proceedings of the Fourteenth International Joint Conference on Artificial Intelligence (Montreal). Morgan Kaufmann., San Mateo, CA (1995) 878-884

The Use of Intermediate Graphical Constructions in Problem Solving with Dynamic, Pixel-Level Diagrams

George Furnas, Yan Qu, Sanjeev Shrivastava, and Gregory Peters

School of Information
University of Michigan
Ann Arbor, MI 48109-1092
furnas@umich.edu

Abstract. Many diagrams can be thought of as graphical representations used to support the solution of problems. This paper discusses how computation based on pixel-level rewrites can produce a rich form of diagrammatic computation making use of intermediate graphical constructions not explicit in the input or output of the problems it is solving.

1 Introduction

One of the more intriguing possibilities which computation offers for diagrammatic representations is the creation of diagrams that actively transform themselves to solve interesting problems. The focus in this paper is on representations that do so without the aid of other non-diagrammatic machinery (unlike, for example [1][2][3] who also use sentential representations to enforce constraints or augment the reasoning). It is a central premise of this paper that one important understanding of computational systems, comes from examining *how* they do the things they do. In the diagrammatic case of interest here, a rich and subtle space of computations is produced through the use of dynamic intermediate graphical constructions. These constructions are temporary features of the 2-D representation that are explicit neither in the initial presentation of the problem (the input) nor the final result (the output). They are ephemera, created on the fly as needed to advance the computation, and then discarded.

The research explores these constructions with a Pixel Rewrite System (generically PRS, in this case a reincarnation of BITPICT [4][5][6]). A PRS comprises a set of rules and a shared blackboard, essentially a pixmap, called the *field*. The left-hand side of each rule is a pixel pattern, which is searched for in the current state of the field. If a match is found, it is rewritten as the pattern in the right-hand side of the rule. Only one rule is allowed to fire at a time; if there is more than one match, any of various conflict resolution schemes is used to pick a winner. That winner performs its rewrite, and the system iterates at high speed. The rewrites, by transforming the 2D patterns in the field, do the computations and solve the problems, and in so doing often make use of non-trivial intermediate patterns.

M. Anderson, P. Cheng, and V. Haarslev (Eds.): Diagrams 2000, LNAI 1889, pp. 314–329, 2000.

Note that rules here use only local, 2D rewrites on rectilinear grids of pixels. There is a large space of related systems as yet unexplored, including variants with multiple layers and non-local pixel rewrites (i.e., with arbitrary separation between subpatterns). Some of these variations will surely alter the space of problems easily addressed – and presumably, the intermediate constructions they support. We have been exploring this particular initial class of PRSs because they are quite spatially intense and computationally rich, yet their simplicity and theoretical tractability should provide good grounding for the larger class. Note also that since the fundamental representations used in the computation are pixmaps, the various shapes involved (input, output and along the way), can have arbitrary form. This is quite different from systems where much work is done sententially, e.g., with analytic geometry, and lines and polygons tend to dominate.   We are exploiting this easy, computationally accessible representation of a richer set of shapes to try to open up the design space of possible human/computer interactions to new forms, behaviors and problems. Here, however we focus on intermediate constructions in computation.

2   On Intermediate Constructions

Structures used in the process of solving a problem, but which are not explicit in the input or output, are not uncommon. There are at least two familiar categories. First are those graphical constructions used in more traditional diagrammatic contexts. In mechanical drawing, for example, it is common to construct auxiliary lines to aid the placement of various features in the final figure: an extra guide-line helps align a series of elements that are supposed to be collinear. More interestingly, parallelogram construction is often used to compute important positions. Given three points forming an angle, $\angle bac,$ an engineer might wish to locate a forth point that will complete a parallelogram, *bacd.* She first draws an auxiliary line parallel to the segment <u>ab</u> through point *c*. Then she draws a second line parallel to <u>ac</u> through *b*. Where these two auxiliary lines intersect, she places her fourth point. This trick is used, for example, to position points in perspective drawings. It is also how a physics student working on a statics problem graphically "computes" a resultant vector, adding two other vectors (e.g., *ab* and *ac*) using the parallelogram rule (the resultant is the vector *ad*).

Such intermediate constructions use space quite intimately to solve the problem. These auxiliary lines reify certain geometric constraints that are easily captured by graphical constructions. The graphical intersection of such constructions yields the locus of points jointly satisfying the corresponding constraints. A straight-edge and compass allow the reification of certain such constraints, and the resulting limited sort of computation made them a mainstay of construction in Euclidean geometry (e.g., [4]) In another strategy, used in proving geometric congruence, special marks (like arcs inside of angles, and single and double tick marks on segments) serve as a special, spatially located record-keeping device to propagate congruence relations throughout a diagram according to simple rules (e.g., opposite angles are equal.) Locating annotations right in the diagram allows easy matching of the graphical rules for propagation of congruence.

Other diagrammatic formalisms proposed for solving problems (e.g., Venn diagrams for syllogistic reasoning [7][8]) similarly use relations of adjacency, inside/outside, overlap, and connection to represent logical constraints, and invoke intermediate constructions to propagate those constraints.

The techniques, while making good use of properties of space, have typically been much less dynamic than one might hope in a computational environment. The methods historically have been developed for manual operation, and as such would rarely be invoked more than a half dozen times in a problem. Also, presumably following from the incidental property of paper that it is easier to draw neatly than to erase neatly, they have tended to be approximately monotonic, in the sense that they keep adding to a diagram.

The second, and rather different, class of intermediate constructions are the local variables and data structures in usual programming languages. It is standard practice in writing code to create temporary variables for numerous purposes in the operation of a procedure. Integer counter variables are created to control iteration. Booleans are used to store state information for later use in execution flow branching. Variables of arbitrary types are used to accumulate intermediate results in extended calculations. Pointers are used to keep track of the focus of operations in arrays. Whole data-structures may be created to support the computation of final results. All of these familiar ephemera are created in the process of computing outputs given inputs, yet may have no explicit description in the problem statement, and are thrown away when computation is complete. They arguably serve a richer variety of uses than the geometric constructions above, and they are certainly more computationally intense: iteratively altered, created and discarded as needed, often used in great numbers.

An underlying assertion here is that various computational paradigms are characterized by the way they create and use such ephemera. The two-dimensional surfaces and drafting tools for engineering drawings support certain such constructions in service of geometric computation. Those of standard procedural sentential programming languages support a well-explored arsenal of intermediate datastructures. Coming to understand each of these computational paradigms in part involves understanding such constructions. In that spirit, to understand the possibilities of diagrammatic computation as it might be possible in more radical ways, for example with pixel level rewrites, we must start to look at the kinds of intermediate constructions such systems use to solve problems. In particular, we want to use space as intimately as the geometric construction examples, but more intensely with a greater variety and fluidity of intermediate constructions, comparable to what is seen in sentential programming.

3   Examples of Intermediate Constructions in a PRS

In this section we examine in detail four examples of pixel-rewrite rule-sets that compute solutions to different spatial problems. These have been selected, from the dozens we have created, to illustrate the use of intermediate graphical states (usually constructions) in the computation. We take them to be representative the kind of richer use of space that computation with diagrams or other picture-like representations may afford in the future.[1] For each example we will first describe the problem and the general way the rewrites solve the problem. Then we will comment preliminarily on how intermediate graphical states were used in the algorithms. The subsequent discussion session will make some general observations on these examples, and comment on some related work.

In these examples, note that the different steps of computation are sequenced by having rules of high priority that fire first, then lower priority rules for the next phase, etc.

3.1 Example 1: Blob Move

We begin with a simple example. Here the "problem" is to shift a green blob-shape (any arbitrary set of 4-connected, green pixels) to the right by one pixel, when a red pixel is placed next to it. This problem arises in pixel-rewrite-based human-computer interaction systems we are developing, where a user needs to be able to push around arbitrarily shaped objects in their workspace. The problem is non-trivial because the rules can only rewrite local pixel patterns of fixed size and shape, and yet must somehow move blobs of arbitrary size and shape. The algorithm shifts the blob shape by adding a column of green pixels to its leading edge and deleting a column of pixels from its trailing edge – with the result that the shape is preserved but shifted right. (For simplicity we show the rules adequate for any *convex* blob. A few rules for single-pixel-wide features can extend the algorithm for arbitrary shapes.) A sample initial problem state is shown in Figure 1.

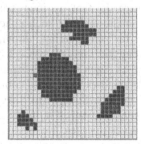

Fig. 1. Sample initial state: a red "pushing" pixel touches a green blob

Step 1. Identify the extent of the blob. All of, and only, the red-touched blob must be shifted, so its full extent is first identified by re-coloring all its dark green pixels ivory. The first rule of Figure 2 starts the re-coloring at the red marker. The second propagates the ivory in all compass directions where there is adjacent dark green. Intermediate states of the re-coloring are shown in Figure 3.

---

[1] Note that we have explicitly excluded here what might be called simulation examples, where PRSs mimics physical processes (e.g., gas models, balls and hoppers, Rube-Goldberg type configurations). There the "problem" of determining some later or final state is "solved" by simulating it, and intermediate graphical states for solving the "problem" are approximations of the intermediate physical states being simulated. These are interesting in their own right, but bear more on the notions of physical computation and simulation.

(all rotations)

Fig. 2. Rules for re-coloring the red-touched blob

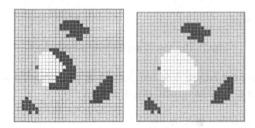

Fig. 3. Extent of the connected blob identified by propagation

Step 2: Mark the leading and trailing edges. Once the blob is identified, its leading (right) edge is recognizable as an ivory-to-gray left-to-right boundary. The gray pixel at any such new frontier is re-painted blue. The opposite transition identifies a trailing edge, and the ivory pixel is repainted black (Figure 4).

Fig. 4. Marking leading frontiers and trailing edges

As a result of these rules the new leading frontier has been identified and marked blue, and the old trailing edge marked black (Figure 5).

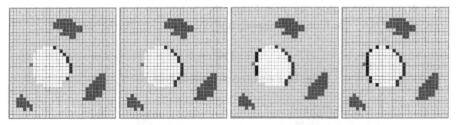

Fig. 5. The new leading frontier and old trailing edge are marked

Step 3. Repaint the blob an intermediate color. We want to reset the body of the blob to dark green, but cannot do so directly. The first ivory pixel we would try to revert to dark green would immediately be repainted ivory by the higher priority rules of Figure 2. Thus we must first repaint the ivory to an intermediate color, here orange, as shown in Figures 6 and 7.

Fig. 6. Rule to repaint blob an intermediate orange color to prevent an infinite loop

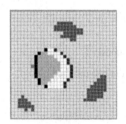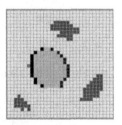

Fig. 7. The rest of the blob repainted orange

Step 4. Make the new edges. The old trailing edge is deleted and the new leading frontier is repainted to original green. (Figures 8 and 9)

Fig. 8. Rules to incorporate the new frontier and delete the old trailing edge

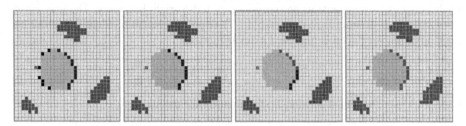

Fig. 9. Steps to incorporate the new frontier and delete the old rear edge

Step 5. Revert blob color. As a final step, the orange pixels are repainted to their original dark green – the blob shape has shifted one pixel (Figures 10 and 11).

Fig. 10. Rule to revert the blob color

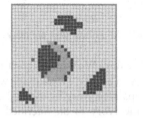

Fig. 11. The blob ends up shifted right by one pixel

Preliminary comments on intermediate graphical constructions in the blob move example.
Several intermediate graphical constructions were used here. First, the propagating ivory color implemented the transitive closure of the adjacency relationship, marking the extent of the blob. Then, leading and trailing edges were recognized and marked. An intermediate orange color was used to avoid unwanted execution looping. The final solution was achieved as these intermediate graphical constructions were removed: leading fontier appended in green, trailing edge deleted, and blob body color restored.

3.2   Example 2: From Source to Sink around an Obstacle

Here the problem is for rules to draw the shortest possible path from a "source" object to a "sink" while avoiding the "obstacle" by some minimal buffering distance. This sort of problem is familiar in robot motion planning contexts. The initial status of a sample field is shown in Figure 12. The obstacle is light lavender while the background is purple. A reddish-black-colored dot in object **A** designates it as a source, while a turquoise dot in object **B** designates it as a sink.

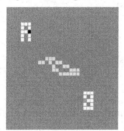

Fig. 12. Initial state

Step 1: Change objects' colors and grow buffer zone. First the reddish-black and turquoise diffuse through **A** and **B** separately, identifying them in their entireties as source and sink, respectively. Next, a set of rules grows a buffer zone, one pixel layer at a time, around the obstacle (Figure 13).

Fig. 13. Buffer zone grown in successively darker shades of red.

Step 2: Grow contours of a distance field around the sink. Next, a similar set of rules starts at the sink, **B**, and lays out successive iso-distance contours in the remaining space, one pixel layer at a time (Figure 14). Cycling through three colors (dark green, blue, olive) is sufficient to preserve shape and down-hill direction in this temporary distance field.

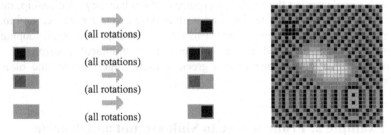

Fig. 14. Iso-distance contours grown, mod3

Step 3: Walking down the field gradient. Next, a set of rules implements a "down-hill walker" who begins at source **A** and travels locally downhill across the field lines (Figure 15), leaving a brighter "chalk line" behind – the shortest path (in city-block metric).

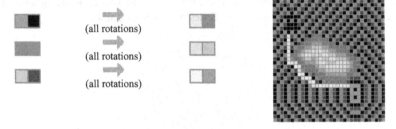

Fig. 15. Downhill path traced through the contours

Step 4: Cleanup. After the shortest path is constructed, the temporary colorings are removed, leaving just the shortest path curve, well clear of the obstacle (Figure 16).

Fig. 16. Shortest path from A to B around obstacle

Note that if there are two or more sinks (Figure 17), each grows its own set of contours, meeting at Voronoi curves equidistant from their respective sources, dividing the space up into separate attractor regions. Each source finds a shortest path to the nearest sink.

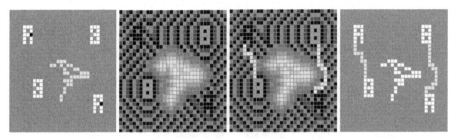

Fig. 17. Find shortest paths from multiple sources to multiple sinks

Preliminary comments on the shortest path example.
At least three sorts of intermediate constructions were used here. As in the blob moving application, we first used special colors to temporarily identify source and sink objects in their entirety (reddish-black and turquoise, respectively). Before the iso-distance contours grew, a temporary buffer zone expanded around the obstacle. This intermediate construction guaranteed that the subsequently generated path would get no closer than five pixels to the obstacle. Perhaps the most interesting intermediate construction in this example is the set of iso-distance contours. These expand to the whole field, transferring the distance information throughout, so the chalk-walker, wherever she starts, knows where to step next.

3.3   Example 3: Reducing Connected Components to Dots

The third example reduces arbitrary 4-connected patterns of pixels (separate topological "components") to individual dots. A simple version of this rule-set appeared in the first part of an example in [4], which ended up collecting and arranging the dots in the lower right corner of the field, finally converting them to a roman numeral count of the initial components. Here we focus on more general rules for reducing the components (they handle blobs and cycles), doing so in a number of steps involving intermediate graphical states. Making use of the ability to erase as well as to write, this kind of transformation creates an intermediate graphical form (which we bastardize the language a bit to call a "construction") that is a departure from the tradition inherited from paper systems. The algorithm first thins the components to single pixel width. Long stringy ends are then reduced by successively nibbling their tips. Next any cycles are determined and broken, leaving stringy ends which are further nibbled. At each step the rules neither create nor destroy connected components, merely reducing each to a single pixel. Figure 18 shows a sample initial state.

Fig. 18. Initial state – six components

Step 1: Thin. Wide parts of the components are eaten away (Figure 19(a)&(b)).

Step 2: Nibble tips. Then tips of single pixel wide components are nibbled (Figure 19(c) and Figure 20. Note that non-rectangular pixel rewrites are allowed, as in (a) and (c).)

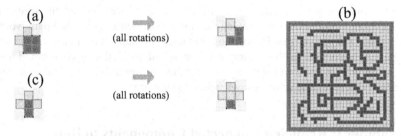

Fig. 19. (a) Thinning rule, (b) thinned components, (c) tip nibbling rule

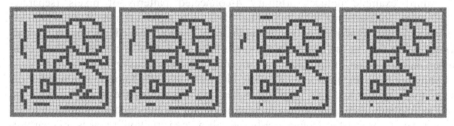

Fig. 20. Intermediate structures for tip nibbling

At this point all components without cycles have been reduced to single pixels. Those with cycles have been thinned, and had all extraneous branches eliminated leaving only the cycles

Step 3: Break cycles. Any cycles will have at least three adjacent green pixels, and a random one such (on the whole field) is turned yellow. It is then converted to blue which propagates in both directions from that point. When the propagating blue meets itself, it has rounded a cycle, and the green pixel with blue on either side is deleted, breaking the cycle. The process iterates until all the cycles are broken. Then the free tips remaining from the broken cycles are nibbled back to a single pixel and restored to green. From there they can be collected and counted as in [4]. The rules for handling cycles are shown in Figure 21, with intermediate structures shown in Figure 22.

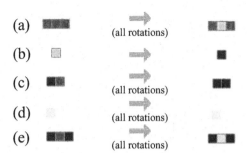

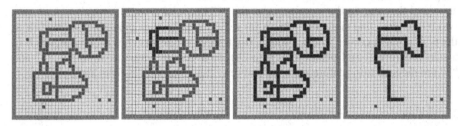

Fig. 21. Cycle-breaking rules. (a) One cycle picked, then at higher priority (b) blue propagation starts, (c) continuing until (d) & (e) it almost meets itself, whereupon the cycle is broken

Fig. 22. Cycle breaking and tip nibbling (eventually reducing the blue component to a dot.)

Preliminary comments on the reducing components example.
In this application, the intermediate graphical states include the thinned versions of the original components, and then the successively nibbled versions. This use of erasure is an important kind of intermediate state. It results in a simplification analogous to the simplification of algebraic expressions. Here however the rules preserve a topological invariant (connectedness) instead of an algebraic one (equality). Later, propagating blue along the green pixels identifies cycles so they can be broken. This is similar to Example 1, where a blob is painted to an intermediate color to find its extent. Note that in the larger problem of counting these components [3], the final dots are intermediate constructs canonicalized for easier counting.

3.4   Example 4: Digit Recognition

The final example is a digit recognition application, recognizing different digits (0-9) of any size and location, rewriting them in a unified format. We complete this task by shrinking the characters, marking the significant features of the shrunken structures, and matching them with fixed patterns. To simplify our example algorithm, the original digits consist of only horizontal and vertical lines.

Figure 23 gives the original digits to be recognized. They have different sizes and locations in the yellow rectangles. The actual rules for this example are rather involved, and so only a few are shown in detail. Instead their actions are described, and shown in the figures.

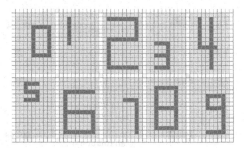

Fig. 23. Digits in varied formats

Step 1: Shrink. First the digits are shrunk as much as possible while preserving the structural features. In the simplest case, the "1" becomes a short vertical stick of only two red pixels – the signature pattern for a "1". Simple rules move this short stick to the bottom-right corner where a single, digit-specific rule matches it and rewrites it in a standard format and position (Figure 24). In a similar way, the digits "3","4" and "7" are reduced to signature patterns, and rewritten in canonical form and color.

Fig. 24. Recognition and rewriting of the digit "1"

Step 2: Substitute special marks for some structural features. For certain digits, we substitute significant features with pixels of special colors. For example, as shown in Figure 25, an upward turn is represented by a green pixel, while a downward turn is represented by an orange one. After this substitution and further shrinking, we can recognize and rewrite many of the other digits. For example, the recognition and rewritings of "2" and "5" are shown in Figure 26.

Fig. 25. Marking structural features

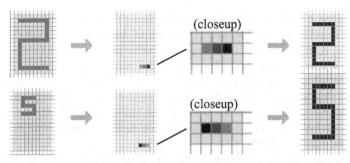

Fig. 26. Recognition and rewriting of digits "2" and "5"

The final result of all the digit recognition rules is shown in Figure 27.

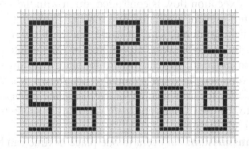

Fig. 27. The result of the digit recognition applied to the field in Figure 23

Preliminary comments on the digit recognition example.
In this application, in a spirit similar to Example 3, we shrank the skeletons of the characters to produce simpler, canonical versions. Color-coded pixels then served to flag compactly the presence of distinctive structural features. Further shrinking made the set of identified features all local for additional rules to then recognize as signature patterns of the various digits For example, a row of adjacent orange, red and green pixels represented a "2". The final recognized digits were rewritten in a new color to prevent execution cycling.

4   Discussion

An exhaustive taxonomy of uses for intermediate graphical forms in this pixel-rewrite system is certainly premature, and given human ingenuity probably unattainable. Several common themes have emerged, however:
- **Reifying spatial relationships:** Certain spatial relationships are rendered graphically, making them available for further computation, including simple iteration. The most common example was the adjacency relation, where neighboring pixels were iteratively identified to compute connectedness, the transitive closure of adjacency – in coloring the blob, in identifying the entire source or sink, in laying out successive contours of the distance field, in finding

and breaking cycles. Note that this is analogous to the construction of parallel lines, or circular loci in Euclidean construction, but involves topological, not a geometric, relationship, and is iterated many more times.

- **Marking features:** Graphical features of the input configuration, critical to subsequent actions, are distinctively marked: the leading and trailing edges of the blob, the structural features of the digits. This is a kind of parsing activity, and the marks are analogous to non-terminals in string grammar parsing, but here they are marked right on the 2d data-structure itself, with all spatial information maintained for future processing.

- **Regularization:** Shapes are brought into canonical form or position for simpler subsequent processing -- digit recognition, counting components (often not "constructions" in the usual sense but alterations, motions, or even "destructions")

- **Execution flow**: Execution flow is controlled, often with the simple use of color to preserve form but prevent unwanted rule match. They control which rules apply *where* (distinguishing source, sink, obstacle, and buffer zone; or blob vs. red-dot pusher; or blue for cycle-breaking algorithm), or *when* - intermediate orange was used in the blob move to prevent infinite looping, final digits were changed so they were not reparsed. In other examples, not shown, we use color to control turn taking in tic-tac toe, or signal an interactive game is over. These various uses are analogous to the use of flags and state variables in sentential programming, or tag variables in rule systems.

- **Information transmission:** An event in one place or time can be coordinated with an event at another by specifically designed intermediate graphical structures.
  Transmission across time: Persistence – of blob color throughout feature marking, of source, sink and barrier throughout field growth, of field throughout chalk-walk. In other examples (not shown) balls bounce randomly off pins to accumulate a sample binomial distribution (a kind of intermediate structure to preserve partial results), tallies accumulate to keep score in games, or produce counts of connected components (cf. [4]).
  Transmission across space: The growth of the successive contours in Example 2 transmitted distance and direction information across the field. In other examples not shown, virtual wires transmit voltages in an active logic diagram, or propagate a signal from where defeat is detected to where a "Game Over" message is displayed.
  Transmission across space & time: Color propagated to mark a whole (space) blob for the duration (time) of move, and set up a distance field to last for the duration of the chalk walk.

This list of high level techniques using intermediate graphical forms can be understood in terms of the nature of a pixel rewrite system. Specifically, consider

- the representational dominance of a 2D-graphical pixellated state,
- the persistence of that state until changed by rules,
- the finite number of rules,
- the fundamental role of local patterns of color and shape for defining matches,
- the local character of the pixel rewrites, and
- the ease of rapid iteration of such rewrites.

These suggest that: (1) state can/must be encoded in local patterns of color or shape, (2)propagation of any such state information across space can/must be mediated via chains of local rewrites, and (3) recognition of global properties can/must be mediated by making them local. Together they explain items in the preliminary taxonomy given above: the need for regularization, the way information is transmitted, the local marking of features, etc.

Similar constructions are being noted in other spatially dense computation situations. In cellular automata, various projectiles (like gliders in Conway's game of LIFE) are used to send signals around through space [10]. In spatial simulations of multi-agent processes (e.g., even the simple cases of StarLogo), evaporating and diffusing residue trails left by agents allow the creation of hexagonal honeycombs, efficient foraging algorithms by ants, and culmination of slime molds [11].

It seems, then, that in future work, more active dynamic 2D representations (in PRSs and elsewhere) will engender rich use and understanding of intermediate forms that will increasingly be inherently spatial, exploiting geometric and topological properties of the medium, yet varied, dynamic, and prolifically expended like the data structures in sentential programming languages.

References

1.   Gardin, F. and Meltzer, B., Analogical representations of naïve physics, *Artificial Intelligence,* 38, 1989, pp 139-159.
2.   Narayanan, N. H., Suwa, M., and Motoda, H. Hypothesizing behaviors from device diagrams, In Glasgow, J., Narayanan, N., Chandrasekaran, B. (Eds.) *Diagrammatic Reasoning: Cognitive and Computational Perspectives.* Menlo Park, CA: AAAI Press, 1995, pp 501-534.
3.   Novak, Gordon S., Diagrams for solving physical problems. In Glasgow, J., Narayanan, N., Chandrasekaran, B. (Eds.) *Diagrammatic Reasoning: Cognitive and Computational Perspectives.* Menlo Park, CA: AAAI Press, 1995. pp 753-777.
4.   Furnas, George W. Formal models for imaginal deduction. *Proceedings of the Twelfth Annual Conference of the Cognitive Science Society;* Cambridge, Mass. Hillsdale, NJ: Lawrence Erlbaum; 1990: pp 662-669.
5.   ---. New Graphical Reasoning Models for Understanding Graphical Interfaces . Human Factors in Computing Systems CHI '91 Conference Proceedings; New Orleans. Hillsdale, NJ: Lawrence Erlbaum; 1991: pp 71-78.
6.   ---. Reasoning with Diagrams Only. *Proceedings of the AAAI Symposium on Reasoning with Diagrammatic Representations;* Stanford, CA . AAAI; 1992.
7.   Shin, Sun-Joo, A situation-theoretic account of valid reasoning with Venn diagrams. In Jon Barwise et al.(Ed), *Situation Theory and Its Applications*, Vol 2., CSLI Lecture Notes No 26, CSLI, Stanford CA, 1991, pp 581-605.
8.   Wang, D., Lee, J., Zeevat. Reasoning with Diagrammatic Representations. In Glasgow, J., Narayanan, N., Chandrasekaran, B. (Eds.) *Diagrammatic Reasoning: Cognitive and Computational Perspectives.* Menlo Park, CA: AAAI Press, 1995. pp 339-393.

9. Kostovskii, A. N. *Geometrical Constructions using Compasses Only* NY: Blaisdell 1961.
10. Hordijk, W.; Crutchfield, J. P., and Mitchell, M. Mechanisms of emergent computation in cellular automata, *Proceedings of conference on Parallel Problem Solving from Nature*, 1998; (Springer: *Lecture Notes in Computer Science*) pp. 613-622.
11. Resnick, M. Beyond the Centralized Mindset. *J Learning Sciences.* 1996; 5(1):pp 1-22.

Universal Arrow Foundations for Visual Modeling

Zinovy Diskin[1], Boris Kadish[1], Frank Piessens[2], and Michael Johnson[3]

[1] Lab for Database Design, *Frame Inform Systems*, Ltd.
Riga, Latvia
[2] Postdoctoral Fellow of the Belgian National Fund for Scientific Research
Dept. of Computer Science, K. U. Leuven, Belgium
[3] School of Mathematics and Computing, Macquarie University
Sydney, Australia

Abstract. The goal of the paper is to explicate some common formal logic underlying various notational systems used in visual modeling. The idea is to treat the notational diversity as the diversity of visualizations of the same basic specificational format. It is argued that the task can be well approached in the arrow-diagram logic framework where specifications are directed graphs carrying a structure of diagram predicates and operations.

1 Introduction

Compact graphical images, diagrams, are very often nothing but *visual models* presenting various aspects of the universe of discourse in a comprehensible and easy to communicate way. Proper modeling is a key to a proper design and indeed, diagrams have proved their practical helpfulness in a wide range of design activities: from thinking out how to erect a shack (where a drawing is useful yet optional) to the design of high-rise buildings of business/software systems (impossible without diagrams). In general, the history of graphical notations invented in various scientific and engineering disciplines is rich and instructive but it is software engineering (SE) where during recent years one can observe a Babylonian diversity of visual modeling languages and methods: entity-relationship (ER) diagrams and a lot of their dialects, object-oriented (OO) analysis schemas in a million of versions and, at last, the recent Unified Modeling Language (UML) which itself comprises a host of various notations.

An important peculiarity of diagrams used in SE is the intention to provide them with very precise, or even formal, meaning. That is, the meaning $\mathcal{M}(D)$ of a diagram D is considered to be described in precise (ideally mathematical) terms. This latter description forms some precise specification S_D possessing precise (formal) semantics $M(S_D)$. Thus, $M(S_D)$ is a formal abstraction of the intuitive meaning $\mathcal{M}(D)$ and we consider S_D as some *(internal) logical specification hidden in* D. It seems the pattern just described is typical (or, at least, is desired to be typical) for diagram usage in SE.

M. Anderson, P. Cheng, and V. Haarslev (Eds.): Diagrams 2000, LNAI 1889, pp. 345– , 2000.
© Springer-Verlag Berlin Heidelberg 2000

The diversity of domains where diagrams are used is vast (even inside SE), correspondingly, we have a vast diversity of intuitive universes from which meanings $\mathcal{M}$ have to be taken. However, as soon as we begin to speak about their formal reducts M, the diversity is compressed into only a few universes of mathematical constructs. Each of the latter can be described in its own specific language, say, the language of set theory, or type theory, or higher order logic, or categorical logic. In fact, it is proven that all these languages are of equal, and universal, expressive power so that any formal semantics can be expressed in any of them. So, S_D above can be taken to be a specification (theory) in any of the formal languages mentioned. However, one of them is of special interest for us because specifications in this language also appear in a diagrammatic graphic form. We mean categorical logic where specifications are nothing but graphs of nodes and arrows, in which some fragments are marked by labels denoting predicates taken from a predefined signature; this graphical format is called a *sketch*. In fact, we always deal with Π-sketches where Π denotes a signature of predicate labels (markers).

So, if the formal model for the meaning of D is also specified in a graphical language $\mathbf{Ske}(\Pi)$, then one can hope for some useful correspondence, even some visual similarity, between the external visual appearance D and the internal logical specification, sketch S_D, behind it. In this framework, the diagram D appears as a visualization, most often, a special abbreviation, of the precise and detailed logical specification S_D. For example, the language of ER-diagrams determines a sketch signature Π_{ER} such that any ER-diagram D can be presented as a special visualization of the corresponding Π_{ER}-sketch S_D (section).

This gives rise to a general thesis that *any diagram with precise semantics (to be described in mathematical terms) actually hides a sketch in a suitable signature of markers*. Then, any diagrammatic notation (language) $\mathcal{L}$ appears as a special visualization superstructure $Vis_{\mathcal{L}}$ over a certain basic specification sketch language $\mathbf{Ske}(\Pi_{\mathcal{L}})$ where $(\Pi_{\mathcal{L}})$ is a signature of markers corresponding to $\mathcal{L}$. This means that any $\mathcal{L}$-specification, a diagram D, can be presented as $D = Vis_{\mathcal{L}}(S_D)$ with S_D a $\Pi_{\mathcal{L}}$-sketch specifying the meaning of D in precise terms and $Vis_{\mathcal{L}}$ a mapping sending $\Pi_{\mathcal{L}}$-sketches into their visual presentations. In this way the *diversity* of diagrammatic notations can be transformed into a *variety* of sketch models in different signatures. Indeed, sketches in different signatures are nevertheless sketches, and they can be uniformly compared via relating/integrating their signatures. Though the latter task is far from being trivial, it is precisely formulated and can be approached by methods developed in category theory.

The view outlined above opens wide and tempting possibilities for the uniform treatment of many diagrammatic notations and for putting them on precise mathematical foundations, refining their vocabularies and providing them with

---

[1] *Categorical logic* is a discipline of viewing and studying logic within the *category theory* (CT) framework. The latter is a branch of modern algebra where mathematical structures are considered in a graph-based specification methodology; a standard reference suitable for computer science is [ ].

formal semantics. It brings much more discipline to the art of designing new diagrammatic notations and provides a base for the systematic treatment of the inter-language issues: translation and integration of diagrams in different graphical languages. In addition, it opens the door for incorporating powerful algebraic techniques developed in CT into the field of visual modeling. In particular, diagram transformation is reduced to a graph-based counterpart of algebraic term rewriting – the so called *diagram chasing* (see [ ] for an example of useful application) elaborated in CT to a great extent.

Of course, the realization of this idea is a wide research program: the sketch treatment of each particular diagrammatic notation needs careful elaboration. Indeed, a diagram language really used in a scientific/engineering domain accumulates a lot of useful "notational tips", habits and traditions. A correct sketch approach to such a language means putting it on precise semantic foundations and then explaining/specifying the notational peculiarities as visualizations of the underlying logical constructs rather than removing them entirely as "non-logical". Of course, while the sketch language as such is a mathematical phenomenon, visualization of sketches in one or another diagrammatic notation easy to use and communicate is a highly non-trivial issue of cognitive, and even more generally, cultural nature far beyond formal mathematics and logics. Thus, the research program in question is essentially interdisciplinary.

The plan of the paper is as follows. Section 2 presents general foundations of the sketch specification machinery. In section 3 several concrete applications in the so called semantic (conceptual) data modeling are considered: we demonstrate the sketch treatment of classic ER-diagrams and the recently fashioned UML. Though these applications are taken from a particular field, they illustrate an approach of (we emphasize once again) quite general nature and applicability.

2   Basics of Arrow Thinking and Arrow Logic

It is important for understanding the purposes of the present paper to recognize the distinction between string-based and graph-based logics ([, ,]). The point is that any specification – as it is presented to its reader – is actually a *visual presentation* of a certain underlying *logical specification* as such. In general, there are possible linear string visualizations of graph-based specifications and, conversely, graphical visualizations of string-based logical specifications. Examples of the latter are the well-known Euler-Venn diagrams for visual presentation of propositional logic statements, or the graphic representation of first-order logic sentences by conceptual graphs [ ] or the visual presentation of first-order logic used in Barwise and Etchemendy's *Hyperproof* [ ], or even the graphical interfaces to relational database schemas used in many design tools. Conversely, graph-based logic specifications can be presented in a linear plain form: after all, a graph is an ordinary two-sorted mathematical structure well specified by formulas.

So, in considering graphic notational systems one should carefully distinguish between specification and visualization, and graph-based logics should be carefully distinguished from graphical interfaces to string-based logics. Any logic where arities of predicate/operation symbols are strings is a string-based logic, no matter how graphical its visualization looks.

2.1 Arrow Diagram Logic: Simple Examples

By definition, one deals with a *graph-based logic* if arity shapes of predicates are graphs (of place-holders). In other words, a predicate arity consists of node place-holders and arrow place-holders organized into a directed graph. A typical case when such a logic naturally arises is when one deals with specifying systems of sets and functions as shown in Table .

The predicate of *set inclusion* (the 1st row) is actually a property of a diagram consisting of two nodes (sets) A, B and an arrow (mapping) between them (2nd column). If this property is declared for a diagram of the shape just described (3rd column) – this is visualized by the block-body figure of the arrow – it means that the source set is a subset of the target and the mapping is their inclusion.

A collection of subsets may be disjoint; this property is specified by an arrow diagram predicate as shown in the 2nd row of the table.

A collection of mappings into a common target set may have the *covering* property: each element in the target set is a value of at least one of the mappings. This predicate is expressed as shown in 3rd row.

The predicate of *separating* arrow family (the 4rd row) is somewhat dual to the covering predicate (in category theory, this duality can be expressed in precise terms). It can be declared for a diagram consisting of a source node and a family of arrows going out of it as it is shown in the 2nd column. The formal meaning is as follows:

(1-1)   for any $x, x' \in X$, $x \neq x'$ implies $f_i(x) \neq f_i(x')$ for some i

Note however that in such a case the tuple-function $f = \langle f_1 \ldots f_n \rangle$ into the Cartesian product of domains,

$$f = \langle f_1 \ldots f_n \rangle \colon X \longrightarrow D_1 \times \cdots \times D_n, \quad fx \overset{\text{def}}{=} \langle f_1 x, \ldots, f_n x \rangle$$

is injective (one-one) so that elements of X can be considered as unique names for tuples from a certain subset of $D_1 \times \ldots \times D_n$, namely, the image of f. In fact, elements of X can be identified with these tuples so that X is a relation up to isomorphism. In the classical ER-terminology [ ], if the domains D_i's are *entity* sets then f_i's are *roles* and any $x \in X$ is a *relationship* between entities $f_1(x), \ldots, f_n(x)$. Then, for example, the right-most diagram in the row means that *Married*-objects are nothing but pairs $(wife, husb)$ whose components are taken from classes *Woman*, *Man* respectively. Thus, we have precisely expressed the internal tuple-structure of X-objects by declaring the corresponding property for some arrow diagram adjoint to the node X.

---

[2] In general, there are richer graph-based shapes, *eg*, *2-graphs* with 2-arrows between arrows, *3-graphs* with 3-arrows between 2-arrows and so on.

Table 1. Sets-and-functions predicates in a graph-based notation

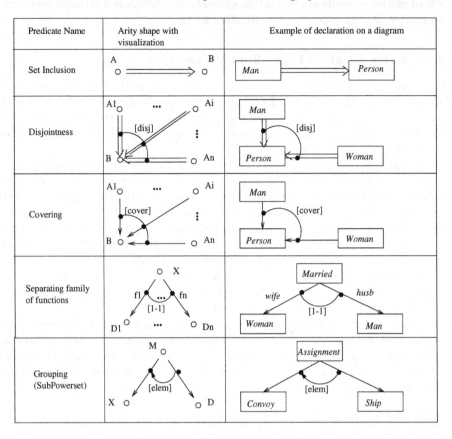

Predicate Name	Arity shape with visualization	Example of declaration on a diagram
Set Inclusion		
Disjointness		
Covering		
Separating family of functions		
Grouping (SubPowerset)		

We may proceed in a similar way for specifying other types of structures. For example, the intended meaning of the Grouping (SubPowerset) predicate (5th row) is to state that the set X consists of subsets of the set D, in other words, of groups of D-objects (e.g., objects of class *Convoy* are groups of *Ship*-objects). Again, in the arrow diagram logic this structure of the X-elements is expressed by declaring a certain predicate (denoted by, say, [elem]) for a certain diagram adjoint to the node X. We omit the technical description.

2.2 Arrow Diagram Logic: General Formulation

Examples considered above may be summarized as follows. Any given diagram property has a predefined shape, that is, a configuration of nodes and arrows for which the property makes sense. So, in the graph-based logic a predicate is specified by its name and a graph (of place-holders) called the *(logical) arity shape* of the predicate (see the middle column of Table). In addition, the arity shape should be supplied with auxiliary graphic means like arcs or double-body arrows for visualizing predicate declarations on schemas. Of course, these auxiliary means do not occur in the logical arity as such – together with the latter they form the *visual arity shape* providing the visualization mechanism.

In order to declare a predicate P with some arity shape G_P for a system S of sets and functions, one must assign S-sets to nodes in G_P and S-functions to arrows in G_P in such a way that adjointness conditions between nodes and arrows are respected. This can be considered as (i) arranging (names of) items from S into a graph so that the system S can be identified with this graph, and (ii) setting up a graph mapping $d\colon G_P \to S$ from the arity graph into the graph of names. Such a mapping can be visualized by labeling nodes and arrows in the shape by names (of, respectively, nodes and arrows in S, see the rightmost column in Table). The couple (predicate_marker, its shape labeled by names), *ie*, (P, d) in our notation, is a predicate declaration for the collection of items whose names were used as labels. Such a pair is called a *marked diagram* and so, in the graph-based logic, a predicate declaration is nothing but a marked diagram. Thus, a graph-based logical specification is a graph of names, in which some diagrams are labeled with markers from some predefined signature. As it was said, we call such specifications *sketches* or, more precisely, Π-sketches where Π is the name of the signature.

An important peculiarity of the machinery is a hierarchical structure of sketch signatures. The point is that the arity shape of a predicate P may itself contain predicate markers defined prior to P, e.g., the disjoint predicate in 2nd row of the table uses inclusion. In such a case a P-predicate declaration in a schema, that is, a P-diagrams $d\colon G_P \to S$ must fit these markers in G_P. For example, if an arrow f in G_P is marked as inclusion, its image, function $d(f)$ in S, must be also an inclusion. Thus, in general, G_P is a sketch rather than graph and d is a sketch mapping, that is, a graph mapping compatible with markers.

The last two examples in Table   demonstrate a key feature of the arrow logic: nodes are always considered homogeneous and their internal structure (the structure of their elements) is specified by declaring the corresponding property of the corresponding arrow diagram adjoint to the node. Actually, this gives rise to a special way of thinking out specifications, the so called *arrow thinking*, and in the next section we will briefly manifest the general foundations of the approach.

2.3   Arrow Thinking and Category Theory: A Brief Manifesto

Category theory (CT) is a modern branch of abstract algebra. It was invented in the forties and since that time has achieved a great success in providing a uniform structural framework for different branches of mathematics, including metamathematics and logic, and stating foundations of mathematics as a whole

---

[3]   So, in the sketch framework, the term 'diagram' is used in some narrow technical sense as above. In contrast, in engineering disciplines the term 'diagram' often refers to a graphical specification considered as a whole; to avoid misunderstanding, in the latter case we will use the term 'schema'.

as well. CT should be also of great interest for scientific and engineering disciplines, in particular, software engineering, since it offers a general methodology and machinery for specifying complex structures of very different kinds. In a wider context, the 20th century has been the age of structural patterns, *structuralism*, as opposed to *evolutionism* of the previous century, and CT is a precise mathematical response to the structural request of our time.

The basic idea underlying the approach consists in specifying any universe of discourse as a collection of *objects* and their *morphisms* which normally are, in function of context, mappings, or references, or transformations or the like between objects. As a result, the universe is specified by a directed graph whose nodes are objects and arrows are morphisms. Objects have no internal structure: everything one wishes to say about them one has to say in terms of arrows. This feature of CT can be formulated in the OO programming terms as that object structure and behaviour are encapsulated and accessible only through the arrow interface. Thus, objects in the CT sense and objects in the OO sense have much in common.

A surprising result discovered in CT is that the arrow specification language is absolutely expressible: any construction having a formal semantic meaning can be described in the arrow language as well. Moreover, the arrow language is proven to be a powerful conceptual means: if basic objects of interest are described by arrows then normally it turns out that many derived objects of interest can be also described by arrows in a quite natural way. The main lesson of CT is thus that to define properly the universe we are going to deal with it is necessary and sufficient to define morphisms (mappings) between objects of the universe. In other words, as it was formulated by a founder of categorical logic Bill Lawvere, *to objectify means to mappify.*

The arrow language is extremely flexible. In function of context, objects and arrows can be interpreted by: sets and functions (in *data modeling*), object classes and references (*OO analysis and design*), data types and procedures (*functional programing*), propositions and proofs (*logic/logic programing*), interfaces and processes (*process modeling*), states and transactions (*transaction modeling*), specifications and their mappings (*meta-modeling*). Moreover, all this semantic diversity is managed within the same syntactical framework of *sketch* specifications.

3   Example of Sketching a Visual Modeling Discipline: Semantic Data Modeling via Sketches

Often, within the same field of applying visual modeling (VM) there are many different VM-methodologies that have resulted in different notational systems. For example, in semantic (conceptual) data modeling there are different ER-based notations, and a lot of OO-based, and many "this-vendor-tool-based" ones. In fact, many specialists in semantic modeling and DB design use their

---

[4] that is, expressible in a formal set theory

own graphic specification languages which they find more suitable and convenient for their purposes. This is natural and reasonable but with the current trend to cooperative/federal information systems, one necessarily encounters severe problems with integrating specifications describing the same universe (or its overlapping fragments) in different notational systems. Several attempts to manage the semantic models heterogeneity were made both in academia (*eg*, [ ]) and industry where the famous UML was recently adopted as a standard in visual modeling and design. However, while UML is indeed a significant industrial achievement towards unification, from the logical view point it is just another (monstrous) notational system rather than a framework to manage the heterogeneity problem.

3.1   General Methodology

The main idea underlying the sketch approach to data modeling is to consider object classes as plain sets consisting of internally unstructured (homogeneous) elements whereas all the information about their type is moved into certain arrow (reference) structures adjoint to classes. Examples of how it can be done were demonstrated in section 2.2. Associations (relationships) between classes are also treated as sets (of tuples) with outgoing projection mappings and thus a semantic data schema in any notation can be considered as a specification of a collection of set-like and mapping-like objects subjected to certain constraints. A natural way to structure such data is to organize these objects into a graph, the *underlying graph* of the specification, and to convert constraints into predicates declared for the corresponding diagrams in the underlying graph. Before such structuring can be performed, the collection of diagram predicates which may be declared for set-and-mappings diagrams is organized into a *signature*, say, Π, and then a semantic specification is converted into a structure which we have called Π-*sketch*: it is a graph in which some diagrams are marked by predicate symbols from Π. In this way the diversity of semantic data models can be considered as the diversity of visualizations over the same basic sketch format and thus we have a particular case of the general approach described in the introduction.

3.2   Interpretation of Arrows

The key point in the semantics of sketches is how to interpret arrows. In the former considerations we interpreted arrows by ordinary functions, *ie*, totally defined single-valued functions. This is the standard category theory setting. In contrast, in semantic modeling it is convenient (and common) to use optional and multivalued attributes/references, and so two additional different interpretations

---

[5]   Indeed, since a formal semantic meaning of a majority of the UML constructs is absent, one has no precise means to relate and compare one's particular model with UML. Moreover, for the same reason, UML specifications can be treated differently by different users and different tool vendors, and so, while solving some problems UML creates new ones.

of arrows arise: by partially defined functions (p-functions) and by multivalued functions (m-functions). Their accurate sketch treatment needs a special explanation.

M-functions and p-functions are not ordinary functions subjected to some special constraints. Just the opposite, a single-valued function is a special m-function $f : A \longrightarrow\!\!\!\!\!\rightarrow B$ when for any $a \in A$ the set $f(a) \subset B$ consists of a single element. Similarly, a totally defined function is a special p-function $f : A \circ\!\!-\!\!\rightarrow B$ whose domain $D_f \subset A$ coincides with the entire source set A. So, to manage optional multi-valued attributes and references in the sketch framework we assume that

(i) all arrows are by default interpreted by pm-functions,
(ii) there is an arrow predicate (marker) of being a single-valued function,
(iii) there is an arrow predicate (marker) of being a totally defined function.

It is convenient to visualize constraintless arrows (without markers) by $\circ\!\!-\!\!\!\!\rightarrow$ whereas $\longrightarrow\!\!\!\!\!\rightarrow$ and $\circ\!\!-\!\!\rightarrow$ are denotations of arrows on which markers are hung: the ordinary tail is the marker of being totally defined and the ordinary head is the marker of being single-valued. Of course, superposition of these markers is also legitimate and it is natural to visualize it by the arrow $\longrightarrow$. Thus, visualization of predicate superposition equals superposition of visualizations: here we have a simple instance of applying a useful general principle that reasonable graphic notation should follow.

3.3 Sketches vs. ER-Diagrams: Simple Example

The semantic situation we wish to model is as follows. Suppose, the user is interested in information about men, married couples and married women of some community for some time period; here "married women" means women who are or have been married during the period.

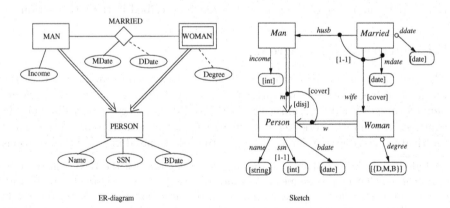

ER-diagram   Sketch

Fig. 1. Sketches vs. ER diagrams

A rough view on the universe is described by the ER-diagram on the left side of Fig. in a conventional ER-notation. The semantics of nodes and attributes is hopefully clear from their names, *MDate,DDate* are dates of marriage and divorce (the latter is optional). The domain of the optional attribute *Degree* is a set consisting of three values *D, M, B*. The class *Woman* is of weak entity type since users are assumed to be interested only in women who are or have been married.

Note, the ER-diagram does not reflect reality exactly: the diamond means that the pair *(wife,husb)* is an identifier (key, in database terms) whereas, in general, this is not the case. Indeed, it is well possible that a married couple got divorced and then got married again. Thus, the correct identifier is the triple *(wife,husb,mdate)*. The diagram is drawn as shown since in the notational ER-standard there is no graphical means to express the required semantics. In contrast, this can be easily done with the sketch machinery: the sketch specifying the situation is depicted on the right side of the figure. Note the [1-1] marker on the key attribute *ssn* and the cover marker on the arrow *wife*.

Rectangle nodes are *abstract classes* whose extensions should be stored in the DB. Oval nodes are predefined *value domains* whose semantics are *a priori* known to the DBMS. For the sketch approach, [int] and [{D,M,B}] are *markers* (in our precise sense) hung on corresponding nodes, that is, constraints imposed on their intended semantic interpretations. For example, if a node is marked by [int] its intended semantics is the predefined set of integers. At the same time, 'Person' is a *name* labeling nodes without imposing any constraints. Generally speaking, we should also give names to the nodes labeled by [int] and [{D,M,B}], say, *'Number'* and *'Label'*. However, since the intended semantics of so marked nodes is fixed and remains the same for all schemas (independently of names of these nodes : *'Label'*, or *'Tag'*, or *'Attribute'* etc.), we adopt the convention of using such markers instead of names (they will be printed in the roman font). So, abstract class nodes are labeled by names, and value domain (printable) nodes are labeled by markers expressing their intended semantics (while their names become redundant and can be omitted).

3.4   Sketches vs. UML: Simple Examples

Let us consider the UML-diagram D_1 in Fig. (i), which models an association class (see, e.g., [ ]). It follows from the informal explanation in [ ] that *Job*-objects are nothing but pairs (c, p) with c a *Company*-object and p a *Person*-object, which themselves can have properties, *eg, salary, dateHired* and the like.

The corresponding sketch specification is presented on the right. Tails and heads of the horizontal arrows declare the same constraints as the left and right superscripts over the association edge on the UML-diagram (see the previous section). Marker [inv] hung on the couple of horizontal arrows denotes the predicate of being mutually inverse functions. Marker [graph] means that $[\![Job]\!]$ is the graph of (each of) these functions, that is, if $(c, p) \in [\![Job]\!]$ then

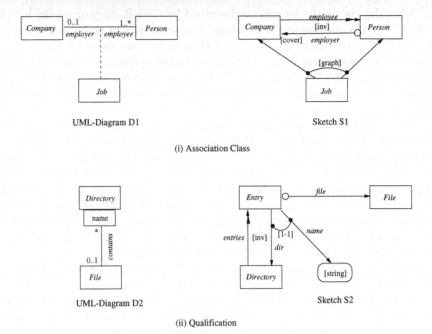

(i) Association Class

(ii) Qualification

Fig. 2. Sketches vs. UML: two examples

$c = [\![\, employer \,]\!]\,(p)$ and $p \in [\![\, employee \,]\!]\,(c)$. Here $[\![\,\,]\!]$ denotes semantic mapping assigning sets to nodes and functions to arrows.

One more example is presented in Fig. (ii). The UML-diagram D_2 models the so called *qualification* construct when to one end of an association edge two coupled nodes are attached: this means that in order to select an object on the other end of the association one should point out a value of the qualifier. In the example, *name* (smaller rectangle) is a qualifier and thus for a given *Directory*-object d, while $d.contains$ is the set of files in d, $d[\text{foo}].contains$ is the single *File*-object having name *foo* in the directory d.

A sketch specification of the same data semantics is presented on the right.

3.5 Sketching Visual Models: Visualization Aspects

In comparing given notational samples between themselves, the question of which one is "more right" is incorrect: any notation with unambiguously specified semantics can be used. However, the question of which notation is more clear and transparent w.r.t. its intended semantic meaning is quite reasonable.

Visualization Aspects of Sketching ER-Diagrams. The sketch specification has a few noteworthy advantages over the ER-diagram.

In the first place, in the ER-diagram, labeling of a node sometimes indicates a set (e.g. MAN), and sometimes it indicates a mapping (e.g. Income is actually a mapping from men to integers). In the sketch specifications, nodes always specify sets, and arrows always specify mappings.

Secondly, the ER-diagram improperly visualizes diagram predicates. For example, a major part of the relationship semantics attributed to the node Married is nothing but the [1-1] property of the corresponding diagram (see sketch). In the ER-diagram this *diagram* property is attributed to the node. It's evidently a poor notation since normally it's assumed that a diamond node denotes relationship, that is, [1-1]-property, for entity nodes connected to it but actually it's not necessarily the case. In the given example, the triple-arrow span possesses [1-1] (as it's shown on the sketch) not the two-arrow span that is suggested by the ER-diagram.

The situation with declaring the node Woman a weak entity is even worse. Suppose, for example, that the same node occurs in another relationship (say, Employment between Woman and Company) where the corresponding role arrow is not a cover, that is, not all Woman-entities are employed. What should be the shape of the Woman node now: double-framed or ordinary? It is clear that the problem has occured because an arrow property is misleadingly attributed to the arrow source.

Finally, these improper visualizations give rise to artificial heterogeneity of nodes in ER-diagrams (relationships, entities, weak entities). As a consequence, since different users may view, and correspondingly specify, the same node in semantically different ways (e.g. one user may consider marriages to be a relationship, whereas another user sees it as an entity type in its own right), integration of different ER-diagrams with overlapping semantics becomes extremely difficult (see, e.g., [ ]). In contrast, within the sketch framework the integration problem can be approached in a quite natural and effective way [ ].

Visualization Aspects of Sketching UML-Diagrams. Compare the UML-diagram D_1 and the sketch S_1 on Fig. (i), which express the same semantics. On the whole, the sketch presents a more detailed specification – all functions involved are shown explicitly – while the UML-diagram can be considered as a special abbreviation. In fact, in D_1 we have two different abbreviations. One is the presentation of two mutually inverse functions by one undirected edge each of whose ends carries the corresponding name and multiplicity constraint. It is a reasonable abbreviation, it makes graphical image more compact and multiplicity expressions are as good as special arrow heads and tails, or even better w.r.t. mnemonic efforts. Note, however, when one considers compositions of many references, undirected visualization can lead to mistakes. As for abbreviating the arrow span outgoing node *Job* in the sketch S_2 by an edge going into another edge in D_2, it is a purely syntactical trick hardly clarifying semantic meaning of the association class *Job*. In addition, such a way of presenting binary associations does not work for more than two-ary associations. In contrast, arbitrary n-ary association are presented by sketches by n-arrow spans in a uniform way.

Diagram D_2 in Fig. (ii) is an even more powerful abbreviation: four nodes and four arrows of the actual semantic picture (specified by sketch S_2 in detail) are compressed in a graphical image with only three nodes and one edge. However, the distance between visual schema D_2 and its semantic meaning ("congruently" specified by sketch S_2) is so large and meandering that diagram D_2 hardly can be considered as presenting a good visualization mechanism.

Summary. Of course, the issue of visualization is of complex cognitive nature and such culture-dependent points as notational habits and preferences, notions of convenience and elegance, and many others, can play a significant role; analysis of such things goes far beyond the sketch formalism as such. Nevertheless, we believe that clear logical structure of sketch specifications and the presence of well-defined semantics for them make the sketch format a proper foundation for building a really good graphic notational system. In particular, it even allows the setting of a formal framework for the problem, for example, one may formulate the *commutativity principle*: visualization of conjunction of diagram predicates should be conjunction of visualizations. In general, we may think of organizing collections of specification and visualization constructs into similar mathematical structures and then visualization appears as a structure-preserving mapping between these collections, that is, their morphism. Actually, it gives rise to a consistent mathematical framework for building graphic notations. Close ideas were developed by Goguen and his group in their studies of reasonable graphical interfaces [ ].

Finally, concerning the visualization of sketches, it must be emphasized that visualization on a computer (where one can use colour and dynamic highlighting of marked diagrams) is much clearer than visualization on paper.

4   Conclusion

The main question of the approach above is whether the specificational tools of a given visual model $\mathcal{M}$ can be converted into a sketch signature. In other words, whether the expressive power of the sketch language is sufficient to cover a wide class of visual models. The answer is positive and somewhat surprising: it follows from general results of category theory that any specification having a formal semantics can be replaced by an equivalent sketch in some signature. In fact, for one remarkable signature called the *topos signature* a universal modeling power was proven: any specification having a formal semantics can be replaced by a sketch in the topos signature (see, *eg*, [ ] where the equivalence of toposes and higher-order logic is proven in a very transparent way). In other words, anything that can be specified formally can be specified by sketches as well. So, sketches do provide a real possibility to handle heterogeneity in a consistent way.

At the same time, by suggesting sketches we do not wish to force everyone to use the same universal graphical language – let one use that collection of graphical constructs one likes. What we actually suggest concerns *what should be specified* in graphical schemas (if, of course, one is concerned about their semantics, which we believe is almost always necessary). After that, the question is *how to visualize* the specification. An accurate distinction between logical specification and its syntactic-sugar visualization, and simultaneously their parallelism, are the two key advantages of the sketch approach from the practical view point.

A clear distinction between logical specification and its visualization is provided by the presence of formal semantics for sketches, and makes the sketch notation favorable in comparison with many notational systems used in software engineering. Of course, logic itself does not decide all of the problems: design of a good notational system is an art rather than technology. Nevertheless, the strict sketch pattern makes it possible to designate the formalizable part of the design task and state a precise basic framework for its consideration (section   ).In the paper we have occasionally touched on this issue of visualization but, of course, the problem needs a special elaboration; we leave it for future research.

Parallelism of specification and graphical visualization is provided by the graph-based nature of the sketch logic, and sharply distinguishes sketches from those visual models which are externally graphical, yet internally are based on predicate-calculus-oriented string logics. The repertoire of graphical constructs used in these models has to be bulky since all kinds of logical formulas require its special visualization. Configurations/shapes of these visualization constructs can be rather arbitrary because there are no evident natural correlations between graphics and logical string-based formulas. In particular, this problem is actual for the modern CASE-tools whose underlying logic is relational, that is, string-based.

Of course, there are situations when something can be easily described by a logical formula but hardly by a graphical image, and there are inverse situations as well. A good language has to combine graphical and string logics into a flexible notational mechanism.

On the whole, the sketch view we suggest gives rise to a whole program of refining the vocabulary of visual modeling, making it precise and consistent, and unified. In the sketch framework, *any particular diagrammatic notation appears as a particular visualization of the same common specificational format* – the format of sketches. Besides this unifying function, an essential advantage of the sketch format is the extreme brevity of its basic vocabulary exhausted by the three kinds of items: nodes, arrows and marked diagrams. Nevertheless, as it was discussed above, the sketch language is absolutely expressive and possesses a great flexibility.

Well, one might argue that given some particular discipline of VM, the virtual possibility of expressing its modeling constructs in the sketch language has nothing to do with effective usage of sketches in the field. This is very much true, and to build a discipline on the sketch ground one should thoroughly explain and specify its modeling constructs in the sketch framework. In effect, this is a research program to be fulfilled for each given VM-discipline but our current experience of applying sketches to real specificational problems in software engineering is promising: in cases where we applied sketches they appeared as a precise logical refinement of the existing notation rather than an external imposition upon it. In particular, in semantic data modeling sketches can be seen as a far reaching yet natural generalization of functional data schemas, ER-diagrams and OOA&D-schemas [,], in meta-specification modeling they appear as essential generalization of schema grid developed by the Italian school (cf. [ ] and [ ]) and we expect that in process modeling sketches will be a natural development of interaction diagrams [ ].

References

1. S. Abramsky. Interaction categories and communicating sequential processes. In A.W. Roscoe, editor, *A Classical Mind: Essays in honour of C.A.R.Hoare*, pages 1–15. Prentice Hall Int., 1994.
2. P. Atzeni and R. Torlone. Management of multiple models in an extensible database design tool. In *Advances in Database Technology – EDBT'96*, 5th Int.Conf. on Extending Database Technology, Springer LNCS'1057, 1996.
3. A. Bagchi and C. Wells. Graph-based logic and sketches. In *10th Int.Congress of Logic,Methodology and Philosophy of Science, Florence, 1995*, Kluwer Acad.Publ., 1997.
4. M. Barr and C. Wells. *Category Theory for Computing Science*. Prentice Hall International Series in Computer Science, 1990.
5. J. Barwise and J. Etchemendy. A computational architecture for heterogeneous reasoning. In *Theoretical Aspects of Rationality and Knowledge*, pages 1–27. Morgan Kaufmann, 1998.
6. C. Batini, G. Battista, and G. Santucci. Structuring primitives for a dictionary of entity relationship data schemas. *IEEE Trans. SE*, 19(4):344–365, 1993.
7. B. Cadish and Z. Diskin. Heterogenious view integration via sketches and equations. In *Foundations of Intelligent Systems*, Proc. 9th Int.Symposium, *ISMIS'96*, Springer LNAI'1079, pages 603–612, 1996.   ,
8. P. P. Chen. The entity-relationship model – Towards a unified view of data. *ACM Trans.Database Syst.*, 1(1):9–36, 1976.
9. Z. Diskin. Formalization of graphical schemas: General sketch-based logic vs. heuristic pictures. In *10th Int.Congress of Logic,Methodology and Philosophy of Science, Florence, 1995*. Kluwer Acad.Publ., 1997.
10. Z. Diskin. The arrow logic of meta-specifications: a formalized graph-based framework for structuring schema repositories. In B. Rumpe H. Kilov and I. Simmonds, editors, *Seventh OOPSLA Workshop on Behavioral Semantics of OO Business and System Specifications)*, TUM-I9820, Technische Universitaet Muenchen, 1998.
11. Z. Diskin and B. Kadish. Variable set semantics for generalized sketches: Why ER is more object-oriented than OO. To appear in *Data and Knowledge Engineering*
12. Z. Diskin, B. Kadish, and F. Piessens. What vs. how of visual modeling: The arrow logic of graphic notations. In *Behavioral specifications in businesses and systems*. Kluwer, 1999.
13. J. Goguen. Semiotic morphisms. Technical report, University of California at San Diego, 1997. TR-CS97-553.
14. M. Johnson and C. N. G. Dampney. On the value of commutative diagrams in information modeling. In *Algebraic Methodology and Software Technology, AMAST'93*. Springer, 1993.
15. J. Lambek and P. Scott. *Introduction to higher order categorical logic*. Cambridge University Press, 1986.

16. G. W. Mineau, B. Moulin, and J. F. Sowa, editors. *Conceptual graphs for knowledge representation*. Number 699 in LNAI. Springer, 1993.

17. J. Rumbaugh, I. Jacobson, and G. Booch. *The Unified Modeling Language Reference Manual*. Addison-Wesley, 1999.

18. S. Spaccapietra, C. Parent, and Y. Dupont. View integration: a step forward in solving structural conflicts. *IEEE Trans. KDE*, 1992.

Diagrammatic Acquisition of Functional Knowledge for Product Configuration Systems with the Unified Modeling Language

Alexander Felfernig and Markus Zanker

Institut für Wirtschaftsinformatik und Anwendungssysteme,
Universität Klagenfurt, A-9020, Austria
{felfernig,zanker}@ifi.uni-klu.ac.at

Abstract. Shorter product cycles, lower prices of products, and the production of goods that are tailored to the customers needs made knowledge based product configuration systems a great success of AI technology. However, configuration knowledge bases tend to become large and complex. Therefore, knowledge acquisition and maintenance are crucial phases in the life-cycle of a configuration system. We will show how to meet this challenge by extending a standard design language from the area of Software Engineering with classical description concepts for expressing configuration knowledge. We automatically translate this graphical depiction into logical sentences which can be exploited by a general inference engine to solve the configuration task. In order to cope with usability restrictions of diagrammatic notations for large applications, we introduce the usage of contextual diagrams. This mechanism makes the conceptual model more readable and understandable and supports intuitively the acquisition of functional configuration knowledge.

1 Introduction

Following the paradigm of mass-customization [ ], products are nowadays sold in many different variants according to the customer's demands and specific needs. To convey the product information to the customer, the technical expert or the sales representative is supported by knowledge-based product configuration systems (or configurators) to find a configuration of the product which conforms to technical and marketing restrictions and fulfills all the requirements of the customer. Product configuration systems have been one of the most successful applications of AI technology in industrial environments (e.g., telecommunication industry, automotive industry, computer industry, etc).

Due to shorter product cycles, lower prices of products, higher customer demands, and the fact that configuration systems have to be developed concurrently with the configurable product, the construction and maintenance of the configuration knowledge bases have become a critical task. Additionally, the high complexity of the products and configuration problems lead to an increased size and complexity of knowledge bases.

M. Anderson, P. Cheng, and V. Haarslev (Eds.): Diagrams 2000, LNAI 1889, pp. 361– , 2000.
© Springer-Verlag Berlin Heidelberg 2000

However, in many cases the computational complexity of configuration problems is not the key issue. The main problems lie - as it is for many knowledge-based systems - in the knowledge acquisition and maintenance phase. In the case of product configuration, the available knowledge is distributed between many different people and organizational units (e.g., technical restrictions or marketing constraints) and changes very rapidly. Additionally, existing configuration systems often employ a special, proprietary terminology to represent the knowledge, which can not be formulated directly by the technical expert. Therefore, we propose in [ ] a framework to support the task of knowledge acquisition and maintenance by semantically extending the description concepts of a standard design language from the area of Software Engineering, i.e. the Unified Modeling Language (UML) [ ]. The graphically represented configuration knowledge is human-readable and can be easily communicated to domain experts for the purpose of validation and maintenance. However, when modeling highly variant products (e.g., automobiles) which offer the customer a vast amount of interdependent options and choices we approach expressiveness restrictions of a single (partitioned) diagram. When acquiring configuration knowledge we have to cope with an intermingled and interdependent structural and functional product architecture. Part of the configuration knowledge applies only in context with a specific functionality which could be provided by a configured artifact, e.g., some motorization choices are only offered, if it is a *luxury* car. This makes additional graphical representation concepts necessary. In this paper we put the emphasis on extending the usability of this diagrammatic representation by introducing contextual diagrams as an additional structuring mechanism to allow the acquisition of knowledge which is only relevant in a specific context. Further we extent the translation rules from [ ] which allow the automatic translation into an executable knowledge representation.

The rest of the paper is organized as follows. In Section 2, we shortly recall the technique to model configuration knowledge bases using UML presented in [ ]. In Section 3 we discuss the role of functions in configuration knowledge representation. In Section 4 we motivate the use of contextual diagrams for modelling functional knowledge. Section 5 sketches how we derive a declarative configuration knowledge base from the diagrammatic representation and gives a formal exposition of the introduced concepts. The paper is concluded with a description of our prototype environment (Section 6) and a summarization of related work (Section 7).

2   Construction of Knowledge Bases Using UML

In practice configurations are built from a predefined catalog of component types of a given application domain. These are characterized by a set of attributes and ports representing physical or logical connections to other components. In [ ] we show, how a general high level modeling language from the area of Software Engineering (Unified Modeling Language - UML) [ ] can be employed to model such product configuration knowledge bases. UML is widely applied in industrial software development processes and the resulting conceptual models alleviate the communication processes between the people involved in the development process. The approach of [ ] employs the extensibility features of UML (stereotypes) to express the domain specific modeling concepts for the product configuration domain, but does not use a proprietary modeling language. The terminology for configuration problems (component port model [ ]) is commonly accepted and can be used for automated compilation of knowledge bases, since the semantics of the individual modeling concepts are stated precisely.

The proposed development process for valid configuration knowledge bases is shown in Figure 1. First, a conceptual model of the configurable product is designed (1) using UML. After syntactic checks of the correct usage of the concepts, this model is then non-ambiguously transformed to logical sentences (2) which are exploited by a general configuration engine for computing configurations of products. Consequently, the configurator is based on a declarative, logic based, explicit representation of the configuration knowledge. The resulting knowledge base is validated by the domain expert using test runs on examples (3). In case of unexpected results, the product model can be revised on the conceptual level. If the knowledge base is valid, it can be employed in productive use (4).

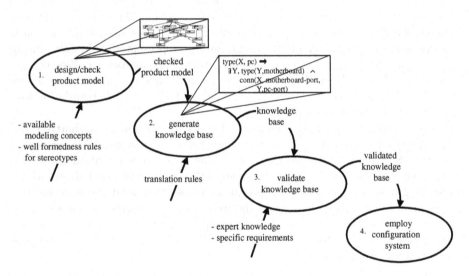

Fig. 1. Constructing a valid configuration knowledge base

Figure 2 shows how we can employ UML to model configurable products using the built-in extension mechanisms of UML. The product structure is only fragmentary modeled through aggregation and generalization between the component types that form the final product.

The following modeling concepts are typical for the product configuration domain [ ] and can be expressed in the graphical model:

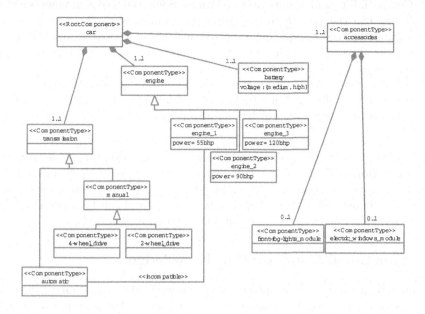

Fig. 2. Product model of a car

- Component types: These are the parts of which the final product is built of. Component types are characterized through attributes (e.g., component *battery* is characterized through attribute *voltage*).
- Function types: They are used to model the functional architecture of an artifact. Similar to component types they can be characterized by attributes (e.g. function type *base_edition* in Fig. 3).
- Generalization: Component types with a similar structure are arranged in a generalization hierarchy and represent choices for the configurable product (e.g., an *engine* can be instantiated either with *engine_1*, *engine_2* or *engine_3*).
- Aggregation: These part-of structures can be compared with classical bill-of-material representations which are semantically enriched by multiplicities, stating a range of how many subparts an aggregate can consist of (e.g., the *accessories* component consists of an optional (multiplicity is 0..1) *front-fog-lights_module*).
- Resources: Parts of a configuration problem can be seen as a resource balancing task, where some of the component types produce some resource and others are consumers (this concept is not shown in our example). In the final product these resources must be balanced.
- Connections and Ports: In addition to the amount and types of the different components also the product topology may be of interest in a final configuration , i.e., how the components are interconnected to each other (e.g., *engine* is connected with *car*).

- Compatibility and requirements relations: Some types of components cannot be used in the same final configuration (*incompatible*). In other cases, the existence of one component type requires the existence of another special type in the configuration (*requires*). In our example, a transmission of type *automatic* is incompatible with *engine_1*.
- Additional constraints: Constraints on the product model, which can not be expressed graphically, may be formulated using the language OCL (Object Constraint Language), which is an integral part of UML.

In [ ] a logical model of a configuration problem in terms of a logic theory is given. In addition, transformation rules are defined, how the concepts of the graphical representation in UML are translated to a declarative logic representation. The resulting knowledge base is then exploited by a general constraint solving software. The knowledge base is therefore represented and can be maintained on a conceptual level and can then be compiled into a processible form.

3   Functional Configuration Knowledge

As already mentioned in our introductory statement the diagrammatic depiction becomes harder to maintain and understand as the number of graphical elements increases. A reduction of the quantity of elements depicted in a diagram, can be easily achieved by splitting it in parts (UML offers the concept of packages [ ] which allows the partitioning of a diagram). A package is simply defined in UML as *a grouping of model elements*. Packages themselves may be nested within other packages. It is therefore straightforward to partition a diagram into packages under the premise of high cohesion among the concepts in the same package and little coupling between different packages. If classes with the same identifier appear in several packages of one graphical representation, they are identical. Therefore the partitioning of a diagram into packages serves solely usability issues and has no influence on the translation. However, the strictly defined hierarchical package structure can not only be used to partition the knowledge, but also to structure the solving process done by the configurator. In some cases, different packages or parts of a configuration problem can be solved more or less independently. This information can be used to partition the solving process which leads to smaller problem sizes and less computational complexity in the solving phase [ ].

Another problem, arises from the nature of the configuration task per se. The configuration knowledge incorporates a structural and a functional product architecture which are interrelated through a mapping from functions to components as described in [ ]. This results in a complexity which can only be partially graphically depicted in the two-dimensional representation space of a single (partitioned) diagram. The structural architecture is represented by interconnected partonomies and taxonomies of physical components each described by a set of physical properties. Additionally, constraints are specified which impose restrictions on the component composition. Similarly, a functional architecture specifies the decomposition of the artifact into functions and constraints on their composition. Further the "many-to-many" mapping between functions and components needs to be modeled and becomes part of the configuration knowledge.

A lot of research has been done in the area of Functional Representation (FR) to describe the function and the structure of a device and identify the processes that are causal for the achievement of the function. See Chandrasekaran [ ] for an overview on the work in this area. However, for the task of configuration design as defined in [ ] the causal processes need not necessarily be modeled. It is sufficient to explain the achievement of functions solely from their structure. [ ] uses the term passive function in cases the function of an object is simply explained by the virtue of its structural properties. In the following we sketch an incomplete functional decomposition of our example car (Figure 3). Note that the structural and the functional model are only two different views on the same artifact (both have the same root node car). These views introduce no new semantics since the underlying model (classes, associations, etc.) is not affected when arranging the classes in different views on the same diagram.

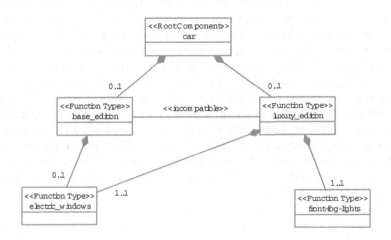

Fig. 3. Functional model of a car

Equivalently to the definition of the physical product structure the functional architecture can be modeled with taxonomies and partonomies of function types. As shown in Fig. 3 modeling concepts like compatibility and requirements relations and additional constraints can also be used. There is no sharp distinction what has to be modeled as function or as component, however we can say that components are those parts the artifact is built of and functions specify the purposes of the artifact.

4 Contextual Diagrams

The third chunk of knowledge which needs to be modeled in the configuration domain is the mapping between functions and components. It describes how the function is realized by implementation of a specific (sub-)structure. Mittal and Frayman introduced the *key component per function* assumption [ ]. Simplified it says that for each function *f* a set of key components can be identified which needs to be part of the final configured artifact in order to implement *f*. A key component therefore is a critical part for the realization of *f*. Further it is possible that several key components alternatively contribute to the implementation of *f*. Additional requirements for the provision of a function can be related to that key component. In our example the implementation of the function *electric_windows* would require that the component *electric_windows_module* is part of the final configuration. As already stated, the functional and structural decomposition are only two different views on the same artifact. It is obvious that the key component assumption can be graphically depicted by introducing a *requires* relation between a function type and each of its key components.

However, we made the experience that the definition of key components for each function has only a limited expressiveness which does not suffice for the definition of the complex functional dependencies which occur in the configuration knowledge of widely known consumer products like a car. As one can see from Fig. 2, the possible product structure is not only constrained by a limited set of components, but also by compatibility and requirements relations, additional constraints etc. Therefore, a function may not only require a specific component but can additionally constrain the product structure. E.g., our example car can consist of several types of transmissions (*automatic, 2-wheel_drive, 4-wheel_drive*) and offers the customer a range of different engines (*engine_1* to *engine_3*) to choose from. However, technically it makes no sense to sell a car combining an *automatic* transmission with an *engine_1*, because the power of the latter is too weak (compare to *incompatible* relation in Fig. 2). If a configured car satisfies the function *luxury_edition* only the most powerful engine (*engine_3*) is allowed to be combined with an *automatic* transmission (see the additional *requires* relation in Fig. 4).

Therefore we define a function in the domain of product configuration as a set of constraints which have to be satisfied if the configured artifact is required to fulfill this specific function. For a diagrammatic representation of more complex mappings from function to structure we introduce contextual diagrams. They graphically depict configuration knowledge which applies only "in context" with the implementation of one or several functions. Fig. 4 contains a graphical representation of the configuration knowledge which additionally applies when the function *luxury_edition* is realized. E.g., the taxonomy of *engine* is redefined and offers less choices and the domain of the attribute *voltage* of component *battery* is restricted to a single value (*voltage = high*).

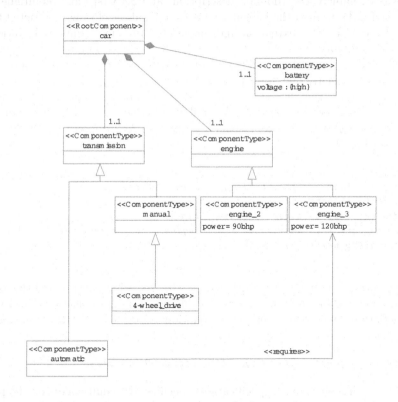

Fig. 4. Contextual diagram for function luxury_edition

5 Knowledge Base Generation

Knowledge base representation

In the following we will shortly present our consistency-based configuration approach and show how we can derive a configuration knowledge base with a declarative representation from the diagrammatic configuration models. The configuration task can be characterized by a limited set of components and functions, a description of their properties, namely attributes and possible attribute values, connection points (ports), and constraints on legal configurations. Given some customer requirements, the result of computing a configuration is a set of components, corresponding attribute valuations, and connections satisfying all constraints and the customer requirements. This model has proven to be simple and powerful to describe general configuration problems and serves as a basis for configuration systems as well as for representing technical systems in general. The formulation of a configuration problem can be based on two sets of logic sentences, namely *DD* (domain description) and *SRS* (specific requirements). We restrict the form of the logical sentences to a subset of range restricted first-order-logic with a set extension and interpreted function symbols. In order to assure decidability, we restrict the term-depth to a fixed number. Additionally, domain specific axioms for configuration are defined, e.g., one port can only be connected to exactly one other port.

DD includes the description of different component and function types (*types*), named ports (*ports*), and attributes (*attributes*) with their domains (*dom*). The following *DD* represents a portion of the complete knowledge base shown in Fig. 2 and Fig. 3.

```
DD:
types = {car, engine, battery, luxury_edition, ...}.
attributes(battery) = {voltage}.
dom(battery, voltage) = {medium, high}.
ports(car) = {transmission_port, engine_port, battery_port, ...}.
ports(engine) = {car_port}
...
```

Additionally, constraints are included in *DD*, reducing the possibilities of allowed combinations of component and function types, connections, and value instantiations, e.g., the *incompatible* relation between *automatic* transmission and *engine_1*.

```
type (ID1, automatic) ∧type (ID2, engine_1) ⇒ false.
```

SRS (system requirements specification) specifies the requirements on the product which should be configured. These requirements are the input for the configuration task. E.g., a customer decides on a car that should provide the function *luxury_edition* and contain an *automatic transmission*.

```
SRS:
SRS = {type(_, luxury_edition), type(_ , automatic)}.
```

The configuration result is described through sets of logical sentences (*COMPS, ATTRS, CONNS*). In these sets, the employed components, the attribute values (parameters), and the established connections of a concrete customized product or service are represented.

- *COMPS* represents sets of literals of the form *type(c, t)*. t is included in the set of types defined in *DD*. The constant *c* represents the identification for a component.
- *CONNS* represents sets of literals of the form *conn(c1, p1, c2, p2)*. c1, c2 are component identifications from *COMPS*. p1 (p2) is a port of the component c1 (c2).

---

[1] We employ a logic programming notation where variable names start with an upper case letter or are written as "_". The variables are all-quantified if not explicitly mentioned.

– *ATTRS* represents sets of literals of the form *val(c, a, v)*, where *c* is a component-identification, *a* is an attribute of that component, and *v* is the actual value of the attribute (selected from the domain of the attribute).

```
Example for a configuration result (COMPS, CONNS, ATTRS):
type(f1, luxury_edition). ... type(c1, car). type(c2,
automatic). ...
conn(c1, luxury_edition-port, f1, car-port). ...
val(c4, voltage, high). ...
```

Based on the definitions given so far, we specify the concept of a consistent configuration:

If (DD, SRS) is a configuration problem and *COMPS*, *CONNS*, and *ATTRS* represents a configuration result, then the configuration is consistent exactly iff $DD \cup SRS \cup COMPS \cup CONNS \cup ATTRS$ can be satisfied.

In order to achieve a complete configuration, we have to additionally specify that *COMPS* includes all required components, *CONNS* describes all required connections, and *ATTRS* includes a complete value assignment to all variables. A more detailed formal exposition is given in [ ].

Diagrammatic knowledge representation

The graphical representation (*GREP*) of configuration knowledge consists of a root model $GREP_{root}$ that contains the knowledge on the structural and functional decomposition of a configurable product. Further simple mappings (e.g., requires relations to key components) from functions to components are also part of $GREP_{root}$. Additionally contextual diagrams $GREP_i$ can be defined which describe more complex mappings between functions and components.

– Let $GREP_i$ be a contextual diagram which corresponds to a logical expression $Expr_i$ consisting of the identifiers of single function types and conjunction operators. Then $Expr_i$ denotes the context in which the model depicted in $GREP_i$ is relevant, e.g., if $Expr_i = type\ (\_,\ luxury\_edition)$ then $GREP_i$ describes the constraints demanded by the function *luxury_edition*. In case $Expr_i = type(\_,\ luxury\_edition) \land type(\_,\ electric\_windows)$ then $GREP_i$ depicts the restrictions that additionally apply when the functions *luxury_edition* and *electric_windows* are jointly implemented.

Fig. 5 illustrates the organization of the diagrams. Based on the root model $GREP_{root}$ several contextual diagrams can be defined that are denoted by a corresponding expression (e.g., $Expr_i$ denotes the context of $GREP_i$). If necessary contexts can be hierarchically organized by conjuncted function types (e.g., $Expr_k$ specifies a subcontext of $Expr_i$ that applies only when both functions *a* and *c* are realized).

Translation rules

In [ ] we give rules for an automatic translation of the graphical concepts into a domain description (*DD*) of a configuration problem consisting of logical sentences. All these rules apply to derive DD_{root} from $GREP_{root}$ and DD_{root} becomes then a part of *DD*.

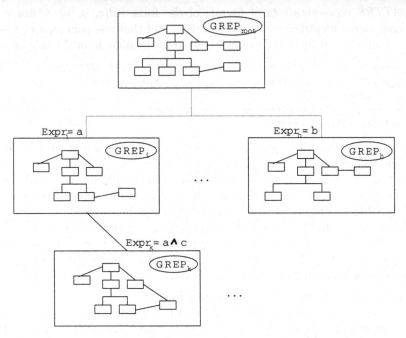

Fig. 5. Organization of diagrams

DD_{root} = the set of logical sentences which can be derived from $GREP_{root}$.

Given DD_{root}: $DD_{root} \subset DD$.

Configuration knowledge is derived from a contextual diagram $GREP_i$ as follows: except classes (component and function types, ressources, ports) and attributes (to avoid a redundant translation) all rules apply as stated in [ ], i.e., *requires*, *incompatible*, *produces*, *consumes* relations are translated analogously. We define the derived logical sentences to be contained in the set DD_i.

DD_i = the set of logical sentences which can be derived from the contextual diagram $GREP_i$.

The logical sentences derived from the contextual diagram apply only in case the function(s) they implement ($Expr_i$) is(are) required.

Given $Expr_i$, $GREP_i$, DD_i: $\{Expr_i \Rightarrow DD_i\} \subset DD$.

In Fig. 4 we depicted an additional *requires* relation between an *automatic* transmission and the engine *engine_3* which applies only for *luxury_edition* cars. An exemplified translation of this piece of knowledge would be the following:

```
type(_, luxury_edition) ⇒
{type(ID1, automatic) ⇒ (∃ID2, type(ID2, engine_3))}.
```

In order to guarantee clear semantics, and the avoidance of inconsistencies and redundant declarations the following restriction applies, when modeling a contextual diagram $GREP_i$:

$$\forall i \ DD_{root} \bigcup DD_i \text{ must be consistent}.$$

This assures that in $GREP_i$ only restrictions that additionally apply are specified and no new classes or attributes are defined which would cause inconsistencies with the model $GREP_{root}$.

6 Prototype Environment

In order to realize the concepts discussed in this paper we have implemented a prototype development environment for the automatic construction of configurators. Since our modeling approach uses standard UML and its built-in extension mechanisms, any tool supporting the UML notation can be used to model the conceptual product model. Our translation tool uses the XMI representation [ ] of an UML modell as input. The computation of final configurations is performed by an industrial-strength constraint solving software (ILOG Configurator). The modeling starts with the definition of a root model (GREP$_{root}$), which contains the functional and structural architecture. Based on the definitions in this diagram, contextual diagrams are derived which specify the mapping from functions to components. After each functional specification syntactic checks are performed.

7 Related Work

There is a long history in developing configuration tools in knowledge-based systems [ ]. Research dealing with knowledge representation regards the declarative constraint representation as the basic representation formalism since the development and maintenance of rule-based systems like R1/XCON [ ] have shown to be error-prone. Using declarative constraint representation is not enough since knowledge bases become more and more complex so that conventional structuring mechanisms do not suffice. Progressing from rule-based systems higher level representation formalisms were developed, i.e., various forms of constraint satisfaction [ ], description logics [ ], or functional reasoning [ ]. [ ] propose a resource-based paradigm of configuration where the number of components of a particular type occurring in the configuration depends on the amount of that resource required. Conforming these paradigms various configuration systems were developed ([ ] [ ] [ ]).

Conventional mechanisms for structuring knowledge bases in the configuration domain are described in Sabin, Freuder [ ] or Peltonen et. al. [ ]. There are many other works dealing with structuring the configuration knowledge. The common point is usage of conventional structuring mechanisms like abstraction, generalization and aggregation.

Feldkamp et. al. [ ] discuss the structuring of the configuration knowledge in terms of reuse, i.e. they describe three different tasks when modeling a configurable product. First, the properties of a product and the corresponding subsystems are modeled. Second the product structure is modeled by using the well known concept of partonomy. The third step is modeling the design logic behind the product. For resolving the problem of defining complex constraints over the product structure they introduce the concept of interfaces. In this context complex constraints are not defined in the components but are part of the interface which describes the connection between several complex components.

Haag [ ] describes the structure of the SAP R/3-configuration engine. Configuration in R/3 means the creation of a concrete bill-of-materials conforming the customer requirements. They start from a bill-of-materials containing the maximum amount of components, these parts are removed from this list, leaving only the actually needed components in the final configuration. The design process is not done on a conceptual level and leaves the maintenance process as a big challenge.

The notion of context as been discussed in several research areas using different interpretations depending on the application area. In [ ] a context is denoted as a higher order conceptual entity, which describes a group of conceptual entities from a particular point of view. Rules and operations are provided to organize the manipulation of contexts. Furthermore, an example for the application of contexts in a cooperative document design environment is given, where documents represent collections of objects. In [ ] the notion of context is employed for supporting cooperative work in hypermedia design providing a set of context operations on hypermedial objects (e.g. editing, inquiry, attribute operations). Compared to these approaches we view our approach as complementary, since our goal is to effectively support the knowledge acquisition for complex (configuration) knowledge bases not regarding any operations for combining different contexts (all contexts are translated into the same DD). In AI research (e.g., [ ] and [ ]) formalizations for a logical concept of *context* are given and application scenarios are presented. [ ] denotes a context as a first class object in logic and it is asserted that propositions can be true in a specific context. We do not use contexts as explicit objects in logic, therefore our notion of context is similar only in a very general way which is as an abstract concept to structure domain knowledge. Our approach applies these concepts in order to support knowledge acquisition for complex knowledge bases on a conceptual (UML) level, especially improving the representation of complex constraints in order to be understandable for domain experts.

8 Conclusions

Our approach proposes the usage of a well-defined conceptual language (UML) to define a product configuration knowledge base. We use the built-in structuring mechanisms of UML as well as newly introduced structuring concepts to design graphical depictions of configuration knowledge that are expressive and human readable and can be transparently communicated to domain experts. We partly exemplified the structural and functional product architecture of a complex consumer product and motivated a new structuring mechanism to model existing functional dependencies which occur in configuration knowledge bases. Further we provide translation rules for an automated generation of an executable knowledge representation. These techniques enhance the phase of acquisition of configuration knowledge and ease the task of maintenance of these knowledge bases.

References

1. V. E. Barker and D. E. O'Connor. Expert systems for configuration at Digital: XCON and beyond. *Communications of the ACM*, 32,3:298–318, 1989.
2. B. Chandrasekaran. Functional Representation: A Brief Historical Perspective. *Applied Artificial Intelligence, Special Issue on Functional Reasoning*, 1993.
3. N. M. Delisle and M. D. Schwartz. Contexts – a partitioning concept for hypertext. *ACM Transactions on Information Systems*, 5,2:168–186, 1987.
4. F. Feldkamp, M. Heinrich, and K. D. Meyer Gramann. SyDeR System Design for Reusability. *AI EDAM Artificial Intelligence for Engineering Design, Analysis and Manufacturing*, 12,4, 1998.
5. A. Felfernig, G. Friedrich, and D. Jannach. UML as domain specific language for the construction of knowledge-based configuration systems. In *Proceedings of SEKE'99*, pages 337–345, Kaiserslautern, Germany, 1999 . , , , ,
6. G. Fleischanderl and A. Haselböck. Thoughts on partitioning large scale configuration problems. *AAAI-96 Fall Symposium on Configuration*, 1996.
7. G. Fleischanderl, A. Haselböck, H. Schreiner, M. Stumptner, and G. Friedrich. Configuring Large Systems Using Generative Constraint Satisfaction. *IEEE Intelligent Systems, Configuration - Getting it right, July/Aug*, 1998.
8. G. Friedrich and M. Stumptner. Consistency-Based Configuration. In *AAAI Workshop on Configuration, Technical Report WS-99-05*, pages 35–40, Orlando, Florida, 1999.
9. R. V. Guha. *Contexts: A Formalization and Some Applications*. PhD thesis, Also published as technical report STAN-CS-91-1399-Thesis, Stanford University, 1991.
10. A. Haag. Sales configuration in Business Processes. *IEEE Intelligent Systems, Configuration - Getting it right, July/Aug*, 1998. ,
11. M. Heinrich and E. W. Jüngst. A resource-based paradigm for the configuring of technical systems from modular components. *Proceedings of the 7th IEEE Conference on AI applications (CAIA)*, pages 257–264, 1991.
12. B. J. Pine II, B. Victor, and A. C. Boynton. Making Mass Customization Work. *Harvard Business Review*, Sep/Oct, 1993.
13. John McCarthy. Notes on Formalizing Context. In *Proceedings of the Thirteenth International Joint Conference on Artificial Intelligence (IJCAI93)*, Chambery, Frankreich, 1993.

---

[2] A revised and extended version of this paper will appear in the International Journal of Software Engineering and Knowledge Engineering (IJSEKE).

14. D. L. McGuinness and J. R. Wright. Conceptual Modelling for Configuration: A Description Logic-based Approach. *Artificial Intelligence for Engineering Design, Analysis and Manufacturing, Special Issue: Configuration Design*, 12,4:333–344, 1998.

15. S. Mittal and F. Frayman. Towards a Generic Model of Configuration Tasks. In *Proceedings of the 11th IJCAI*, pages 1395–1401, Detroit, MI, 1989. , , ,

16. M.Stumptner, A. Haselböck, and G. Friedrich. "Cocos: 1994, A Tool for Constraint-Based, Dynamic Configuration". *Proceedings 10th Conf. AI for Applications*, 10,2:373–380, 1997.

17. Object Management Group (OMG). XMI Specification. *www.omg.org*, 1999.

18. H. Peltonen, T. Männistö, T. Soininen, J. Tiihonen, A. Martio, and R. Sulonen. Concepts for Modeling Configurable Products. In *Proceedings of European Conference Product Data Technology Days*, pages 189–196, Sandhurst, UK, 1998.

19. J. Rumbaugh, I. Jacobson, and G. Booch. *The Unified Modeling Language Reference Manual*. Addison-Wesley, 1998. ,

20. J. T. Runkel, A. Balkany, and W. P. Birmingham. Generating non-brittle Configuration-design Tools. *Proceedings Artificial Intelligence in Design 94*, pages 183–200, 1994.

21. D. Sabin and E. C. Freuder. Configuration as Composite Constraint Satisfaction. *Artificial Intelligence and Manufacturing Research Planning Workshop*, 1999.

22. M. Stumptner. An overview of knowledge-based configuration. *AI Communications*, 10,2:111–126, 1997.

23. Manos Theodorakis, Anastasia Analyti, Panos Constantopoulos, and Nikos Spyratos. Context in information bases. In *Proceedings of the 3rd International Conference on Cooperative Information Systems (CoopIS'98)*, pages 260–270, New York City, NY, USA, August 1998. IEEE Computer Society.

24. Timo Soininen et. al. Towards a General Ontology of Configuration. *AIEDAM*, 12,4:357–372, 1998.

25. B. Yu and H. J. Skovgaard. A configuration tool to increase product competitiveness. *IEEE Intelligent Systems*, 13,1:34–41, 1998.

Evaluating the Intelligibility of Diagrammatic Languages Used in the Specification of Software

Carol Britton, Sara Jones, Maria Kutar, Martin Loomes and Brian Robinson

Department of Computer Science, University of Hertfordshire,
College Lane, Hatfield, Herts. AL10 9AB
M.S.1.Kutar@herts.ac.uk

Abstract. This paper presents an approach to evaluating the intelligibility of diagrammatic languages used in the specification of software. Research suggests that specification languages can be assessed in terms of properties that influence the intelligibility of representations produced using the languages. The paper describes the properties identified and highlights three in particular that have been shown to influence the intelligibility of representations: motivation of symbols in the language; the extent to which the language allows exploitation of human visual perception; and the amount of structure inherent in the language. The paper argues that the first two of these properties are not present to any great extent in diagrammatic languages used in software specification. In order to enhance the intelligibility of software specifications, we suggest that more attention should be paid to ways in which these languages can exploit the amount of structure inherent in the language.

1   Introduction

Today's clients and users of software systems play an increasingly important part in the software development process, particularly during the early stages when requirements for the system are established.   Good communication between all stakeholders is essential if requirements are to be effectively elicited, documented and validated, but this can only take place if there is sufficient shared understanding of the representations of the clients' and users' problems, and the proposed solutions. The specification of software systems is a relatively mature area; specialised diagrammatic languages such as data flow and entity-relationship diagrams have been widely used for many years. However, with the increasing involvement of clients and users who are unfamiliar with such languages, a new problem has arisen for developers of software: how to ensure that representations of the system are readily intelligible to everyone who needs to understand them.

This paper addresses the problem of intelligibility of specifications by focussing on the languages used to produce representations of the software system. The research

M. Anderson, P. Cheng, and V. Haarslev (Eds.): Diagrams 2000, LNAI 1889, pp. 376-391, 2000.

proposes a set of properties of languages used in the specification of software, and investigates the degree to which these properties influence the intelligibility of representations produced using the languages. Four of the properties concern the symbols in a language: number, consistency, discriminability and motivation. The other two properties relate to how the language can be used in a specification; these are the 'perceptual' content and the amount of structure in a language.

Whatever the language used in systems specification, the representations produced using it will only be effective if they can be readily understood by everyone who needs to look at them. However, evaluating such representations directly is complex and difficult. Representations are products of languages used in different ways, by different people, in different situations. This means that there are a large number of factors that influence whether a particular representation can be easily understood or not. The difficulty of evaluating representations directly has been recognised by other authors [33], [37]. Because of these difficulties, the context of the research described in this paper is restricted to focus on the languages that are used to produce the representations, in particular, representations that are based on pen and paper.

It is frequently the case that languages used in the specification of software systems are tightly coupled with a development method. This also complicates the issue of evaluation, since there are so many factors that affect the success or failure of the specification process. All of these factors are important, but they are too complex and numerous to be considered together in detail in one piece of work. The research described here concentrates on individual specification languages; it is beyond the scope of the paper to consider development processes, project contexts, or ways in which the languages may be used in combination.

The subject of the paper is further restricted to the area of requirements validation, during which representations are used to support clients and users of the system in checking that the developer has understood and accurately specified their requirements. The validation process involves a number of activities on the part of clients and users, which can only be performed successfully if the representations to be validated are intelligible to them. The intelligibility of a representation will, of course, be dependent on the ability level and cognitive characteristics of the reader [39]. However, since a wide range of individuals are likely to be involved in checking representations of software systems, we aim here simply to gain a view of the needs of a 'typical' reader who is unfamiliar with languages for specifying software.

There has recently been considerable debate in cognitive psychology about the relationship between internal, or mental, representations, and external representations (see, for example, [33]). Since internal representations are not yet well-understood, it is currently impossible to reach any general conclusions regarding the support which particular forms of external representations provide for working or reasoning with different kinds of internal representations. We agree with Cox and Brna [7] that the debate about internal representations is unlikely to lead directly to practical solutions at this stage. We make no claims here about the links between internal and external representations, focusing simply on the intelligibility of external representations.

There is currently no detailed account of the cognitive processes involved in validating software requirements. We therefore analyse the concept of intelligibility by making a loose analogy with the processes involved in reading [10], in order to focus our attention on the kinds of cognitive tasks that may be involved. We argue that intelligibility in the context of interest to this paper can be characterised in terms of two activities:

- extracting information from the representation;
- checking the correspondence of information in the representation with existing knowledge.

The importance of these activities, which are not externally observable themselves, lies in their role as the basis for three further activities that are essential for effective validation and that can be observed:

- developing and extending ideas about the intended system and possible changes in the work environment;
- suggesting changes and additions to the representation;
- making annotations to the representation.

We assume that, if clients and users can perform these activities effectively, the languages used to produce the representations will be intelligible to them.

A final restriction on the topic of the paper relates to what has been termed *effectiveness:* whether a language 'exploits the capabilities of the output medium and the human visual system', rather than *expressiveness*: whether a language has a semantics that is rich enough to express the desired information [23]. The focus of the work is on what aspects of a language, over and above an appropriate semantics, are likely to facilitate the production of representations which may be easily understood by untrained users.

To summarise: the specific context of this paper is the intelligibility of languages for producing paper-based representations that are used in the validation of requirements for software systems by clients and users. The aim of the research described in the paper is to investigate how representations used in software specification can be made more readily intelligible, particularly to readers who are unfamiliar with the specification languages used. The research is based on the hypothesis that certain properties of specification languages influence the degree to which readers can understand representations produced using the languages.

The rest of this paper is organised as follows: Section 2 discusses diagrammatic languages used to specify software, and Section 3 describes other approaches to evaluating specification languages. Sections 4 and 5 introduce properties of specification languages and highlight the three properties that relate to intelligibility. Section 6 investigates the intelligibility of diagrammatic languages, and finally Section 7 discusses the importance of structure.

2   Diagrammatic Languages Used to Specify Software Systems

Although prototypes or other technology-based representations are widely used during development of a software system, paper-based representations are still needed to elicit and validate requirements in the very early stages, and must be readily understandable to anyone who needs to see and discuss them. The fundamental criterion that a representation should be clearly understood suggests that natural language descriptions are the most appropriate means of representation in this context. Historically, however, descriptions in natural language have not been wholly successful, since such descriptions have tended to be verbose, imprecise and ambiguous. It is not clear whether this has been the fault of the language or of the writers, but the majority of system developers today feel that better communication is achieved through use of languages, particularly those that include diagrams, which have been developed especially for the purpose of specifying software systems [9], [20], [24].

There are so many specialised languages that are used to specify software, that it is not possible to list them all. Families of diagrammatic specification languages include structured techniques, such as data flow and entity-relationship diagrams (see, for example, [12], [35]), object-oriented languages, such as class diagrams and state charts [1], [31] and graphical versions of mathematical specification languages, such as [21]. In addition to these, new languages and variations of existing ones are constantly under development; practically every conference on requirements engineering or software specification includes at least one paper describing a new way of specifying software.

The majority of languages used to specify software include diagrams, and many claim that this produces representations that are easy for untrained users to understand. Within cognitive psychology, a distinction is commonly made between graphical (or pictorial) representations and propositional (or linguistic, textual, sentential) ones [10]. In graphical representations, there are no obvious discrete symbols, but propositional representations, on the other hand, are made up of discrete symbols (words or letters, depending on the level of granularity which is required). Moreover, in graphical representations much of what is represented is implicit, for example the relations 'beside-ness', ordering or causality. In propositional representations, explicit symbols are used for everything that is represented. Finally, graphical representations generally have little defined grammar, whereas propositional representations follow the grammar rules of the language in which they are written.

As we have already said, many different languages may be used in specifying the requirements of a software system. One of the most popular types of language in the early stages of system development is rich pictures [27], which can be characterised as broadly graphical. In rich pictures, the symbols used may be drawn from an infinite set of possibilities, and it can be difficult to make a definite judgement about where one symbol begins and another ends. Rich pictures rely heavily on visual communication, and the location of symbols within the plane of the picture is often used in a significant way. Information can remain implicit in many parts of the representation. Finally, the grammar according to which rich pictures are drawn is only weakly defined.

Another type of language which has been widely used in system specification for may years is the data flow diagram, which, according to the above classification, falls between the two extremes of graphical and propositional. Data flow diagrams are created using a pre-defined set of symbols, and should be constructed following a grammar, but some information (for example, regarding the ordering of processes) is left implicit, and there is quite a strong reliance on the visual modality.

In the following section of this paper, we survey other research that has been carried out into the evaluation of software specification languages and introduce our own approach to evaluating the languages in terms of intelligibility.

3   The Evaluation of Specification Languages

Although there is a considerable amount of research relating to the evaluation of software specifications, most of the approaches focus on the requirements specification itself, rather than on languages used to produce representations of the system [9], [11], [36].   There are, however, four evaluation frameworks that are concerned specifically with languages for software specification [6], [8], [19], [29], as well as the cognitive dimensions of Thomas Green [15], [16], which can be applied to any information structure.

There are two reasons why the research described in this paper is not based on any of the four existing frameworks that were developed for the evaluation of specification languages. First, none of these evaluation frameworks is widely known within the software engineering community, or appears to have been used successfully by researchers other than the authors themselves. Second, and more important, in these approaches, intelligibility is assessed on a purely subjective basis; there is no attempt to break down the concept so that it might be amenable to objective assessment.

Although Green's cognitive dimensions framework [15], [16] is a general approach to evaluating information structures, it is a useful starting point for evaluation of languages for a number of reasons. First, cognitive dimensions are a pragmatic approach; they can be applied to any information structure (in this case, specification languages) and provide a ready-made vocabulary for evaluation. In addition, the dimensions have already been used successfully in a number of cases, both by Green himself, [16], [17], [18], [25], and by other authors, [4], [28], [30]. Cognitive dimensions articulate concepts that are important and relevant in the context of specification languages, thus allowing a better understanding of those concepts and the inter-relationships between them.   Intelligibility is not, itself, part of Green's framework, but several of the cognitive dimensions help to clarify ideas in the cognitive psychology literature that are relevant to intelligibility.

Cognitive dimensions are an effective way of analysing the activities that a reader has to carry out during the validation of representations of software requirements: extracting information from the representation, and checking the correspondence of information in the representation with existing knowledge (see Section 1 of this paper). As an example, the cognitive dimension of abstraction gradient highlights the activity of decomposing a representation into manageable chunks.   This is an important part of extracting information from a representation. A further example relates to the dimensions of closeness of mapping and role expressiveness, which highlight two secondary activities that the reader of a specification has to perform: relating elements of the representation to elements in the domain and inferring the purpose of individual components in a representation. These two activities are part of checking the correspondence of information in the representation with existing knowledge.

A further advantage of cognitive dimensions is that they provide a framework in which to evaluate structure and form, independently of content. This is relevant in the context of this research, since it supports investigation of specification languages separately from particular representations produced using them.   Finally, the dimensions are helpful, in that they provide an accessible route into the cognitive psychology literature, and a framework and vocabulary in which discussion can take place. As an example, the dimensions of visibility and redundant recoding pinpoint relevant parts of the literature on the need for readers to be able to pick out the important parts of a representation [23], [37].

4   Properties of Languages Used to Specify Software

The research reported in this paper is based on the hypothesis that certain properties of specification languages influence the degree to which representations produced using the languages can be understood by readers, particularly those readers who are untrained in languages for software specification.

Properties of languages for investigation were first suggested in   [3] and further developed in [5]. The properties were selected after a critical analysis of the literature in a variety of disciplines: computer science, philosophy of language, linguistics and cognitive psychology. A review of the relevant literature can be found in [5]. The properties of languages that were identified as contributing to the intelligibility of software specifications are described briefly below.

(i) The number of different symbols in the language
Programming languages with a large number of symbols and features appear to be more difficult to learn and understand than languages with fewer features [13]. With regard to languages used in software specification, this implies that representations in a simple diagrammatic language such as storyboard, which has 2 symbols (over and above natural language), will be easier to understand than those in a language such as Z, which has 76 symbols. However, the number of symbols in a notation is a fairly crude measure; other issues have to be considered, such as the amount of information carried by each symbol, and how the organisation of a representation is compromised by using a language with few symbols.

(ii)   Consistency of symbols in the language
It has been pointed out [14] that the number of symbols in a language is less important than consistency in meanings. Inconsistent symbols are difficult to learn and recall, whereas symbols that appear to conform to a pattern are easier to understand. An example of consistency can be seen in the symbols $\subset$ (proper subset) and $\subseteq$ (subset). We should also be aware, however, that it may be difficult for users to discriminate between symbols that are consistent, such as $\subset$ and $\subseteq$. In some contexts there is a trade-off between the properties of clear consistency of symbols and discriminability (see below). Diagrammatic languages tend to have relatively few symbols and consistency is therefore less of an issue.

(iii)   Discriminability of the symbols in the language

Discriminability has been identified by authors in both linguistics and cognitive psychology as having an influence on the intelligibility of a language [32], [39]. It refers to how physically distinct each symbol is from others in the language, and thus the ease with which different symbols in a language can be distinguished from each other. To the untrained eye, for example, the mathematical symbols ↔ (relation) and → (function) look similar, yet they have different meanings and are a potential source of confusion for readers.

Discriminability of symbols is a problem with many mathematical languages. In contrast, diagrammatic languages, such as entity-relationship and data flow diagrams, generally have symbols, such as lines, arrows and boxes, which are clearly distinguishable from each other. However, confusion may arise in the case of someone who is unfamiliar with the languages. An untrained user may well find that the symbol for a process in a data flow diagram is difficult to distinguish from the symbol for a data store (see Figure 1 below).

Process in a data flow diagram   Data store in a data flow diagram

Fig. 1.

(iv)   Degree of motivation of the symbols in the language

A language may be considered to be motivated if there is an intuitive or cultural relationship between the elements of the language and objects that they represent. The use of the symbol of an eye to represent the external view in rich pictures is an example of motivation [27], as is the use of the actor symbol in the Unified Modeling Language [1].   An actor 'Patient' can be seen in Figure 2, below, with the unmotivated symbol used to denote the use case 'Make appointment'.

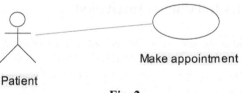

Make appointment

Patient

Fig. 2.

The concept of motivation in languages is found in linguistics [32]. Motivation is an important property of a language, since, by definition, it means that the symbols in the language will be closely related to the concepts that they represent and therefore that their meaning will be clear, even to untrained users. Software specification languages are rarely motivated, but some diagrammatic languages do include motivated symbols, such as a line representing a link (entity-relationship diagrams), and an arrow signifying direction (data flow diagrams).

(v)   Extent to which the language allows exploitation of human visual perception

This property is related both to the notion of effectiveness [23]: whether a language 'exploits the capabilities of the output medium and the human visual system', and to 'computational offloading' [33]. Computational offloading is the extent to which a representation of a problem can reduce the amount of cognitive effort needed to solve the problem by providing the means for direct perceptual recognition of important elements in it. In this paper languages which allow exploitation of human visual perception are referred to as 'perceptual'.

Perceptual and non-perceptual languages each have advantages and disadvantages in software specification. Representations in diagrammatic languages often embody perceptual concepts, such as connectedness (as used in data flow diagrams) and inclusiveness (as found in state diagrams) which most people can readily understand. However, the amount of information these diagrams can contain is restricted. They soon reach a saturation point, after which any extra information only causes confusion. With non-perceptual languages, such as mathematical languages, the problem is that representations are frequently beyond the understanding of untrained users, even though they are generally able to hold much more information than representations that rely heavily on perceptual features.

(vi) Amount of structure inherent in the language

If a representation is clearly structured, it will involve less effort on the part of readers to find, decompose and abstract information, and thus be easier to understand [10], [14], [34], [39]. Decomposition and abstraction are important because a reader can only cope with a small amount of information at a given time. An effective language should encourage decomposition of the representation into 'brain-sized chunks', each of which is intellectually manageable by the reader. Abstraction helps readers of representations to concentrate on the most important elements while ignoring details that are currently irrelevant. Many diagrammatic languages are highly structured; including data flow diagrams, state diagrams and Product/Process Models [26].

5   Key Properties Relating to Intelligibility

Literature reviews and empirical work carried out as part of the research reported in this paper indicate that three of the properties described above have a particularly strong influence on the intelligibility of representations; these are the properties of motivation, 'perceptualness' and structure.

Motivation (see property (iv) above) is identified as an important characteristic by many authors in cognitive psychology (for a review of the relevant literature, see [5]). The claim is often made that graphical representations are 'natural' because they are in some sense analogical to the world which they represent [10], [22], [37]). In this research, empirical work into the effect of motivation on intelligibility was carried out by means of two studies, one involving a group of software developers and the other two groups of students. In the first study, which was in the form of a questionnaire, the software developers were asked to select symbols to match given concepts, as shown in the example below.

Example of question asked of participants.
You will see below a list of words and phrases used in everyday conversation. For each of these, please circle the letter corresponding to the symbol that you think best represents it.

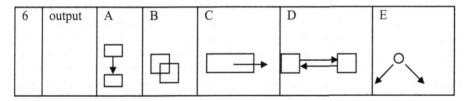

In the example question shown here all the participants chose symbol C to represent the idea of output.

The second study was based partly on the formal specification language, Z. Although Z is a symbolic language, the way in which Z specifications are organised in schema boxes is diagrammatic. Z was chosen as a vehicle for this study partly because it is generally considered to be very difficult for untrained users to understand. It is often the case that more can be learnt from difficult cases than from situations where everything goes according to plan. On pragmatic grounds also, Z was an obvious choice for the study, since two contrasting groups of students were available to act as participants; one group had no experience of any formal language and the other had previous experience of Z and logic.

Each group of students in the study was asked to complete a separate multiple choice questionnaire; this took 10 minutes for Group 1 (inexperienced) and half-an-hour for Group 2 (some experience). All students at each level completed the same questionnaires. The multiple choice questions were all forced choice questions and where students felt they did not know the right answer, they were asked to guess. In the study, two different aspects of motivation were considered: first, whether untrained subjects were able to correctly identify the symbols used in Z to represent particular meanings, and second, whether subjects who had received some training were able to correctly remember the meaning of the same symbols. A separate question in both questionnaires asked subjects to suggest symbols in an imaginary specification language for a given set of concepts, including 'object', 'process' and 'time passing'.

Subjects in Group 1, who were inexperienced, were given meanings and asked to choose what they thought were appropriate symbols to represent those meanings. For example, subjects were given the meaning "there exists" and asked which one of a given set of symbols best represented that meaning. Symbols presented as options were all taken from either Z or logic. The students found this part of the questionnaire very difficult, and it was hard to draw any conclusions from the results. However, in the question where they were asked to suggest symbols, motivation appeared to be an important factor. For example, all the subjects included an analogue clock in their symbols for 'time passing' and several drew a skeleton file shape for 'data store'.

Group 2 had followed a course on formal languages the previous year, in which they were taught Z and logic. These subjects were given the same set of symbols as were shown to Group 1, and were asked to identify their meanings. For example, given the symbol ∃, subjects were asked to say whether this means "and", "for all", "there exists", "if .. then", "or", "if .. and only if", or "not". Performance in this task was generally high, and it is interesting to note that all subjects correctly identified the meanings of ∃ and ∀. From the results of Group 1, the symbols ∃ and ∀ do not appear to be motivated. However, the motivation of these symbols (∃ is a backwards E, as in "Exists", and ∀ is an inverted A, as in "All") had been pointed out to the subjects in Group 2 during their course the previous year. The results form this question suggest that, once attention has been drawn to the motivation of a symbol, its meaning may quite readily be recalled, even if, at first sight, it does not appear to be motivated. In the question where they were asked to suggest symbols for concepts, these subjects also tended to select symbols that had an obvious link with the concept.

Results from both these studies [5], [21] indicated that motivation of symbols (where there is a strong link between the symbol and the object it represents) supports accurate recall of the meanings of the symbols, and thus contributes to intelligibility.

The second property that has a significant influence on intelligibility is the extent to which the language allows exploitation of human visual perception (see Section 4, property (v)). The question of how to exploit human visual perception is one that has long concerned designers of representations in all fields. As long ago as 1935, Jan Tschichold noted that ".... readers want what is important to be clearly laid out. They will not read anything that is troublesome to read, but are pleased with what looks clear and well arranged, for it will make their task of understanding easier", (cited in [38]). More recently, the work of many cognitive psychologists has shown that if a language provides the means for direct perceptual recognition of important elements in a representation, then it follows that the representation will be easier for readers to understand (see, for example, [22], [33], [39]). The related property identified in this research is based on the work of authors such as these and is itself a definition of the level of intelligibility of a language.

Both Larkin and Simon [22] and Winn [39] have identified the importance of spatial layout in facilitating the search for relevant information within a representation. For example, an arrangement of individual symbols clustered together can suggest to the reader that elements represented by the symbols are closely related. Winn has also identified the prominence of symbols within a representation as being an important determinant of how easily meaning can be extracted. He suggests that the visual distinctiveness of symbols determines the order in which readers process components of a representation, their attention being drawn automatically to the most prominent symbols first [39].

Most languages used to specify software contain both perceptual and non-perceptual elements. Diagrammatic languages, such as data flow diagrams, are perceptual, although they contain non-perceptual elements, such as labels for the various components. Representations in languages which are not perceptual in themselves are frequently laid out to show certain aspects perceptually, such as the use of schema boxes in Z specifications.

The third property that contributes to the intelligibility of representations is the amount of structure in a language. The importance of structure in relation to intelligibility of representations has been recognised by several authors in cognitive psychology, such as [22], [34]. Research into the effect of structure has also been carried out by means of an experiment, which is reported in [2], [4]. Findings from the experiment indicate that a change of structure can have a significant impact on the intelligibility of a specification. These findings support the view expressed in [5] that a clear, easily visible structure is an essential component of an intelligible specification, and that the ability to support such a structure is an important property of a specification language.

6   Evaluating the Intelligibility of Diagrammatic Languages

The claim has often been made that 'pictures' are somehow better than 'words' at capturing more 'naturally' what we know about many aspects of the world (see, for example, [22]). We may therefore be led to believe that diagrammatic languages used in software specification must somehow be 'better' than the textual languages. However, this would be too simplistic a view, as the reasons for the supposed superiority of diagrammatic representations are not well-understood [33]. We found, in the research described here, that three properties of specification languages (motivation, 'perceptualness' and structure) have a significant influence on the intelligibility of representations produced using the languages. Diagrammatic languages are frequently used to specify software, since they are assumed to be easier to understand for untrained users. It looks, however, as if many of the widely used diagrammatic languages do not possess two of the relevant properties to any real degree. This means that representations produced using these languages are likely to be much harder to understand than is generally thought.

In the specification of software, the abstract nature of the concepts involved makes it extremely difficult for a developer to exploit motivation in a diagrammatic representation of requirements. Representations used in software development often aim to show, for example, the working of a process or the structure of information. Neither of these correspond to obvious visible objects in the world around us; they are therefore what Larkin and Simon [22] have termed 'artificial diagrams'. Disciplines where the artefact being modelled exists in physical form, such as a bridge or a building in civil engineering, are more likely than software development to have languages with motivated symbols, such as those used in the drawings of an architect or structural engineer. The problem in software systems development is that normally the effects and interface of the artefact are perceived rather than the thing itself. This makes it hard to provide motivated symbols for use in representations: a shape can represent a room and a curve a bridge, but what represents a process or an interaction?

The ability to embody perceptual concepts, such as connectedness (as used in data flow diagrams) and inclusiveness (as used in state diagrams), is often seen as an advantage of diagrammatic languages. However, without elaborate 'extras' the amount of information diagrams can contain is restricted. A further problem with diagrammatic languages is identified in [18]. Functionally-related items may be placed close together in order to emphasise the relationship (this is referred to as 'clustering'); however, this can lead to diagrams which are cluttered and hard to understand. There is also a risk that untrained users may see any adjacent items in a diagram as related, even when this is not the case.

As we have already mentioned above, representations in software specification are frequently laid out to highlight certain aspects perceptually. The main typographic techniques for exploiting the human visual system are summarised in [38] as contrast, repetition, alignment and proximity. One way of measuring the potential of a language to produce perceptual representations is to check whether the language allows each of these typographic techniques to be used in representations. If we consider Figure 3, below, we can examine the extent to which the language of data flow diagrams provides potential for use of the four techniques.

Contrast means that elements of a representation that are the same should be represented in exactly the same way; if they are not the same, then they should be clearly different. The data flow language does not encourage use of contrast: readers who are new to the language may confuse the symbols for process, external entities and data stores, and find it difficult to distinguish between names of data flows, such as 'New_Stock_level ' and 'Updated_Stock_level'.

Repetition means repeating some aspect of the design throughout the representation to give the impression of coherence and organisation. In a set of data flow diagrams this can be achieved by labelling each level in exactly the same way. In individual diagrams, the same font should be repeated for names of the same type of elements, such as processes or data stores. This can cause a problem, however, if one element has a particularly long name and the symbol needs to be made larger, since it may wrongly appear to be particularly important. We can see an example of this in Figure 3; in order to fit its label, the data store PRODUCT_INFORMATION should be drawn larger, but this would give the false impression that it is more significant than the data store ORDERS.

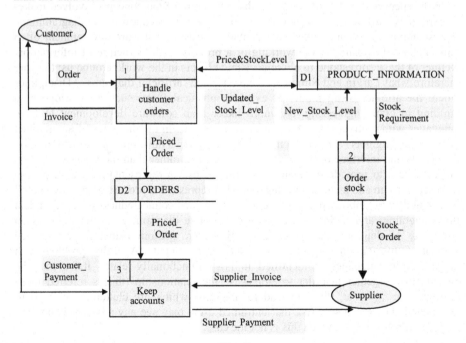

Fig. 3.

The technique of alignment aims to ensure that nothing is placed randomly in a representation. Each element should, where possible, have a strong visual connection with something else in the representation and the overall structure should appear to be a cohesive unit. This is not a problem in the example in Figure 3, but in complex data flow diagrams space is often at a premium. Components of the representation are placed where there is room, rather than in a position which emphasises their relationship to other elements in the representation.

Proximity means that elements of a representation that are related to each other should be grouped close together, as mentioned above. Again, in data flow diagrams, lack of space means that this clustering is often impossible. A further problem arises from the fact that external entities are only shown on the top level diagram; this means that, even if an external entity is very closely related to a lower level process, they do not appear on the same diagram. If the Order stock process in Figure 3 is decomposed at a lower level, we lose the strong visual connection between this process and the Supplier.

Data flow diagrams are so widely considered as easy to understand that they are still the first specification language taught to students on many system development courses. From this brief assessment of the data flow language in the light of the four techniques for making representations more perceptual we can see that, in fact, data flow gives the developer very little freedom to exploit human visual perception.

7   Conclusion: The Importance of Structure

In our previous paper on properties of software languages [5], we suggested that it is important for a specification language to provide a clear, easily visible structure for representations, since, in the absence of a given structure, readers will have to waste time and effort in constructing one for themselves. We argued that any structure is better than none, since it helps the reader to 'chunk' the information presented. Over and above this, a structure will be particularly effective if it in some way reflects the structure of information in the domain, or if it provides the reader with useful abstractions to reason with [37]. Green and Blackwell [17] found that readers' understanding of programs depended more on the structure of the information in the program than on whether the program was written in a visual or text-based language. Green [14] also makes the point that the structure is only useful if it is clearly visible.

Many of the diagrammatic languages used to specify software have a rigorous and highly visible structure. In data flow diagrams, for example, processes can be decomposed hierarchically, affording views of the system at different levels. Abstraction is supported by separation of details about individual elements of the diagram, through data dictionary and process descriptions. Users can see a data store, data flow or process in the context of the system, and then, independently, study the attributes that comprise it. State diagrams, class diagrams and interaction diagrams are further examples of specification languages that produce well-structured representations.

In this paper we describe properties of specification languages and identify three properties, in particular, that influence the intelligibility of representations. We have shown that diagrammatic specification languages are weak in terms of two of the properties (motivation and the exploitation of human visual perception), but strong in terms of the third property, the amount of structure inherent in the language. Unfortunately, however, the property of structure is frequently not exploited to the full. Too often representations are produced which sacrifice a clear, highly visible structure to saturating the representation with information.   We suggest that practitioners and teachers of software specification should place more emphasis on the structure of diagrams, if representations of the system are to be produced that are readily intelligible to everyone who needs to look at them.

References

[1]   Booch, G., Jacobson, I. and Rumbaugh, J. (1999) *The Unified Modeling Language User Guide,* Addison-Wesley.
[2]   Britton, C., Loomes, M. and Mitchell, R. (1993) Formal Specifications as Constructive Diagrams, *Microprocessing and Microprogramming*, Vol.37, pp. 175-178.
[3]   Britton, C. and Jones, S. (1997) Which properties make a modelling notation easy for untrained users to understand? *Proceedings of the International Workshop on Representations,* Queen Mary and Westfield College, London University, pp. 2-10. Available from Department of Computer Science, Queen Mary and Westfield College, London University.
[4]   Britton, C., Jones, S. & Lam, W. (1998). Separating the system interface from its internal state: an alternative structure for Z specifications. *Proceedings of Formal Aspects of the Human Computer Interaction, BCS-FACS Workshop,* 87-102, Sheffield Hallam University .
[5]   Britton, C. and Jones, S. (1999) The Untrained Eye: How Languages for Software Specification Support Understanding in Untrained Users, *Human Computer Interaction,* 14, pp. 191-244.
[6]   Brun, P. and Beaudouin-Lafon, M. (1995) A taxonomy and evaluation of formalisms for the specification of interactive systems, in M. Kirby, A. Dix and J. Finlay (Eds.), *People and Computers X, Proceedings of HCI'95, 197-*212, Cambridge University Press.
[7]   Cox, R. & Brna, P. (1993). Analytical reasoning with external representations. *Proceedings of the AI-ED 93 Workshop on Graphical Representations, Reasoning and Communication.* Edinburgh.
[8]   Davis, A. (1988) A Comparison of Techniques for the Specification of External System Behavior, *Communications of the ACM, 31 (9),* 1098-1115.
[9]   Davis, A. (1993) *Software Requirements: Objects, Functions and States,* Prentice Hall International.
[10] Eysenck, M. & Keane, M. (1990). *Cognitive psychology: A student's handbook..* Lawrence Erlbaum Associates.

[11] Farbey, B. (1993) Software Quality Metrics: Considerations about Requirements and Requirement Specifications, in R. Thayer and A. McGetterick (Eds.), *Software Engineering: a European Perspective,* IEEE Computer Society Press, pp.138-142.

[12] Fertuck, L. (1992) *Systems Analysis and Design,* Wm. C. Brown Publishers.

[13] Green, T. (1980) Programming as a Cognitive Activity, in H. Smith and T. Green (Eds.), *Human Interaction with Computers,* Academic Press.

[14] Green, T. (1983) Learning Big and Little Programming Languages, in A. Wilkinson (Ed.), *Classroom Computers and Cognitive Science,* Academic Press, New York.

[15] Green, T. (1989) Cognitive Dimensions of Notations, in A. Sutcliffe and L. Macaulay (Eds.), *People and Computers V, Proceedings of HCI'89,* Cambridge University Press.

[16] Green, T. (1991) Describing information artefacts with cognitive dimensions and structure maps, in D. Diaper, and N. Hammond (Eds.), *People and Computers VI, Proceedings of HCI'91,* Cambridge University Press.

[17] Green, T. and Blackwell, A. (1996) Thinking about visual programs, in *Thinking with diagrams,* IEE Colloquium Digest No: 96/010, Institute for Electronic Engineers, London.

[18] Green, T., Petre, M. and Bellamy, R. (1991) Comprehensibility of visual and textual programs: A test of superlativism against the "Match Mismatch" conjecture, in J. Koenemann-Belliveau, T. Moher and S. Robertson (Eds.), *Empirical studies of programmers,* 121-146, Norwood NJ, Ablex.

[19] Haywood, E. and Dart, P. (1996) Analysis of Software System Requirements Models, in *Proceedings of Australian Software Engineering Conference,* 131-138, IEEE Computer Society Press.

[20] Johnson, C., McCarthy, J. and Wright, P. (1995) Using a formal language to support natural language in accident reports, *Ergonomics, 38 (6).*

[21] Kutar, M., Britton, C. and Jones, S. (1998) A Graphical Representation for Communicating Sequential Processes, Proceedings *of Formal Aspects of the Human Computer Interaction, BCS-FACS Workshop,* Sheffield Hallam University, pp. 145-162.

[22] Larkin, J. and Simon, H. (1987) Why a diagram is (sometimes) worth ten thousand words, *Cognitive Science, 11,* 65-99.

[23] Mackinlay, J. (1986) Automating the design of graphical presentations of relational information, *ACM Transactions on Graphics, 5 (2),* 110-141.

[24] Meyer, B. (1985) On Formalism in Specifications, *IEEE Software, 2 (1).*

[25] Modugno, F., Green T. and Myers B. (1994) Visual programming in a visual domain: A case study of cognitive dimensions, in *People and Computers IX, Proceedings of HCI'94,* Cambridge University Press.

[26] Myers M., Kaposi A., (1997) *A First Systems Book,* Kaposi Associates

[27] Patching, D. (1990) *Practical Soft Systems Analysis,* Pitman Publishing.

[28] Petre, M. (1995) Why looking isn't always seeing: Readership skills and graphical programming, *Communications of the ACM, 38 (6).*

[29] Pfleeger, S. L.(1998) *Software Engineering : Theory and Practice,* Prentice Hall.

[30] Roast, C. (1997) Formally comparing and informing notation design, in H.Thimblely, B. O'Conaill and P. Thomas (Eds.), *People and Computers XII, Proceedings of HCI'97*, Springer.

[31] Rumbaugh, J., Blaha, M., Premerlani, W., Eddy, F. and Lorensen, W. (1991) *Object-Oriented Modeling and Design*, Prentice Hall.

[32] Sampson, G. (1985) *Writing systems*, Hutchinson.

[33] Scaife, M. and Rogers, Y. (1996) External cognition: How do graphical representations work? *International Journal of Human-Computer Studies, 45*, 185-213.

[34] Sengler, H. (1983) A model of program understanding, in T. Green, S. Payne. and G. van der Veer, (Eds.), *The Psychology of Computer Use*, Academic Press.

[35] Sommerville, I. (1995) *Software Engineering* (5th edn), Addison Wesley.

[36] Sommerville, I. and Sawyer, P. (1997) *Requirements engineering: A good practice guide*, Wiley.

[37] Stenning, K. and Oberlander, J. (1995) A cognitive theory of graphical and linguistic reasoning: Logic and implementation, *Cognitive Science, 19*, 97-140.

[38] Williams, R. (1994) *The Non-Designer's Design Book*, Peachpit Press.

[39] Winn, W. (1993) An account of how readers search for information in diagrams, *Contemporary Educational Psychology, 18*.

Executing Diagram Sequences

Joseph Thurbon

Artificial Intelligence Group
School of Computer Science and Engineering.
University of New South Wales
Sydney, Australia
joet@cse.unsw.edu.au

Abstract: We present a general framework for using diagram sequences as plan specifications. We also present an implemented system based on that framework, which generates imperative programs from diagram sequences similar to those used in teaching programming. The specific notations we use in the system are based closely on the diagrams typically used for teaching introductory programming, but the framework is general enough to account for and express many uses of diagram sequences. The system and the underlying theory highlight some areas where planning, reasoning about action, the refinement calculus and diagrammatic reasoning are synergistic. For example, by framing the definition of algorithms as a type of plan specification, it becomes clear that decomposition of a planning problem into sub-plans is analogous to refinement in the software engineering sense. More importantly, the system gives insight into the underlying structure of the largely informal use of diagrams that is routinely found in the explanation of algorithms. Obvious applications include teaching (since the inspiration for the system is a common method for teaching) and software engineering, where diagrams are often used to specify type systems rigorously (e.g. class diagrams), but specify program dynamics informally.

1   Introduction

While single diagrams can convey large amounts of information compactly, there are many interesting domains where sequences of diagrams are used in every day life. It is desirable to account for these uses in a theory of diagrams[1]. For example, each Lego™ kit is accompanied by instructions that are almost free from text, yet are easily comprehensible to the youngest of children. Programmers routinely appeal to diagrams not only to express the static elements of software systems (eg. class diagrams in UML), but also the dynamics of the topological structure of programs as they execute. When teaching programming to novices, operations on trees, linked lists

M. Anderson, P. Cheng, and V. Haarslev (Eds.): Diagrams 2000, LNAI 1889, pp. 392-406, 2000.

and other data structures are almost uniformly introduced pictorially. In all these situations, sequences of diagrams are used to succinctly and exactly prescribe actions that have quite complex specifications. This paper presents a framework in which these and other uses of diagram sequences can be explained, and to a certain extent mimicked[1]. In this regard, we follow in the spirit of Foo [5], which gives a formal explanation of existing uses of single diagrams in the specification of actions which have local extent. We also present a visual programming environment, built upon the framework, which produces working code from sequences of diagrams similar to those found in text books (e.g. [8]) describing basic abstract data type (ADT) operations. It also highlights the relationships between diagrammatic reasoning, reasoning about action, planning and the refinement calculus.

Throughout the paper, we refer to a simple example, SWAP, which was chosen because it is one of the canonical examples found in the refinement calculus literature, it is simple and familiar enough to be explained quickly, and is complex enough to motivate our approach.

Overview

Informally, our approach is to view sequences of diagrams as plan specifications. Each diagram in a sequence represents constraints on, or properties of, a state in some related domain. Given a sequence of diagrams $<D_1,...,D_n>$ we generate plans such that, given a state satisfying the constraints of (having the properties represented in) D_1, executing the plan beginning in that state will result in a state satisfying the constraints of D_n. Further, during the execution, the plan will have passed through, in order, states satisfying the constraints specified by D_2 through to D_{n-1} (although it may pass through others as well). These intermediate diagrams can be viewed as hints to, or as constraints on, the plan generator. For example, one views the intermediate diagrams in Lego™ instructions as hints, but the intermediate states in an algorithm definition as constraints[2]. While the first interpretation of these intermediate diagrams is fairly obvious, the second is perhaps more interesting, and is the interpretation we use for the rest of this paper.

We introduce an algebraic method to express relationships between diagrams and states. We then show how sequences of diagrams can be gradually refined into a form where they can be easily translated into programs that are correct implementations of the original sequence. Finally, we add some notation for dealing with control constructs so that plans for non-linear sequences of diagrams can be constructed, and programs generated for them.

In this paper, we focus primarily on a programming system we have developed, rather than a detailed discussion of the theory underlying the system. Enough detail is

---

[1] While this is one framework that explains the use of these sequences, we do not make any assertions about the cognitive status of this framework. That said, the process by which our system operates bears many similarities with the processes we have observed when teaching first year programming

[2] That is, a program that sorted an array but did not shuffle the elements in a pair-wise fashion would not be considered to be an implementation of Bubble-Sort, yet clearly there are many implementations of Bubble-Sort. The intermediate states play a crucial role in characterising a particular algorithm.

provided for this paper to stand alone, however we defer a complete exposition of the theory, including a more detailed discussion of the motivations behind it to [13].

Motivating Example

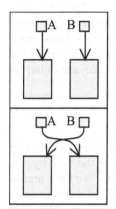

Figure 1: Swap

When teaching algorithms to students, or explaining programs to experienced programmers, one frequently employs the following pattern of explanation. The tutor or co-worker draws an initial diagram detailing the structure of the preconditions of the algorithm, and explains what the diagram means. Then, through repeated rubbings-out, re-drawing and additions to the diagram, some new diagram is reached. This new diagram may be the desired result of the algorithm, or some important sub-goal of it, followed by more erasing, drawing and so on. In the classroom situation, the teacher would then explain how one might capture the essence of what was expressed in the diagrams with imperative code. We will call this informal process *drawing a program*. Figure 1 is the sort of diagram sequence that might typically be used to express the notion of swapping the contents of two variables.

After the sequence has been drawn, it is translated into (usually imperative) code, with explanations of how to reach each successive diagram from the last. Implicit in this task is the assumption that each modification that was made to the diagrams corresponds to some modification one can make to the code state.

Despite the apparent ubiquity of these pictures, and the explanation process, to our knowledge there has been little attempt to capture this process in such a way that it may be rigorous rather than informal, while maintaining the obvious intuitive power of the diagrams that are used. Of course, there are many new problems that need to be addressed in the translation stage: memory management, ensuring pointers do not go astray, keeping references to objects that will be required later in the program, and so on. There are also diagrammatic issues that should be resolved. For example, in the diagrams above, how do we know that the arrows have changed the objects that they point to, rather than the objects having moved? What changes in the diagram are significant? How do the diagrams relate to a specific problem domain?

Our overall goal, then, is to provide answers to the above questions, create a framework in which the translation step is automated, and isolate those elements of this task that are general techniques suitable for diagrammatic reasoning and those that are domain specific.

2   Background

Since this paper draws on ideas from software specification, the refinement calculus and visual programming, we will briefly discuss some of the relevant results from these areas. This section is not intended to be a complete introduction to any of the fields that we discuss, but should provide sufficient context for the reader to place our work.

Software Specification

An algorithm is usually defined along the lines of "a step by step procedure for taking any instance of [a] problem and producing a correct answer for that instance" [8]. The 'problem' is usually expressed in terms of logical pre- and post-conditions, but greater emphasis is often on the steps that are taken. One may define an algorithm more formally as the Hoare triple [7] $\{P\}S\{Q\}$ where P is a precondition, Q a post-condition and S is either a single action/statement or a sequence of 'algorithms', each with their own pre- and post-conditions. Since each sub-algorithm in S modifies the state of the entire program, we may more fully characterise the algorithm as $\{P\}<S>\{Q\}$ where

$$S = \{I_0\} \qquad \{I_1\} \; \ddot{O} \; \ddot{O} \; \ddot{O} \; \{I_{n-1}\} \qquad \{I_n\} \qquad (1)$$
$$\{P_1\}S_1\{Q_1\} \quad \ddot{O} \; \ddot{O} \; \ddot{O} \quad \{P_n\}S_n\{Q_n\}$$

Here, $P = I_0$ and I_n entails Q. Each step S_i of the algorithm is listed explicitly along with its own pre- and post- conditions, and corresponds to a 'step' in the 'step by step procedure'. Each I_i represents an intermediate state in the algorithm. We have argued elsewhere ([13], [14]) that these steps are as much a complete and valid definition of an algorithm as specifying the steps. There is a simple reason to view algorithm definitions as specifications of states to be visited, rather than as specifications of steps to be performed. If a sequence of diagrams can be found such that each diagram corresponds to a state (I) in the specification, then that sequence is a valid representation of that algorithm. By contrast, the step specification of an algorithm has no such natural diagrammatic presentation. This representation is more than a guide or a clue to a programmer - it is a representation which contains exactly the same amount of information as the text specification. We believe that this is exactly the kind of process which occurs during the drawing of a program.

Refinement Calculus

The refinement calculus is another method of software specification, comprising both a system for formal specification of behaviour and a process to translate those specifications to code. One can describe the process of refinement as the gradual behaviour preserving modification of a program[3] from pure specification to pure code. For a full treatment and explanation, see [10]. The process begins with a program specified in the form w: [pre, post] which can be read 'If the initial state is described by the predicate pre, then, by changing only the variables listed in the frame w, establish some final state satisfying the predicate post.' A program is said to be a correct implementation of a specification if it satisfies this criteria. This is analogous to the specification by a Hoare triple $\{P\}S\{Q\}$ with the S being omitted, but constrained by the frame. It is especially similar to our rendering of an algorithm where a sequence of states rather than a sequence of actions are paramount. Specification statements are delimited by square brackets, whereas code is written unadorned, and segments of a program are separated be semi-colons. The overall goal of refinement is to transform all specification statements into code. If one program p_2

---

[3] In the refinement calculus, the notion of a program is a statement of behaviour which may consist of both specification and code components.

refines another p_1, it is written $p_1 ! p_2$. The intuitive meaning of the refinement operator (!) is 'if $p_1 ! p_2$ then any program that is a correct implementation of p_2 will be a correct implementation of p_1.'

Revisiting the SWAP example from above, *initial program* is a program that is a specification of the behaviour of swap. Through a series of refinements (we list the actual refinement rules in italics), one can eventually reach the last (pure code) program, which is a refinement of the initial program. The parts of the program that have changed are underlined for clarity, but are not a part of the usual refinement calculus notation

initial program   (2)
x, y: [x=X ∧ y = Y, x=Y ∧ y = X]   ! *(introduce local block)*
x, y, t: [x=X ∧ y = Y, x=Y ∧ y = X]   ! *(following assignment)*
x, y, t: [x=X ∧ y = Y, t=Y ∧ y = X]; x := t   ! *(following assignment)*
x, y, t: [x=X ∧ y = Y, t=Y ∧ x = X]; y := x; x := t ! *(assignment)*
t := y; y := x; x := t
pure-code program

We will use the notion of refinement to show the correctness of the implementations that we create from sequences of diagrams. We will show that when we view diagram sequences as program specifications, inserting a diagram into a sequence S_1, resulting in a new sequence S_2, always yields a situation where $S_1 ! S_2$.

Visual Programming
There have been numerous visual programming languages (VPL) built over the years. They all seem to diverge from the style of diagrammatic programming that might be implied by the examples given above. Instead they are usually more closely related to flow-charts [12] or spreadsheets [3]. Other approaches to visual programming include Pictorial Janus[6], which has a very declarative nature, with semantics based on description logic. Work by Erwig and Meyer[4], a heterogenous (mixed modal) visual language framework (HVLF), is general enough to incorporate visual segments into existing imperative, functional and declarative languages. While it seems extremely natural to use their HVLF for languages that have declarative semantics (one of the great strengths of diagrams) the corresponding incorporation of diagrams into imperative code is somewhat less persuasive. What is persuasive is their argument for a heterogenous system. In particular, they argue for diagrams to be used to express local information, and text to express non-local and global information. This complements results by Foo [5], which indicate how to determine which types of information and actions are local in nature.

None of the above approaches utilise the same approach as ours: that of letting the diagrams in a sequence implicitly specify the code which would correctly implement that specification. A significant side effect of our research into diagrammatic reasoning has been the VPL we present herein, which at least partially fills this void. To stay true to our aims, one of the major constraints we place on our work is that we maintain a close and obvious relationship with the conventions that already exist in teaching imperative programming.

In summary, a sequence of diagrams can be considered a specification of an algorithm, in the traditional sense. We will provide a method of implementing those diagrammatic specifications correctly by refining a given sequence into one that has a straightforward translation to its implementation. From results in the refinement calculus, we know that this translation is also an implementation of the original sequence. Finally, the system that we have developed based on this technique is a VPL that is unlike any other of which we are aware.

3   Technical Framework

This section is a condensed version of several sections from [13]. It is provided for completeness. In this section we define the structure of a diagram algebraically, and provide semantics for single diagrams and diagram sequences. We then show how to derive programs that implement the sequences.

Single Diagram Specification

A diagram $D_n = $ <$Instances_n$, $Names_n$, d_n, $null_n$, $unknown_n$, $Properties$ > where

- $Instances_n$ is a finite set of constants
- $Names_n$ is a finite set of strings
- d_n, $null_n$, and $unknown_n$ are constants that represents the diagram as a whole, the null value, and that a value is unknown, respectively

For convenience, we define $Values(D_n) = Instances_n$ " $\{null_n, unknown_n\}$. These correspond to the values that properties can have. In Figure 2, we represent null by a black box, and unknown by an uncoloured box.

- $Properties \# \{Instances$ " $d\}$ x $Names$ x $Values(D)$. These correspond closely to the evaluations for each variable reachable within the lexical closure[4] of a program. Each tuple <x, n, v> is read "x has a property called n with a value v". We say the property adorns x. We sometimes say n adorns x, since no two properties adorning the one instance (or diagram) may have the same name.

In Figure 2, the labels i_1 and i_2 are the constants we use to denote the instances. We place them explicitly in the diagram here for exposition only: they would not usually be a part of the diagram's pictorial representation, only the algebraic one. The example diagram could be fully specified as D = <$Instances$, $Names$, d, $null$, $unknown$, $Properties$> where

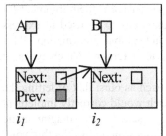

Figure 2

1. $Instances = \{i_1, i_2\}$
2. $Names = \{$ìAî, ìBî, ìNextî, ìPrevî$\}$
3. $Properties = \{$<d, ìAî, i_1>, <d, ìBî, i_2>, <i_1, ìNextî, i_2>,   <i_1, ìPrevî, $null$>, <i_2, ìNextî, $unknown$>$\}$

Since d, $null$, and $unknown$ are constructed in the same way for every diagram, we will simply specify a diagram by as <$Instances_n$, $Names_n$, $Properties_n$>.

---

[4] The lexical closure of a program (used in languages such as Smalltalk and lisp) is essentially the set of variables in scope at a given point, and their values

Diagram Sequences

A diagram sequence $<<d_1,...,d_n>, !>$, is a linear sequence of diagrams, plus a counterpart structure[5]. The counterpart structure can be defined as a partial one to one function $!$ from the first diagram to the second diagram in any pair in the sequence. Considering the initial example of **SWAP** in Figure 1, the counterpart structure is required to indicate that the values of A and B have changed, rather than that the icons representing the values have moved. The function is partial, since elements may be removed from, or added to, a diagram over the length of sequence. In Figure 1, call the first diagram d_1 and the second d_2. Call the instances in the diagrams (subscripted appropriately) *left* and *right*. We can then specify the diagrams as

$$<\{left_1, right_1\}, \{``A", ``B"\}, \{<d_1, ``A", left_1>,<d_1, ``B", right_1>\}> \qquad (3)$$
$$<\{left_2, right_2\}, \{``A", ``B"\}, \{<d_2{}', ``A", right_2>,<d_2, ``B", left_2>\}>$$

The counterpart structure simply maps each constant to another by effectively changing the subscript. So for example $! (right_1) = right_2$. This counterpart structure answers the first question raised in the motivation: it is how we can tell that it was the references that changed in Figure 1, rather than the locations of the icons on the diagram. While we assume that a counterpart structure is provided as a part of the specification, it may be possible to infer the counterpart structures given some domain dependant rules for analysing the diagrams. We do not address this problem in this paper.

Code State Specification

This section defines a simple mechanism for interpreting program states of a running program algebraically. It defines exactly what information is present in the 'data segment' of the program that is executing, after a certain section of that program has been executed. Not surprisingly, not only the language in which we are programming, but also the current lexical scope of the program, define the structure of our program states.

For consistency, we will refer to composite reference types as *classes*, and instances of them as *objects*, regardless of the language which we are discussing; apart from a few implementation details (e.g. pointer addition), the semantics of C pointers and Java references are the same.

A code state $C = <Objects, scope, null, unknown, Labels, Evaluations>$ where
- *Objects* is the set of objects created during the execution of the program. This set will grow and shrink as objects are created and destroyed.
- *null, scope,* and *unknown,* correspond to the value null, the current lexical scope of the program (which is similar to *this* in Java), and a place holder for variables that have not been initialised respectively.

---

[5] We follow Barwise[1] in the use of counterparts. A counterpart structure identifies which icons on different diagrams represent the same 'real-world' object. For example, in Figure 1, $left_1$ and $left_2$ represent the same object during a program's execution. This information cannot always be inferred, so must be represented explicitly. Our situation is somewhat more complex than in [1], since we deal with a domain that necessarily changes over time, whereas the domain of Hyperproof is static.

- *Labels* corresponds to the names of the variables defined in the system. One label is defined for each variable in the lexical closure of the current code state, and one label is defined for each member variable of each composite type in the language.
- *Evaluations* # *Objects*" {*scope*} x *Labels* x *Objects*. Each member of *Evaluations* <*a*, *b*, *c*> can be read as the assertion *a.b* = *c*

Since *scope, null* and *unknown* are the same for every code state, we refer to code states as <*Objects, Labels, Evaluations*>.

Program States As Models Of Diagrams

We are now able to define the conditions under which we can say whether a code state satisfies the constraints defined by a diagram. We introduce a function Π from diagrams to code states, which maps names to labels and instances to objects. We call a code state C a model of a diagram D, written C$D under the following criteria:

> Given a diagram D_n = <*Instances$_n$, Names$_n$, Properties$_n$*> and a code $\quad$ (4)
> state C = <*Objects, Labels, Evaluations*> we say that CD_n$ if there
> exists a one-to-one function ! that satisfies the following constraints:
> 1. ! (d_n) = *scope*;
> 2. ! ($null_n$) = *null*;
> 3. for an instance *i*, if ! (*i*) is not *null*, then for all other instances
>   *j* ! (*i*) does not equal ! (*j*); and
> 4. For every *c* % *Instances$_n$* " {*d*},
>   a.   for every <*c, n, v*>% *Properties* either
>   i. < ! (*c*), ! (*n*), ! (*v*)> % *Evaluations* or
>   ii. *v* is unknown

If we wish to specify a particular function ! under which C is a model of D, we write C$$_!$D and say C is a !-model of D. The third condition means no object can be represented by two instances. This relationship answers the second and third question asked in the motivation section. In answer to the second question raised in the motivation section: a change in a diagram is significant if it changes the states that are models of the diagram. ! is the only relationship our framework requires between diagrams and the problem domain. Thus, although our notations are specific to programming, the framework is easy to generalise [13].

Diagram Sequences and State Traces

We call a pair of code states defined using the same language a *state pair*. We call a sequence of code states defined using the same language a *state trace*. Given a state pair <*C$_1$, C$_2$*>, and a diagram sequence of length two, <<*D$_1$, D$_2$*>, ! >we say that the state pair are a model of the diagram pair if there exist functions ! and " such that $C_1$$$_!$$D_1$ and $C_2$$$_"$$D_2$ and further, for all elements α in D_1 for which ! is defined, ! (α) = " (! (α)). The second condition ensures that ! and " map counterparts to the same 'real-world' object, and is read as ! and " respect the counterpart structure.

Similarly, we say a state trace $<s_1,...,s_n>$ is a model of a diagram sequence $<<d_1,...,d_n>, !>$ if for every $i\%\{1..n\}$ s_i is a $!_i$-model of d_i and for $i\in\{1..n-1\}$ $!_i$ and $!_{i+1}$ respect $!$. We write $<s_1,...,s_n> \$ <<d_1,...,d_n>, !>$.

Code Actions, Diagrammatic Actions

A *diagram action* is an action that has a precondition and effect that is specified diagrammatically. We denote the diagram obtained by the application of a diagram action α on a diagram D as $\alpha(D)$. A *code action* is a program statement that has a precondition and an effect specified in terms of code states. We denote the code state obtained by the application of a code action a to a code state C as $a(C)$. A diagram/code action pair $<\alpha, a>$ is called *homomorphic* if they satisfy the condition that for all diagrams D and all code states C such that $C\$D$

1. If D satisfies the preconditions of α then C satisfies the preconditions for a, and
2. $a(C)\$\alpha(D)$

That is, if α is applicable to a diagram, then the corresponding action a is applicable to all models of that diagram. Further, the new code state obtained by applying a to the C must be a model of the diagram obtained by applying α to D.

Programs as Implementations of Diagram Sequences

A *program* is a sequence of code actions. Executing a program p on a code state s results in a state trace, denoted $p(s)$. Given two state traces t_1 and t_2 we call t_1 a *sub-trace* of t_2 if t_1 is simply t_2 with some states other than the first and last omitted. That is, they begin and end in the same state, but t_1 omits some of the states in t_2.

A program p is an implementation of a diagram sequence $D = <d_1,...,d_n>$ if for any code state s such that $s \$ d_1$, there exists a sub-trace of $p(s)$, *Sub*, such that $Sub\$D$. Informally, the program yields a state trace such that there is a state that is a model of every diagram in the sequence, but there may be additional states.

The next result follows directly from the definitions given above, where $!$ is the refinement operator

$$<d_1,...,d_j, d_b...,d_n> ! <d_1,...,d_j, \underline{d_k}, d_b...,d_n> \qquad (5)$$

That is, inserting a diagram into a sequence results in a new sequence that refines the first.

Planning Correct Programs

To implement a diagram pair $<D_1, D_2>$, one finds a sequence of actions $<\alpha_1,...,\alpha_n>$ that transforms D_1 into D_2 i.e. $\alpha_n(...(\alpha_1(D_1))...) = D_2$. From repeated application of Equation 5, $<D_1, D_2> ! <D_1, \alpha_1(D_1), \alpha_2(\alpha_1(D_1)),...,D_2>$. During this search, we restrict the diagram actions that we use to those with homomorphic code actions. It is easy to show that that sequence of homomorphic actions, say, $<a_1,...,a_n>$ is an implementation of the derived sequence. We know the derived sequence refines $<D_1, D_2>$, therefore $<a_1,...,a_n>$ is an implementation of $<D_1, D_2>$.

The general case of planning programs for arbitrary length sequences of diagrams is handled simply by taking each pair of adjacent diagrams in the sequence and planning programs for them. Again, it is easy to show that the concatenation of each of these plans is an implementation of the whole sequence.

4   The Implemented System

This section presents the system we have implemented based on the framework above. We build on the framework to include control constructs such as iteration and selection. It is interesting that once single linear sequences of diagrams can be translated into source code, we have at our disposal all of the tools required to implement iteration, selection, and function calls to existing code. This last feature provides interoperability with existing Java classes.

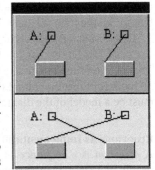

Figure 3

For the remainder of this paper, we will refer to the basic diagrammatic unit requiring translation as a *D-function*. We call the output of the translation process a *C-function*. Each C-function comprises a single Java class. To avoid confusion, we will refer to 'member functions' of a Java class as methods.

Figure 3 shows a simple D-function as it presented to the user. The first (*initial*) diagram in the sequence is adorned by two properties; these are referred to as *D-parameters*. The second diagram is called the *target*.

Code States

As mentioned above, the structure of a code state depends on exactly what type of code one it attempting to build a state of. This in turn has implications generating code to implement D-functions. To simplify these matters, we have defined a Java class called CodeState that has several purposes.

The first is to maintain a mapping between names in a diagram and labels in a code state. Properties adorning a diagram (the D-parameters) are mapped from left to right to the parameters of the C-function. Properties adorning instances are mapped so that their names are mapped to labels that have the same string value. That is, if an instance is adorned by a property named "next", it must be mapped to an object with a data member called "next". This fixes the Π under which we check if the CodeState a Π-model of the initial diagram. An alternative implementation that searched for all Πs was considered, but the resulting system had all of the difficulties associated with Prolog's backtracking, without any of the benefits. Additionally, when one draws a program, one usually has a single intended interpretation of that program. We believe that fixing the ! under which the CodeState operates helps capture that intended interpretation unambiguously.

The second use for a CodeState is that it allows us to simulate pass-by-reference in Java. This is particularly useful for dealing with Function where more than one Property on a Diagram changes value (c.f. SWAP) so more than a single return value is required.

Finally, through use of the Java introspection API, the CodeState class simulates dynamic typing in Java. This allows for the easy generation of C-functions (since every parameter can have Object as its type) without compromising type safety, which is implemented at run time in a manner similar to the structural typing of Smalltalk.

Linear Sequences

To implement a simple sequence, such as Figure 3, the planner searches for a sequence of actions that transform the initial diagram into the target diagram. The user can inspect the plan found, and the plan generated for Figure 3 is shown in Figure 4. From one of these plans, four pieces of code are generated.

The first piece is a translation of the actions in the plan into code; we call this a *plan*. There are four actions available to the planner in our system, namely *IntroduceVariable*, *RemoveVariable*, *Assignment* and *SetNull*, all of which except SetNull are used in the plan for Figure 3. All planning is done by breadth first search, and the search space is exponentially large. However, we

```
public static void executePair1_0(CodeState codeState) {
/*
    The following function is written to the CodeState API.Were
    we writing to a language with dynamic typing, we could use
    the following code
        Object pA0;
        pA0 = A
        A = B
        B = pA0
        Remove pA0from scope;
    however, Java is statically typed, necessitating the following
    code.
*/
    codeState.addVariable("pA0");
    codeState.assign("pA0", "A");
    codeState.assign("A", "B");
    codeState.assign("B", "pA0");
    codeState.removeVarable("pA0");

}
```

(6)

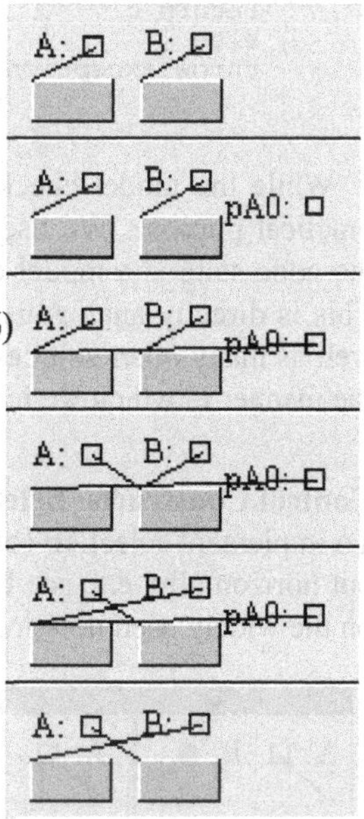

Figure 4

have been able to massively prune the search space by applying several heuristics when selecting actions at each search step. Some are very simple (for example, do not select an action that does not change the diagram, such as assigning a property to itself) but we discuss one of particular interest in a later section.

The generated plan shown above is the exact output of our system (with the exception of white-space that has been edited to make it more suitable for publication). The name of the function is generated automatically.

The second piece of code is a code state *constructor*. This is a method that takes as many parameters as there are D-parameters, and builds a CodeState with appropriate mappings from the D-parameters to the C-parameters, and from names to labels.

The third piece of code is a *model checker*. It is a boolean method based on the initial diagram in the sequence and takes a single CodeState as a parameter. It returns true if the CodeState passed it is a model of the diagram from which the model checker was generated.

The final piece of the code is essentially glue wrapping all of the code together. In the case of a single sequence, the *glue method* is shown in equation 7. With these four sections of code, control free programs can be implemented, and the code generated from Figure 3 works as expected.

```
CodeState doSwap(Object A, Object B) {                           (7)
    c = BuildCodeState(A, B) //constructor
    if (codeStateIsModel(c)) { //model checker
        executePlanFromSequenceOne(c); //plan
        return c;
    } else {
        throw Exception due to unmatchable state
    }
}
```

While the model checking code is technically redundant[6], it serves an important practical purpose. We use the model checker to differentiate between normal (when the code state is a model of the initial diagram) and abnormal execution of the plan. This is directly analogous to the design-by-contract methodology espoused in [9] as well as many other sources. Using a model checker in a Java conditional, as above, is the manner in which we incorporate control flow into our system.

Control Constructs: Selection and Iteration

To implement selection (or if/then statements), we allow multiple sequences to be laid out horizontally, e.g. see Figure 5. This layout, which we call a *split*, is based closely on the widely used notation established by in Nassi and Schneiderman in [11], and can be of arbitrary width.

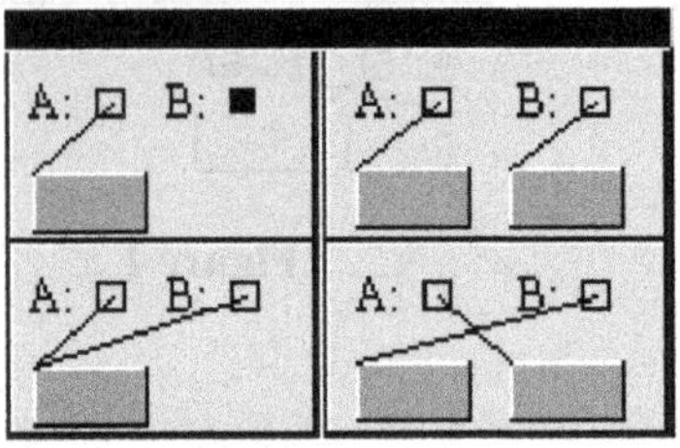

Figure 5

A model checker and plan are generated for each sequence in the split, but only a single translator and glue method are required. The glue method (as shown below in equation 8 for a split of width two) tests each of the model checkers, from left to right, and executes the first plan whose model checker returns true. Assuming that the C-function is called SwapIfNotNull, and sequences are numbered left to right, the following glue code would be generated. A wider split simply corresponds to a deeper nesting of if/else/if statements.

```
CodeState doSwapIfNotNull(Object A, Object B) {                  (8)
    c = BuildCodeState(A, B)
    if (modelCheckSequenceOne(c)) {
        executeSequenceOne(c);
        return c;
    } else if (modelCheckSequenceTwo(c)) {
```

---

[6] The theory only guarantees any particular behaviour when the code state is a model of the first diagram, and makes no guarantees otherwise

```
        executeSequenceTwo(c);
        return c;
    } else {
        throw Exception due to unmatchable state
    }
}
```

The entire framework used to generate this C-function was already in place to deal with linear sequences.

Similarly, Iteration (or looping) is dealt with by providing an appropriate glue method. Figure 6 is an example, with the corresponding glue code in Equation 9, which is a method for finding the last element in a linked list.

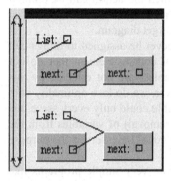

Figure 6: Find End

```
public static CodeState findEnd(A, B) {       (9)
    c = BuildCodeState(A, B)
    somethingExecuted = true
    while(somethingExecuted) {
        somethingExecuted = false;
        if (modelCheckSequenceOne(c)) {
            //List = List.next
            executeSequenceOne(c);
            somethingExecuted = true;
        }
    }
}
```

A *loop* contains a single Split. The glue method generated for a loop will iterate as long as any of the plans in the split execute normally. Again, the diagrammatic notation we use for loops is based closely on Nassi-Schneiderman diagrams. A loop will never throw an exception, since termination of a loop is a completely normal condition in programming.

Mixed Modal Programming

We have only presented a few simple examples thus far, but we have tested our system some considerably larger examples. These examples highlight the desirability of mixed modal programming. We have already alluded to the simplest form of this, which is calling existing text functions from within a D-function. Our

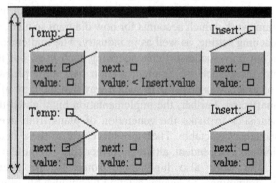

Figure 7

system also allows properties of the initial diagram to have text expressions as values, such as in Figure 7. These text expressions are translated into Java boolean expressions that the model checker evaluates at run time. The inner loop of insertion sort (which inserts a single element into a sorted list) demonstrates the technique that we use to mix text and icons in a single diagram. This loop will iterate while

Temp.next.value is less than Insert.value. That is, by the time the loop terminates, Insert will need to be inserted 'between' Temp and Temp.next.

Heuristics and the Planner

Foo defines a method by which one can describe the elements of a diagram which are required to define the pre- and post-conditions of an action[5]. This *action window* is essentially a boundary around the elements of a diagram that are needed to describe it. Using the counterpart structure, we are able to translate this window to the diagram that the planner it targeting. If the objects described by the translated window are not also local (related topologically) in the last diagram, then the action can be omitted from the plan. An example of such an action would be an assignment that assigned an instance to a property that it was not assigned to in the target diagram.

One result of such a heuristic is that variables can never be assigned to things that they will not be assigned to in the target diagram. Of course, element that do not appear in the target diagram (e.g. planner added variables) are not effected by this heuristic, so we do not lose any expressiveness. However, we do lose optimality with respect to plan length, since under this heuristic a variable could only every serve one purpose. Thus we often require the introduction (and removal) of variables from the scope of a program, when perhaps an existing variable could have served the purpose of the added variable.

On our current range of examples, which range from very simple programs like swap above, to more complex ones such as insertion sort, the above heuristic alone reduces the search space by up to two orders of magnitude.

5   Conclusion

Our main goal for this paper was to provide an implementation of our theoretical framework, which accounts for how diagram sequences are used for programming in a teaching setting, as well as in industry. The implementation shows that the framework is not only successful characterising such uses, but is also feasible to implement. In providing such the implementation, we were pleased that the addition of control constructs was possible in such a way that the core of the theory remained completely unaffected. Further, the implementation highlighted that using our approach relies on heuristics to make the generation of plans from even moderately complex diagram sequences tractable. The most successful of these heuristics was one that seems to be domain independent, although further study in these heuristics is warranted.

The work also highlighted some of the similarities between diagrammatic programming and several existing techniques of software specification. This indicates that our approach could be a useful addition to these existing techniques, as well as being interesting to the diagrammatic reasoning community.

6   Future Work

We are currently implementing a system which is the complement of this one; it takes simple code and creates from it a diagrammatic specification. Such a tool, in conjunction with this one, could make a powerful tool for students to explore the relationship between the diagrams that they draw, and the code which is required to implement it. Secondly, we would also like to implement a second system based on our framework to validate our ideas about which of our techniques are truly domain independent.

7   Acknowledgments

This work could not have happened without the help of Norman Foo, and other my colleagues at the Knowledge Systems group at the University of New South Wales. Early portions of this work were supported by an APA scholarship.

References

1. Anderson, M. & McCartney, R. Diagrammatic Reasoning and Cases. *AAAI/IAAI, Vol 2, 1996.*
2. Barwise, J. & Etchemendy, J. *Heterogeneous Logic.* in *Diagrammatic Reasoning* Glasgow, J. Narayanan N.H., Chandrasekaran B. eds pp 211-234.
3. Burnett, M. & Gottfried, H. Graphical definitions: Expanding Spreadsheet Languages through Direct Manipulation and Gestures. *ACM Transactions on Human Computer Interactions 5(1).* March 1998.
4. Erwig, M. & Meyer, B. Heterogenous Visual Languages – Integrating Visual and Textual Programming. *Proceedings, 11th Symposium on Visual Languages, V. Haarslev (ed.), Darmstadt, Germany, Sep. 5-9, 1995, IEEE Press, 1995, pp. 156-163pp 318-325.*
5. Foo, N. Local Extent in Diagrams. *Formalizing Reasoning with Visual and Diagammatic Representations.* Technical Report FS-98-04. AAAI Press, 1998
6. Haarslev, V. Formal Semantics of Visual Languages using Spatial Reasoning *Proceedings, 11th Symposium on Visual Languages, V. Haarslev (ed.), Darmstadt, Germany, Sep. 5-9, 1995, IEEE Press, 1995, pp. 156-163*
7. Hoare, C. A. R. An Axiomatic Basis for Computer Programming *Communications of the ACM 12(10), 576-580, 583 1969*
8. Kingston, J. *Algorithms and Data Structures: Design, Correctness, Analysis.* Addison Wesley Publishing Company, Sydney 1990.
9. Meyer, B. *Object-Oriented Software Construction* (2nd Ed) Prentice Hall 1997
10. Morgan, C. *Programming from Specifications* Prentice Hall, U.K. 1989
11. Nassi, I. & Schneiderman, B. Flowchart Techniques for Structured Programming. *SIGPLAN Notices 8(8):12-26.* 1973
12. *Prograph* © Pictorius Inc. http://www.pictorius.com
13. Thurbon, J. & Foo, N.Y *Planning with Diagram Sequences* [under review].
14. Thurbon, J., & Foo N.Y Programming with Pictures *proceedings of Pan-Sydney Workshop on Area Visual Information Processing* Jin, J., Eades, P., Yan, H., & Feng, D. eds. November 1998

MetaBuilder: The Diagrammer's Diagrammer

Robert Ian Ferguson, Andrew Hunter and Colin Hardy

University of Sunderland, School of Computing, Engineering and Technology, Informatics Centre, St. Peter's Campus, Sunderland, SR6 0DD, UK. Tel. +44 (0)191-515-2754, Fax. +44 (0)191-515-2781, Email. {ian.ferguson, andrew.hunter, colin.hardy}@sunderland.ac.uk

Keywords: *MetaCASE*

Abstract

A software tool named MetaBuilder is described. MetaBuilder's purpose is to enable the rapid creation of computerised diagram editing tools for structured diagrammatic notations. At its heart is an object-oriented, graphical meta-modelling technique - a diagrammatic notation for describing other diagrammatic notations.

The notation is based upon the concept of a mathematical graph consisting of nodes and edges. Construction of a "target tool" proceeds by drawing a meta-model of the target notation. Items in the target notation are modelled as "classes" and the syntax of the target notation such as connectivity between elements are expressed as "relationships" between the classes. Once the meta-model is complete, a new tool can be generated automatically. Thus the time to develop such notation specific drawing tools can be dramatically reduced. As the design of a piece of software can be expressed diagrammatically, the MetaBuilder software can be used to build itself!

1 Introduction

The need to express information diagrammatically is common to many disciplines from project management to electrical design. The use of software to support this activity can greatly ease the task of creating and maintaining diagrams, particularly when the diagrams are of a highly structured nature, corresponding to formal rules. The development of such software is however a complex, time-consuming and therefore expensive task.

One possible solution to this problem is to automate some or all of the process of creating such software. In order to do this, some means of describing diagrammatic notations in a machine readable form must be employed. One possible approach is to use a "meta-diagramming" notation: a diagram to describe other diagrams. A system, known as MetaBuilder and its associated notation have been developed at the University of Sunderland and used to build several structured diagram editing tools.

M. Anderson, P. Cheng, and V. Haarslev (Eds.): Diagrams 2000, LNAI 1889, pp. 407-421, 2000.

2 Problem - The Construction of Diagramming Software is Complex

The production of structured diagramming software is a complex and therefore time-consuming task [1]. It thus follows that it is expensive to develop a tool to support a new notation and the rapid provision of support for ad-hoc or experimental notations is impossible. One domain where this situation is of profound significance is the area of software engineering. Structured diagrammers (known as CASE - Computer Aided Software Engineering tools or, more specifically, upperCASE tools) to support various notations used during software development projects have existed since the mid 1980's. Uptake of these tools has however been notoriously poor with probably no more than 20% [2] of all software projects employing them. Chief among the reasons for this poor uptake is the fact that CASE tools compel their users to adhere to a set of rigid processes and procedures known as a method. Unfortunately many software engineers prefer (for quite valid reasons) to use ad-hoc or custom methods for each project. For reasons of cost given above, CASE tool manufacturers have been unwilling or unable to produce tools for a limited market (i.e. tools that may only be used on one project.)Due to the difficulties in building diagramming software a classic "Catch-22" situation arises whereby tool builders will not support new (and better) methods until engineers are using them and software engineers will not adopt a new method until it is supported by CASE tools.

3 Solution - Use Software to Build other Diagramming Software

One solution to the problem of building diagramming tools is to (partially) automate the process. The approach taken has been to examine several diagramming tools, identify the common components of such tools and to use these components to build a generic tool. This tool is then parameterised by the addition of the description of a given notation to make a tool to support that notation. This concept is illustrated in Figure 1.

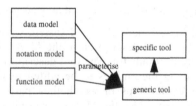

Figure 1. Generic and specific tools

When applied to the production of CASE tools, this approach is known as MetaCASE and the description of the notation as the metamodel. MetaBuilder has its origins in the field of software engineering where it is used to generate diagrammatic notation editors for CASE tools.

3.1 MetaModelling

The notation specific part of a diagramming tool is, in essence, a description of the notation that the tool will support. It specifies the things that can appear on a diagram, the relationships between them and their appearance. That description can be textual (as in systems such as TBK[10] and MetaEdit [12]) but an interesting possibility is the use of a diagram to provide the description. By building a tool which supports this Metamodelling notation, and combining it with a suitable set of software components to provide the generic editing capabilities, the construction of diagramming tools can be reduced to the task of building a metamodel of a target notation and single mouse click to automatically generate the new tool. This approach can dramatically speed up the task of constructing such diagramming tools. In addition, it removes the need for skilled software engineers to construct such tools.

4 Other Related Work

Various approaches to easing the task of building diagrammers have been attempted. This sections reviews some of the most influential systems.

4.1 Other MetaCASE Systems

MetaBuilder (described in section 5) is part of a "tradition" in MetaCASE which includes tools such as Ipsys TBK[10], Lincoln Software's Toolbuilder[11], MetaEdit[12], and KOGGE[4]. The common approach to MetaCASE taken by these tools is characterised by the MetaCASE system consisting of one or more generic editors which are customised (into specific tools) by the addition of a tool description. The tool description contains a model of the underlying data repository expressed in some form of Entity-Relationship(ER) notation.

TBK/Toolbuilder TBK uses a three part textual description of tools. The underlying PCTE repository is described using Data Description Language (DDL) the graphical appearance of entities using Graphics Description Language (GDL) and the tools interface using Format Description Language (FDL). These are combined with a generic editor tool partly at compile time and partly at run time. A graphical front end to TBK known as Toolbuilder exists and obviates the need for much coding. Toolbuilder allow the creation of a metamodel using an extended ER diagram notation. Drawing tools are used to define the graphical appearance of entities.

MetaEdit MetaEdit is a Smalltalk based system which consists of an editor tool which parameterised by a tool description in the MOF language. It is not clear from documentation what MOF stands for. The tool description is similar to that of TBK is that it involves describing the entities to be edited in terms of the their attributes, appearances and the relationships in which they participate The latest version of MetaEdit (MetaEdit+) again hides the MOF files behind graphical interface. This to relies on an extended ER notation to build the metamodel

KOGGE KOGGE's tool description is in three parts concepts (described by extended ER and GRAL - graph specification languages), menus (described by the use of Directed Acyclic Graphs - DAGs) and user interactions (specified using Harel's statechart notation). Three corresponding diagramming tools allow for the input and editing of KOGGE tool specifications.

Relationship to MetaBuilder MetaBuilder is identified as belonging to this group of tools by its use of a generic editor which is customised by the addition of a method model MetaBuilder differs from its predecessors in that it uses an entirely object-oriented approach: The generic part of the tool is provided in the form of an abstract class, the repository is described using an object modelling technique and the views of the repository are defined in terms of graphics primitive classes.

4.2. Toolkits and Component Based Approaches

Whilst the previous systems come form a MetaCASE background, MetaBuilder has similarities with systems from a diverse range of areas. One important class of system are toolkits. These systems are characterised by having no one executable program which the diagrammer builder uses but instead provide kits of software components which can be assembled to form diagram editors. In general these systems operate at a lower level than MetaBuilder, requiring (in some cases substantial) knowledge of the techniques of software development in order to produce results.

JComposer/Jviews The JComposer[8] system is a Java based system for building multi-user graphical interfaced editors which allows simultaneous editing of objects by a group of cooperating users. Jcomposer is a tool for building tools based on the Jviews components. It provides a visual language for defining repositories, the graphical views of the repository, the user interface of the built tool and the propagation of events describing changes to items in the repository.

JComposer/JViews are based up an architecture known as change propagation and response graphs (CPRG) [9] which is designed to support consistency and versioning operations upon data in a repository. The CPRG architecture uses an ER modelling technique to define the repository, but the architecture itself is implemented as a set of OO classes. This contrasts with the MetaBuilder persistent object database which uses an OO approach both for repository design and implementation.

Xanth The Xanth system is library of software components used to support interactive structured graphics. Whilst it is certainly more advanced than a library of graphics primitives and some of the classes provided (particularly those for representing hierarchical structures and networks) are useful in constructing CASE tools, there are no data modelling or repository tools included in the system.

Escalante Escalante[14] is a system for creating structured graphics editors. It can almost be considered a MetaCASE system in that allows a model of a diagram to be expressed in entity-relationship-attribute terms but there is only limited support for building tools which do anything with the diagram once it has been drawn. Adding facilities such as report writers and code generators would be difficult with such a system.

Hotdraw Hotdraw is a collections of classes implemented in the Smalltalk language that can be assembled to produce diagram editors. No specific support is provided for representing the structure of the target notation.

Relationships to MetaBuilder The Xanth system is too low a level to be considered a MetaCASE tool. The classes that it provides are however useful to toolbuilders and would be useful should MetaBuilder be ported to a C++ environment. Escalante is in many ways similar to MetaBuilder and the capabilities of the two system intersect. The lack of access to the underlying repository however precludes its use for some task in tool building.

4.3 Visual Programming Languages

One of the design objectives of MetaBuilder was to build a graphically rich environment. As experience in using MetaBuilder increased, it became clear that some of the graphical notations that it was being used to develop tools for were complex enough to be considered visual programming languages. A number of visual languages were reviewed but in the majority of cases, the environment associated with the language has been built either "from scratch" or with the aid of a graphics toolkit rather than using any dedicated automated support. A notable exception to this was the Vampire environment.

Vampire Vampire[13] is a system for developing visual programming languages. Languages built with Vampire are defined using a rule-based approach. A form of pattern matching is used to compare input in the form of "iconic" diagrams with graphically expressed rules. The pattern matcher decides whether a rule "fires" or not and the output is expressed as a diagram which has been transformed by the rule.

Elements of the language defined with Vampire become part of Vampire class hierarchy but the concept of class in Vampire is somewhat unusual in that the classes have attributes which can be either values or rules rather than data members and function members familiar from (say) C++.

Relationships to MetaBuilder Although like MetaBuilder, Vampire is based upon an OO approach to building OO language the similarity soon ends. Vampire concept of class and use of rules is significantly different to that of MetaBuilder. Vampire is aimed principally at producing executable visual languages rather than the structured diagram editors familiar with MetaCASE. It would be an interesting exercise to attempt the production of CASEtools using Vampire.

4.4 Generic Diagram Editing Systems

A class of systems dedicated to producing diagram editors exists. Typical examples of such systems are VLCC, GenEd and DiaGen. These systems use a variety of means for the description of graphical notations including hypergraph grammars and positional grammars.

Relationships to MetaBuilder One of MetaBuilder's design goals was that it should produce tools that were more than "mere" diagram editors but were capable of being extended and integrated into other systems. As such much development effort has gone into mechanisms (such as the persistent object repository) for achieving this rather than concentrating on the expressive power of the notation. MetaBuilder owes more to CASE and metaCASE systems than to the generic editors of this section and as such, it's notational description is not directly comparable. These systems take a variety of approaches to defining notations (including hypergraph grammars, positional grammars and entity-relationship modelling) but none use the visual object model of MetaBuilder.

5 MetaBuilder

This section provides a description of the MetaBuilder notation and a simple example of its use.

5.1 History

MetaBuilder has its origins in a series of diagram editors created to support a software engineering method called MOOSE (Method for Object Oriented Software Engineering) [6]. These were engineered in an object-oriented (OO) fashion and as such it soon became obvious that the tools had many of their classes (or components) in common. This set of classes was gradually refined and became known as the MetaMOOSE OO component library. As MetaMOOSE [5] was used in more projects, it became apparent that much of the code needed to combine these generic components into specific tools could be generated automatically from a suitable graphical notation. That notation and its associated tool become known as MetaBuilder.

5.2 The Metabuilder Metamodelling Notation

The metamodelling technique is best introduced by example. Consider, the diagram shown in Figure 2. This represents a simple type of structured diagram (which we name box-o-method) consisting solely of labelled nodes joined by edges. In this section, the metamodel representing this kind of diagram is developed, introducing the various modelling concepts in the process.

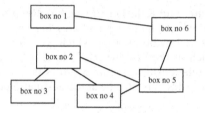

Figure 2. An example box-o-method diagram

Before describing the MetaBuilder notation, it is important to distinguish between the metamodelling notation (that used to describe the specific component of a diagrammer and the target notation (that which is being modelled in this case box-o-method).

The MetaBuilder metamodelling notation is based upon the concepts of object orientation (OO). Any OO modelling notation could have been used as the basis of MetaBuilder, but it was convenient to derive the MetaBuilder tool from an existing object diagrammer built as part of the MOOSE notation.

Classes The basic modelling concept in MetaBuilder is therefore the class. The concept is typically introduced (informally) to undergraduates by saying that any "blobs" in the target notation, must be represented by classes in the metamodel.

In the box-O-method example, there is clearly only one type of "blob" - the node -

and would be modelled as shown in Figure 3

Figure 3. Metamodel of box-o-method - Stage 1

Relationships The edges in box-o-method are modelled using relationships. Informally, any lines in the target notation become relationships in the metamodel. Stage 2 of the metamodel of box-o-method is shown in Figure 4

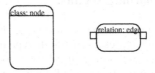

Figure 4. Metamodel of box-o-method - Stage 2

Class->Relationship and Relationship->Class Relationships It is necessary to model the manner in which the nodes and arcs may be connected to form a "legal" diagram. This is achieved by designating which type of class maybe at the source of a relationship and which at the destination. This is achieved using two types of link between classes and relationships. Figure 5 shows that nodes may be at the source of edges and shows that nodes may also be at the destination of edges.

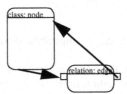

Figure 5. Metamodel of box-o-method - Stage 3

Cardinality Rules Restrictions may be placed on the number of edges that may originate or terminate at a node. The restriction is modelled as an integer value (representing the maximum permissible number of edges) placed near the appropriate link. A value of 0 represents an unlimited number of edges. Generalising this, the number of any instances of a given relationship that a class may participate in, can be restricted by the cardinality of the class->relationship and relationship->class links.

Figure 6 shows that an unspecified number of edges may originate and terminate at

any node. i.e there is no restriction

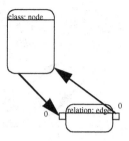

Figure 6. Metamodel of box-o-method - Stage 4

Data Members The existence of labels on nodes or edges can be modelled using data members. By adding a data member to a class or relationship, a data carrying facility is added to the underlying representation allowing the storage of textual information. Figure 7 shows the "name" data member added to the node class.

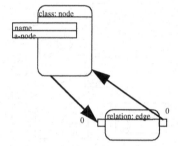

Figure 7. Metamodel of box-o-method - Stage 5

Has_a relationships A second example box-o-method diagram (Figure 8) shows that nodes may have "spots" associated with them. This can be designated in the metamodel by the use of the has_a relationship. This implies that a instance of one class maybe directly associated with an instance of another without the mediation of a line or relationship. The metamodel shown in Figure 9 demonstrates this feature.

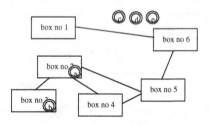

Figure 8. Example box-o-method diagram with "spots"

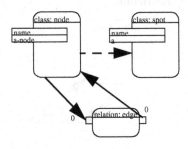

Figure 9. Metamodel of box-o-method - Stage 5

Inheritance - Variants of nodes can be created as shown in Figure 10 where two types of node are present: nodes and coded nodes which have an extra numerical label. Other than this label, coded nodes are identical to ordinary nodes, i.e they participate in the same relationships and have identical data members.

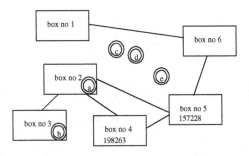

Figure 10. Example box-o-method diagram with coded nodes

In this situation, the coded node could be modelled as an entirely new class, and have the data members and relationships duplicated. This can rapidly lead to complex, hard to maintain metamodels. A much cleaner solution is to introduce the concept of inheritance (or the is_a_kind_of) relationship. The destination class of such a relationship is a special kind of the source class and inherits all its attributes. Figure 11 shows how this can be modelled. Generally, inheritance is used to create a "special kind" of an already defined class

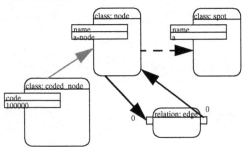

Figure 11. Completed box-o-method metamodel

Symbols The appearance of each class and relationship must be specified to complete the metamodel. The MetaBuilder system includes a simple drawing editor for creating symbols.

Textual Annotations The symbol editor allows each data member to be associated with a user interface "widget" which allows the contents of the data member to be edited. Three types of widget are currently supported:*textitems* which allow single lines of text to edited, *textwidgets* which support multiple lines of text and *popup_editors* which allow more sophisticated editing operations to be performed. In addition, simple non-editable labels maybe added to a symbol for non-varying textual annotations.

Other notation items Various other items of notation can appear on a metamodel such as "tools" and "root objects", but these are "housekeeping" items imposed by the method of implementation. They have no bearing on the metamodelling technique and as such are beyond the scope of this paper.

A screenshot of MetaBuilder being used to prepare the metamodel of box-o-method is shown in Figure 12

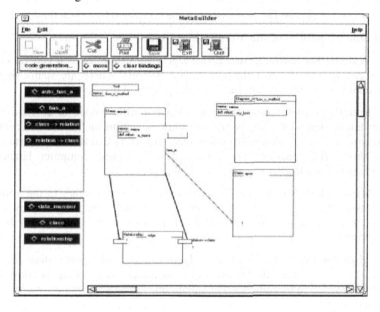

Figure 12. Screenshot of the completed metamodel

5.3 The MetaBuilder Tool

Generating a tool Once the metamodel is complete, building the target tool is simply a matter of pressing a button. Figure 13 shows the completed box-o-method tool in use.

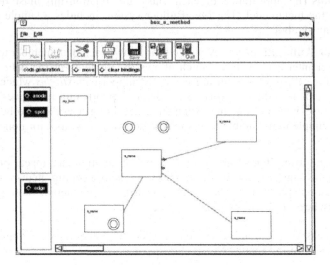

Figure 13. Screenshot of the generated box-o-method tool

5.4 Other Facilities of the MetaBuilder System

Widgets The textitems and text widgets are examples of a general class of items that can be placed on a diagram called widgets. These are items such a buttons, checkboxes and menus familiar from the graphical user interface.

Functions Allow computational behaviour to be added to classes. Triggered by user interaction with widgets or some drawing action such as mouse clicking, symbol movement or dragging and dropping these functions can access the data stored in data members. The functions are expressed in the same computer language used to implement the MetaBuilder system - Itcl[15] [16].

HyperCASE Navigation The term hyperCASE [3] refers to the ability to easily access related diagrams (or different views of the same diagram) by simple mouse operations on a diagram. MetaBuilder facilitates such navigation by the use of functions.

Multimedia A set of special widgets can be placed on a diagram which allow access to the multimedia facilities of a computer. Video widgets allow short clips of digitised video to be embedded in a diagram. Audio widgets perform the same task for digital sounds.

Report generation - code generation A common requirement having drawn a diagram (at least in the field of computing) is to have a textual report generated automatically from it. This kind of facility is the basis for code generation systems in CASE tools. A default code generation function is present in all classes in the metamodel. This default function prints out the values of each of the class's data members and then calls the code generation function of any related classes. This default function can easily be customised to provide the desired format of generated report.

Active diagrams The facilities described in section 5.4 allow the construction of "active diagrams" - diagrams with some associated behaviour.

5.5 The Advantages of the MetaBuilder Approach

The MetaBuilder approach is characterised by its use of a graphical notation based on OO techniques for metamodelling mapped directly onto an OO implementation of the system. Several practical advantages accrue from using this approach to diagrammer construction that are not all present in other approaches:

Rapid Development - Complex systems of hyperlinked diagrammers can be built in minutes rather than days. The OO metamodelling approach means that notation specifications seamlessly translate into the implementations. This allows a rapid prototyping approach to be taken to tool construction. If an end user requires changes to a tool or notation, they can be effected rapidly.

Ease of Tool Construction Experience in teaching the MetaBuilder system suggests that the task of implementing such diagramming tools can be brought within reach of undergraduate computer scientists in a matter off hours. This is due to the fact that MetaBuilder abstracts all of the underlying implementational complexity allowing the toolbuilder to concentrate on the target notation.

Ease of Enhancement Some advantages of the approach described above pertain to the construction of MetaBuilder itself rather than to the building of diagrammers or to end users. The OO approach to the system allows MetaBuilder to be extended and improved more easily than would otherwise be the case. If some extra facility is required in a tool (e.g. a new type of "widget") it can be rapidly added by using MetaBuilder to add the new facility to itself, thus making it instantly available for use in all tools built with the MetaBuilder system.

HyperCASE and Multimedia The HyperCASE facility of MetaBuilder is "Internet aware" in that it allows hyperlinks to diagrams stored on another computer entirely. When combined with the multimedia capabilities, this creates a diagram publishing facility, the possibilities of which have only begun to be explored.

6 MetaBuilder in Use

MetaBuilder has been used extensively within its original intended application domain of CASE tools and to a lesser extent to build diagramming tools in other areas. This section describes some of the systems built with MetaBuilder.

6.1 MOOSE Toolset

Although MetaBuilder was originally derived from the MOOSE toolset (see section 5.1) MOOSE has subsequently been rebuilt using MetaBuilder. MOOSE consists of a variety of diagramming and text editing tools which make use of the full range of MetaBuilder's capabilities. Screenshots of and output from some of the MOOSE tools show the variety of supported notations.

Analysts sketchpad tool The analysts sketchpad tool was designed to allow the creating of "back of the envelope" sketches to be created in a more formal manner. It has many of the attributes of a free form drawing tool but differs in two respects: Firstly, audio and video clips maybe embedded in the diagram allowing the software engineer to capture aspects of interviews with clients or site visits to serve as an original reference point. Secondly, any of the items on the drawing can be promoted to the status of a class in the MOOSE class diagrammer via a hyperCASE link.

MOOSE class diagrammer The MOOSE class diagrammer shown in Figure 14 is a typical "node and edge" diagram, supporting a variety of nodes and relationships between them.

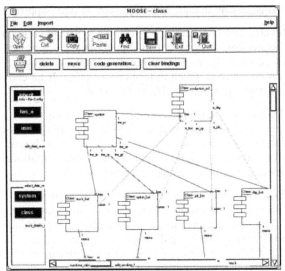

Figure 14. Class diagrammer

MOOSE object diagram The MOOSE object diagram notation (Figure 15) is composed entirely of classes with has_a relationships between them - no link relationships are present. The diagram is built up by simply dragging one node on to another. This action is sufficient to link the nodes via the appropriate has_a relationship.

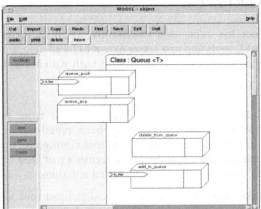

Figure 15. Object diagrammer

The VisualMOOSE tool is a user interface design tool. It allows the various screens, forms and dialog boxes of a program's user interface to be created by dragging and dropping the components from a palette of available controls. The controls that appear on the diagram are not simply icons but are actual working controls. This means that as soon as the control is placed onto the design diagram it is fully functional and will behave as if it were part of a completed program.

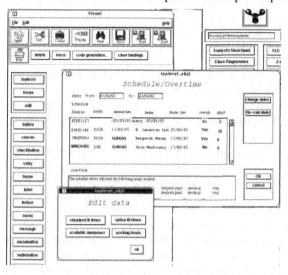

Figure 16. VisualMOOSE

6.2 MetaBuilder is Self-defining

Possibly the most interesting use to which MetaBuilder has been put is re-implementing itself. The MetaBuilder notation is "complete" in the sense that it can describe itself. The current version of MetaBuilder is capable of making amendments and improvements to itself. This greatly speeds up development work!

7 Conclusion

MetaBuilder has mainly been deployed to support diagrams originating in the field of computing. As such the metamodelling notation cannot be considered to be a universal notation capable of describing every other diagrammatic notation. Whilst the notation is however sufficiently general to describe most graphical notations that can be expressed in terms of the classes of object upon the diagram and the manner in which those items are related, the MetaBuilder tool itself has evolved to build a certain kind of diagrammer; i.e. CASE tools.

Within its own domain, MetaBuilder has been proven to be a versatile, rapid and powerful solution to the problems it was designed to address. Its wider applicability has been hinted at by its use in constructing tools such as Rezip and the Mobile Agent Manager. Future work on MetaBuilder will include investigating and attempting to remove the limitations of the notation as well as general improvements to the facilities offered to the diagramming tool builder.

The MetaBuilder system can be downloaded (under a shareware license) from the MetaBuilder home page [17] which also contains more complex examples of metamodelling.

References

1   Alderson A., 1991, Lecture Notes. In:*European Symposium on Software Development Environments and CASE technology, Konigswinter, Germany, June 1991* Lecture Notes in Computer Science 509 Springer-Verlag

2   Fisher, A.S., CASE: Using Software Development Tools, 2nd Ed., John Wiley & Sons, New York, NY, 1992

3   Cybulski, J. L. & Reed, K., 1992.*A Hypertext BASED Software-Engineering Environment. IEEE Software*, March 1992

4   Ebert, J., Süttenbach, I.U. (1997) Meta-CASE in Practice: A Case for KOGGE, In A. Olive, J.A. Pastor: Advanced Information Systems Engineering, Proceedings of the 9th International Conference, CaiSE97, 203-216

5   Ferguson, R.I., Parrington N.F., Dunne, P., Archibald, J.M., & Thompson, J.B., *MetaMOOSE - an Object-Oriented Framework for the construction of CASE tools*, in Journal of Information and Software Technology, Feb, 2000

6   Ferguson R.I., Parrington N.F. & Dunne P., 1994. MOOSE: A Method Designed for Ease of Maintenance. In:*Proceedings of the International Conference on Quality Software Production 1994 (ICQSP 94)*, Hong Kong: IFIP

7   Ferguson, R.I., Parrington N.F., Dunne, P., Archibald, J.M. & Thompson, J.B., MetaMOOSE - an Object-Oriented Framework for the construction of CASE tools In: Proceedings of International Symposium on Constructing Software Engineering Tools (CoSET'99) Los Angeles, May 1999

8   Grundy, J.C., Mugridge, W.B., Hosking, J.G. (1998) Visual Specification of Multi-View Visual Environments, IEEE Symposium on Visual Languages

9   Grundy,J.C., Hosking, J.G. & Mugridge, W.B., Supporting Flexible Consistency Management via Discrete Change Description Propagation, Software- Practice and Experience. Vol26, No. 9, pp 1053-1083, September 1996

10   IPSYS Software, 1991. *TBK Reference Manual*. Macclesfield, UK: Macclesfield, UK: IPSYS

11   IPSYS Software, 1991. *Toolbuilder Reference Manual*. Macclesfield, UK: Macclesfield, UK: IPSYS

12   Kelly, S., Lyytinen, K., Rossi, M. (1996) MetaEdit+ A Fully Configurable Multi-User and Multi-Tool CASE and CAME Environment, Lecture Notes in Computer Science, Vol. 1080, pp. 1-21

13   McIntyre, D.W. (1995) Design and Implementation with Vampire In: Burnett, M.M., Goldberg, A., Lewis, T.G. Visual Object-Oriented Programming Concepts and Environments, Prentice Hall, Chapter 7, pp.129-159

14   McWhirter, J.D. & Nutt, G.J., Escalante: An Environment for the Rapid Construction of Visual Language Applications, IEEE Symposium on Visual Languages (VL94),pp 15-22,1991

15   Ousterhout, J.K.,*Tcl and the Tk Toolkit*, Addison-Wesley, Reading MA, 1994

16   Harrison, M., *Tcl/Tk Tools*, OReilly, 1997

17   Ferguson, R.I., The MetaBuilder Project, online at http://www.cet.sunderland.ac.uk/rif/metabuilder/welcome.html, Jan. 2000

Diagrammatic Control of Diagrammatic Structure Generation

Stefan Gruner[1] and Murat Kurt[2]

[1] LaBRI, Université Bordeaux 1
`stefan@@labri.u-bordeaux.fr`
[2] Infonie GmbH Berlin
`murat@@infonie.de`

Abstract. Labeled graphs are a subclass of the class of all diagrams which are widely used in various disciplines. In consequence, graph grammars have been developed as powerful and intuitive means for the generation and manipulation of such kind of diagrams. As the operations of graph grammars are local and non-deterministic in principle, additional concepts of control are required. While textual control structures for the application of graph grammars are well-known already, more intuitive diagrammatic control mechanisms are still missing. We use a combination of the UML Activity Diagrams and the earlier Dijkstra Schemas for this purpose. We present a control flow editor which the user can use to build up diagrammatic control structures for the execution of graph-grammatical diagram constructions. The control flow diagrams are interpreted and animated such that the the user can observe an automatic diagram construction running along the specified control flow diagram.

1 Introduction

Diagrams can be found at almost every place in a modern society, but the meaning of the term "diagram" is really vague. So, first of all, we have to declare what kind of diagrams we are working with, before we can start with the discussion of our ideas. In **Fig.** , three familiar examples of different kinds of diagrams are given. The example given in Fig. **a**, which we may observe at a street crossing, belongs to the family of *sign*-like diagrams. The example given in Fig. **b**, which we might watch on TV after a political election, belongs to the family of *function*-like diagrams. Those (and other possible) families of diagrams do *not* belong to the scope of our considerations in the following. Instead, we are interested in *graph*-like diagrams as depicted in Fig. **c**, which may for example appear on public transportation maps. Here in the figure, the Central Station is connected with the University via the Main Street by tram line T1 and with the Air Port via the Music Hall by underground line U2. The Main Street and the Music Hall are additionally connected by bus line B12.

M. Anderson, P. Cheng, and V. Haarslev (Eds.): Diagrams 2000, LNAI 1889, pp. 422–   , 2000.
© Springer-Verlag Berlin Heidelberg 2000

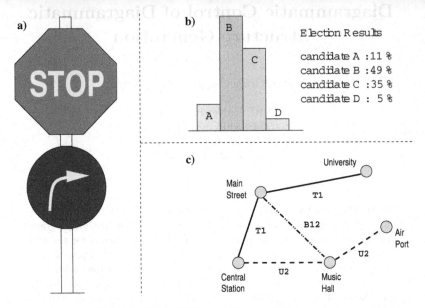

Fig. 1. Three different kinds of diagrams

Universe of Discourse

The diagram given in Fig. c is an example of a *labeled graph*. The labeled graphs constitute a subclass of the class of all diagrams and they are widely used in various disciplines of computer science and electrical engineering. For example, the well-known Entity-Relationship diagrams [ ] for database schema design, Structured-Analysis diagrams [   ] as well as module- and class architecture diagrams [ ][ ] in Software Engineering, Petri nets [ ] in Cybernetics, Automata- and Switching Theory, Semantic Networks [ ] in Artificial Intelligence, or certain types of Circuit Maps [ ] in electronic Hardware Engineering can sensibly be viewed as labeled graphs. Because of the technical relevance of such labeled graph diagrams, *graph grammars* (see below) have been developed as powerful and intuitive means for the generation and manipulation of such labeled graphs. Unfortunately, however, the operations of graph grammars are local and non-deterministic in principle so that not all the desired manipulation tasks can be fulfilled by graph grammars in an adequate manner. Thus, additional concepts of *control* are required.

Contribution of this Paper

While textual control structures for the application of graph grammars are well-known already, more intuitive diagrammatic control mechanisms are still missing. We propose a combination of the UML *Activity Diagrams* and the earlier *Dijkstra Schemas* for this purpose. Our main practical contribution is an (almost) already implemented *control flow editor* which the user can use to build up those diagrammatic control structures for the execution of graph grammar derivations. Moreover, our control flow diagrams are *interpreted and animated* (by different colors) such that the user can observe a diagram construction (or manipulation) running along the specified control flow diagram.

Organization and Addressed Readers of this Paper

The rest of our paper is structured as follows: In section 2, the notion of graph grammars is briefly introduced. (Readers who are familiar with graph grammars may omit that "auxiliary" section.) In section 3, we compare some related approaches to imposing control onto graph-grammatical diagram constructions. Section 4 explains the genealogy of our control flow diagrams which are described in section 5. Small examples are given in all these sections. In section 6, a bigger example is exercised for deeper intuition. Section 7 concludes our results.

Due to the interdisciplinary nature of the 1st DIAGRAMS2000 Conference, we try to give all explanations as intuitive as possible; thus, any reader having a minimum of philosophical or mathematical experience should be able to understand the contents of our contribution. Especially interested readers can obtain a slightly more technical "background version" of this paper from one of the authors on demand [ ].

2 Graph Grammars

Though the theory of graph grammars can be quite difficult in detail [ ], the principle of graph grammars is so easy to understand that they are meanwhile applied as powerful tools supporting the solutions of several relevant problems in recent computer science [ ][ ]. In this section, the principle of *graph replacement* is briefly sketched which is necessary for understanding the context of our approach.

Example

Please imagine that the underground line U2 between the Music Hall and the Air Port (Fig. c) has to be closed for the next season due to repair and modernization activities. For compensation, the transportation bureau of the town decides to prolong the bus line B12 from the Main Street to the Air Port. As a consequence, the transportation map must be adapted accordingly. Given the *graph replacement rule* of **Fig.** , this adaption can easily be performed as follows:

First, we have to find a *graph-homomorphic match* from the left-hand-side graph **L** (Fig.) into the input graph (Fig. c). We can match the *node variable* *stop-1* onto the node Main Street, the node variable *stop-2* onto the node Music Hall, and the node variable *stop-3* onto the node Airport. Then, we can match the *edge variable* (of bus line type) *b-line* onto the edge B12 and the edge type (of underground line type) *u-line* onto the edge U2 between Music Hall and Airport. Following the right-hand-side graph **R** of the rule (Fig.), the matched underground line is *deleted*, and a new bus line is *created* which carries the same name as the bus line found in the pre-condition of **L**. In **Fig.** , the application

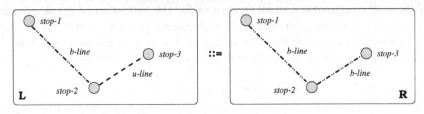

Fig. 2. A graph replacement rule

of the rule (from Fig.) to the input graph (from Fig.) is shown. Whilst the
match from **L** to the input graph is the *pre-condition* of the graph replacement,
an according match from **R** to the result graph is the *post-condition* of it.

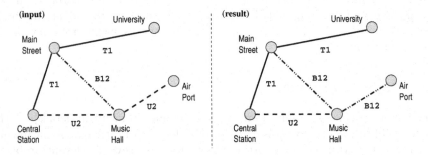

Fig. 3. An input and a result of a rule application

Comment

The graph-homomorphic match described above is not the only possible one,
because you could also match *stop-3* to Central Station and *u-line* to U2 between
Music Hall and Central Station. The advantages and disadvantages of *that* kind of
inherent non-determinism in the application of graph grammars are discussed in
the literature and do not belong to the scope of this contribution, but a second
kind of graph-grammatical non-determinism will be discussed below. Finally we
have to state that, using graph grammars, it is not only possible to replace single
edges (as done in our simple example), but also single nodes as well as rather
complex subgraphs and patterns at once.

Concluding this "auxiliary" section we can state that a graph grammar is
a *finite structure* $\mathcal{G} = \{G, r_1, \ldots, r_n\}$ whereby the start graph G is the axiom
and the graph replacement patterns $r_1, \ldots, r_n$ are the rules of the calculus which
is a generalization of the well-known textual Chomsky grammars. The different
possible semantics of graph grammars are described in the literature [ ] and do
not belong to the scope of this paper.

3 Comparison of Related Control Approaches

After having briefly sketched the general idea of graph grammars in the previous section, we can now discuss some more specific issues which will directly lead us to the key issue of our paper. The question of non-determinism in the graph-grammatical generation of diagrams will be answered with special kinds of control diagrams for better support of diagrammatic structure generation.

3.1 Motivation of the Problem

As described in [ ][ ], graph grammars have turned out to be applicable for solving various problems of recent computer science. It is also known, however, that *pure* graph grammars are *no* suitable solutions for practical problems in several cases, because

i. graph grammar derivations (as any grammar derivations) are of non-deterministic nature whilst deterministic and predictable behavior is often required (see section 6 for example),

ii. graph grammar derivations are *local* operations on some data graph (input graph) whilst global operations are often required for reasons of consistency between different parts of the data graph which do not belong to the same neighborhood.

   Example

   For better readability of the public transportation map (Fig. c), the transportation bureau decides to emphasize all junction stations where the passengers can step over from one transportation line to another one. This can be done with the graph replacement rule of **Fig.** , but neither the rule itself nor the graph grammar which the rule belongs to is able to *ensure* that the rule is applied in *both* the neighborhoods of the station nodes Main Street *and* Music Hall for the sake of map-global consistency.

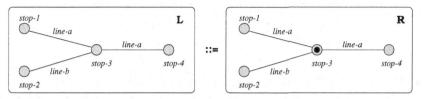

Fig. 4. Redrawing a junction of different lines: global application not ensured

3.2 Solutions Found in the Literature

For this reason, different concepts of *control* have been introduced in the recent graph grammar literature — in particular, we can mention the PROGRES system [ ] and the GRRR system [ ] which both are promising attempts to make the use of graph grammars feasible for various application tasks. Further approaches to imposing control on graph-grammatical diagram handling are, for example, described in [ ] and [ ]. Both the first approaches offer already implemented and usable tools, whilst both the latter approaches are still in a purely theoretical stadium as far as we know at the moment. A similar but more detailed discussion of this issue can be found in [ ].

Approaches without practical Evidence

The *transformation units* of [ ] allow the specification of rule application sequences with ";" and application loops with "!". For example the expression $r_1!;(r_2;r3)!;r_4$ means to apply r_1 as often as possible, then to apply the sequence of r_2 and r_3 as often as possible, and finally to apply r_4 exactly once. Thus, the presented control flow language is only textual but not diagrammatic and, even worse, no concept of branching or case distinction is supported.

In the *rule expression* approach of [ ], diagrammatic control structures are proposed the syntax of which is quite similar to the one of our approach. In the related approach, however, the semantics of the diagrammatic control symbols is slightly different. In contrast to the transformation unit approach of above, no concept of loop iteration is mentioned in the rule expression approach.

Approaches with implemented Tools

Comparing the control techniques of the relevant graph replacement tools PRO-GRES [ ] and GRRR [ ] we can observe that they are orthogonal to each other as shown in **Fig.** :

PROGRES	GRRR	New Approach
explicit	*implicit*	*explicit*
textual	*diagrammatic*	*diagrammatic*

Fig. 5. A comparison of control concepts supported by relevant tools

- Whilst the *explicit* control concepts of PROGRES are powerful they occur only in a *textual* notation which is rather inconvenient from a non-expert user's point of view. Explicit control in PROGRES means that the user can write program which executes the applicable graph transformation rules in the required manner.

- The opposite is true in GRRR: there we can find *diagrammatic* control concepts the handling of which is, therefore, quite user-friendly, but their control behavior is built-in, thus: *implicit*, and can not be modified by the user according to his requirements. Implicit control in GRRR means that the user can attach certain *flags* on certain nodes which invoke certain built-in graph replacement strategies. The replacement strategies themselves cannot be modified from outside.

3.3 Our Approach

Starting from this discussion we wanted to combine the advantages and avoid the disadvantages of both the control concepts of PROGRES and GRRR which means that we required a controlling system of as well explicit and diagrammatic control structures. As a result, we can present an (almost already) implemented diagrammatic *control flow editor* for the attributed graph grammar system AGG [ ] as the main contribution of this paper. The control structures specified with our editor can be interpreted and executed, and the user can watch the system running on the *control flow monitor* [ ]. The control flow monitor is another operational view of the control flow editor, and the conceptual basis of its *control flow diagrams* stem syntactically from a new combination of the widespread UML [ ] and the earlier Dijkstra Schemas [ ]. From a semantical point of view, however, the UML is left even farther behind in the interpretation of our special control flow diagrams. Finally it is worth mentioning that the process of constructing our control flow diagrams in the control flow editor can be described by graph grammars again [ ] such that —back to the beginning— the "methodological circle" is closed.

4 Watching out for Suitable Control Structures

We have already mentioned above that, manipulating some data graph G by a graph grammar $\mathcal{G} = \{G, r_1, \ldots, r_n\}$, we have to face *two* different kinds of non-determinism, namely

- i. the choice of *what* rule r_i shall be applied to G,
- ii. the choice *where* a selected rule r_i shall be applied *within* the data graph G.

In the following, we are concerned with the first kind of non-determinism but *not* with the second one (the handling of which is sufficiently treated by the latest literature [ ][ ]). Thus: we want to be able to determine an application sequence $r(1); r(2); \cdots; r(m)$ which we call a *graph transformation* on G, where $r(i) \in \mathcal{G}$ for all $i = 1, \ldots, m$ and $0 \le m \le \infty$.

4.1 Dijkstra Schemas

The well-known *Dijkstra Schemas* [ ] would allow us to specify the three most important control concepts in a diagrammatic fashion. These are: concatenation (*sequence*), conditioned decision (*if-then-else*), and conditioned iteration (*while*).

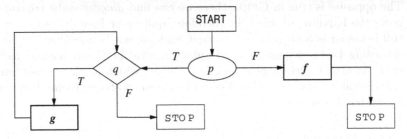

Fig. 6. A Dijkstra Schema

Example

The Dijkstra Schema depicted in **Fig.** is interpreted in the following way: after the system has started, its global state is checked via some predicate p. If the result is False, some function f is executed and the system halts. Otherwise, some function g is executed until some predicate q results in False.

Comment

Because of some obstacles explained in [ ], we are not able to apply the elegant concept of Dijkstra Schemas immediately to our problem of diagrammatic graph transformation control [ ]. The reason is that we are concerned with semi-deterministic operations on graphs whilst those *program schemas* are valid only for deterministic functions on ordinary global variables. Later we will see, however, what parts of the Dijkstra notion are re-used for which purpose in our hybrid control flow approach.

4.2 Activity Diagrams

Also the more recent *Activity Diagrams* [ ] seem to offer us interesting diagrammatic features for the specification of control. Conditioned jumps are supported as well as the possibility of splitting the control flow into several concurrent branches.

Example

In **Fig.** , a small example of an Activity Diagram is shown. After the system has started, some activity A is performed, followed by an activity B. If some condition *cond* is evaluated to False, the system jumps back to A and iterates. Otherwise, the both the activities C and D are performed in an asynchronous manner, this means: without respect to any application order in time. After both activities C and D have been applied, activity E is performed and the system halts.

Comment

In contrast to the Dijkstra Schemas, Activity Diagrams are not restricted with respect to data types or function applications. Thus: as their original semantics is vague anyway, Activity Diagrams can easily be re-interpreted in such a way that our control requirements are fulfilled. Especially the *fork* and *join* items, the concurrent-asynchronous interpretation of which makes no sense in our case, will get a different meaning as shown in the following section. Please note further that some "legal" kind of non-determinism is re-introduced by the *fork* concept, which is not part of the strictly deterministic Dijkstra terminology. *Well*-structured loops (*while*), however, are *not enforced* by the Activity Diagrams. Instead, the user is able to simulate not only *while* loops but arbitrary GOTO loops by (ab)using the concept of conditioned decisions (*if-then-else*) which Dijkstra has "considered harmful" for well-known reasons [ ]. Agreeing with Dijkstra at this point of discussion, we will avoid such "spaghetti" facilities in our approach as well.

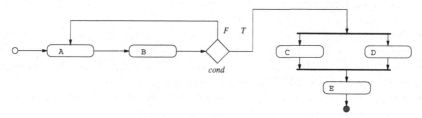

Fig. 7. An Activity Diagram

5 Control Flow Diagrams and Their Interpretation

In this section, we present the concepts of our solution to the problem how graph grammar operations on labeled data graphs can be controlled by diagrammatic hence intuitive and user-friendly means. We merge the most prominent concepts of Dijkstra Schemas and Activity Diagrams such that their advantages are increased and their disadvantages are decreased. Then we explain how our hybrid control flow diagrams are interpreted in a graph transformation environment.

Structure

Our control flow diagrams (which belong to the diagram subclass of labeled graphs, too) are constructed according to [ ] from the following set of syntactical items:

- All the nodes of the control flow diagram are connected by arrows → representing the control flow through the diagram.

- The arrows may (but need not always) be attached with the usual truth symbols T and F.
- There must be exactly one *start* symbol $\bigcirc$ and at least one *stop* symbol $\odot$. The start symbol has exactly one outgoing arrow, and the stop symbol has exactly one incoming arrow.
- The *rule application* symbol $\square$ must have exactly one incoming arrow and may have one (unlabeled) outgoing arrow; it may also have two outgoing arrows (which are labeled with T and F in this case).
- The *if-then-else* symbol $\triangle$ has exactly one incoming arrow and exactly two outgoing arrows which are labeled with T and F.
- Loops may be constructed only with the *while* symbol $\diamond$ (whilst the emulation of GOTO jumps via the *if-then-else* symbol is forbidden). The loop symbol must have exactly two incoming arrows and two outgoing arrows. One of the outgoing arrows is the exit. Every path beginning with the other outgoing arrow must eventually lead back to the $\diamond$ symbol such there is no other possibility of "illegally" leaving the loop without using the well-defined exit. The exit is labeled with F, the other outgoing arrow is labeled with T. Correspondingly, one of the loop inputs must come from outside of the loop and must not be a branch of it.
- Every rule application symbol $\square$ must be attached with exactly one parameter r_i $(i = 1, \ldots, n)$.
- The decision symbols $\triangle$ and $\diamond$ must be attached with proposition-logical expressions (AND, OR, NOT) over some parameters r_i.
- Finally we may *fork* and *join* the control flow (without any condition) by using the according symbols $\top\!\top$ and $\bot\!\bot$, respectively.

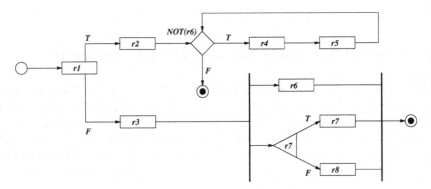

Fig. 8. A control flow diagram $\mathcal{C}$ for some graph transformation

Example

Fig. shows an example of how all these symbols may be used. The interpretation of our example of the control flow diagram $\mathcal{C}$ given in Fig. is explained informally as follows:

Given a graph grammar $\mathcal{G} = \{G, r_1, \ldots, r_8\}$ the rules r_i of which shall operate on the data graph G. Given further an interpreter $\mathfrak{I}$ which is able to operate on $\mathcal{C}$ and $\mathcal{G}$ by reading in $\mathcal{C}$ and writing in G according to the rules of $\mathcal{G}$. After having started in $\bigcirc$, the interpreter *tries* to apply r_1. If this trial *has been* successful (*post-condition T*), $\mathfrak{I}$ tries to apply r_2. Whether successful or not, $\mathfrak{I}$ enters the *while* structure $\Diamond$. *As soon as* r_6 *is* applicable (*pre-condition F*), $\mathfrak{I}$ stops at $\odot$, otherwise the application of r_4 and r_5 is tried sequentially in the loop (pre-condition T), no matter if with success or without. If the trial of r_1 has not been successful however (post-condition F), $\mathfrak{I}$ tries to apply r_3 and enters the fork $\mathbin{\text{TT}}$ afterwards in any case. At this point, $\mathfrak{I}$ makes a non-deterministic choice by chance. *Either* rule r_6 is tried or the *if-then-else* structure $\triangle$á is entered: if r_7 is applicable (pre-condition T) this is done, otherwise (pre-condition F) the application of r_8 is tried. Finally, $\mathfrak{I}$ joins at $\mathbin{\text{$\bot\!\!\bot$}}$ and stops the interpretation of $\mathcal{C}$ at $\odot$.

Comment

In general, the decision symbols $\triangle$ and $\Diamond$ may be attached with proposition-logical expressions on propositions r_i, whereby a proposition r_i is interpreted as "rule r_i is applicable in the next step". In these cases, a label r_i appears as abbreviation of a boolean *predicate* expression $p(r_i)$ which is asking the applicability question. In the case of $\Box$, however, the label r_i appears as abbreviation of a boolean *procedure call* expression $c(r_i)$ which returns true or false, and which has the possible *side effect* of a graph-transformative application of r_i. Please note further that making decisions on post-conditions ($\Box$) is *not equivalent* to making decisions on pre-conditions ($\triangle$, $\Diamond$). In Fig. , for example, if r_5 has been applied we are sure that r_1 has been applied as well, but we are not sure that r_3 has been applied, too, if we only know that r_7 has been applied. A complete run of $\mathfrak{I}$ by $\mathcal{G}$ through $\mathcal{C}$ from $\bigcirc$ to $\odot$ is called a *graph transformation* on G. Non-terminating loops over $\Diamond$ are not regarded as graph transformations, thus: graph transformations are *finite per definitionem* here.

The application environment of our diagrammatic control flow component is the attributed graph grammar tool AGG [ ] which does not offer the possibility of concurrent application of graph replacement rules — nor do PROGRES or GRRR. Under these circumstances it does not make sense to interpret the $\mathbin{\text{TT}}$ and $\mathbin{\text{$\bot\!\!\bot$}}$ symbols in an asynchronous-concurrent manner as proposed by the UML. For this reason we have decided to interpret the *fork* construct as a non-deterministic (random) exclusive disjunction (XOR) which re-introduces the sometimes useful concept of "restricted anarchy". Further remarks on the interpretation as well as on other notions of graph transformation can be found in [ ].

Implementation

At the moment, our control flow editor together with a control flow interpreter $\mathfrak{I}$ is almost completely implemented, except of unconditioned fork and join concepts $\mathbin{\text{TT}}$ and $\mathbin{\text{$\bot\!\!\bot$}}$ [ ]. For reasons of re-usability, the implementation language is JAVA. The integration and maintenance of our control flow editor with the

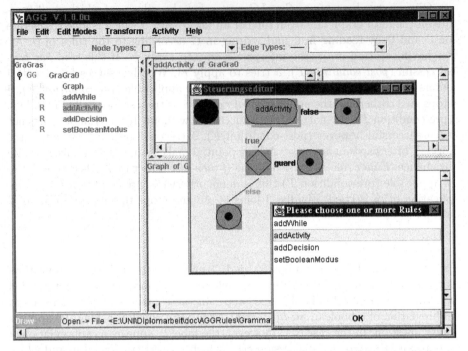

Fig. 9. Surface of the control flow editor with AGG in the background

already existing graph grammar tool AGG [ ] is partially done and partially ongoing work. The GUI operations of the editor are *syntax-directed* such that the user is protected from making syntax errors while designing a control flow diagram C. The editor operations which the user can apply are formally specified by means of graph grammars again. The rule names r_i from a given graph grammar G can be inserted into C via "mouse & menu" only, which also supports a consistent control flow design. The control flow diagrams designed by the user are pretty-layouted automatically by the control flow editor, but the tool also allows the user to manipulate the layout via "drag & drop" for more sophisticated results. **Fig.** shows the screen surface of our control editor in front of the AGG surface in the background.

6 Example

In this section, a bigger example of our concepts is given for deeper intuition. We are constructing a public transportation map by use of some very simple graph replacement rules, and we show the control flow diagram by which the construction process is managed. Our example graph grammar $G = \{G, r_1, r_2, r_3, r_4, r_5\}$, which is depicted in **Fig.** , consists of the following components:

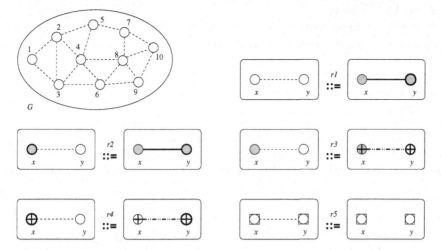

Fig. 10. A graph grammar $\mathcal{G}$ with start graph G and five simple rules

G is the start graph. In contains a template of a town with 10 possible stations and several possible connections.

r_1 is a graph replacement rule which declares the destination and the first piece of tunnel of an underground line.

r_2 prolongs an underground line by one piece of tunnel between to stations.

r_3 creates a junction by connecting an underground station with a bus line.

r_4 prolongs a bus line into an empty area.

r_5 wipes out a non-activated line pattern, (whereby the boxed circles serve as pattern for stations of any type, i.e. underground, bus, or junction).

If we want to construct a useful transportation map with this graph grammar it is obvious that we must not apply the rules in a pure grammatical, i.e. non-deterministic manner. Otherwise for example, rule r_5 could destroy our map before any bus line or underground rail is drawn. Instead, we have to impose deterministic control onto the diagram construction process. Let's now assume that the public transportation bureau of our town has to satisfy the following planning constraints due to economic reasons:

α. There must be five underground stations.

β. The number of junctions shall be "small".

γ. Every possible place must be connected.

Provided with $\mathcal{G}$ and the control flow diagram $\mathcal{C}$ depicted in **Fig.** a, we can construct a whole *family* of possible transportation maps which all satisfy our economic constraints (the second of which is "soft"). The control flow diagram $\mathcal{C}$ implements the following diagram construction algorithm:

1. apply r_1 once;
2. apply $_2$ three times;
3. while r_3 or r_4 are applicable do:
   if r_4 is applicable apply r_4 once;
   else apply r_3 once;;
4. while r_5 is applicable do:
   apply r_5 once;;

The transportation map shown in Fig. b is one possible result of our diagram construction procedure starting from G by C via $\mathcal{G}$. We hold the number of junctions comparatively small by giving r_4 a higher priority than r_3 in step 3.

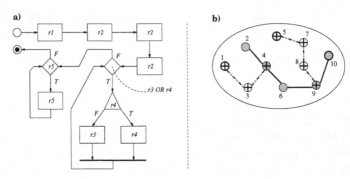

Fig. 11. A control flow diagram C and a transformation result from G

7 Summary

The class of labeled graphs is a relevant subclass of the class of all diagrams. Graph grammars are a powerful means for constructing and manipulating such graph-like diagram structures. However, pure graph grammars need to be enhanced with sophisticated control structures for several application purposes. Whilst other graph transformation systems offer either diagrammatic *or* explicit control structures, our approach of hybrid UML-Dijkstra-based control flow diagrams supports the user in both diagrammatic *and* explicit control flow design. Our tool operates fully syntax-directed in order to protect the user from making avoidable mistakes. An operative run through a given control flow diagram with respect to a given graph grammar and a given data graph is called a graph transformation.

References

1. C. Albayrak, *Die WHILE-Hierarchie für Programmschemata*. Doct.-Diss., Fac. of Comp. Sc., RWTH Aachen, 1998. Shaker, Aachen/Maastricht, 1998

2. G. Booch et al., *The UML User Guide.* Addison Wesley, Reading 1998   ,   ,

3. P. Chen, *The Entity-Relationship Model: Towards a Unified View of Data.* ACM Transact. on Databs. Syst. 1/1, pp.9-36, 1976

4. E. Dijkstra, *GOTO Statement considered Harmful.* Comm. of the ACM 11/3, pp.147-148, 1968   ,

5. F. Drewes et al., *Graph Transf. Modules and their Composition.* In [   ]

6. H. Ehrig et al. (Eds.), *TAGT'98: 6th Internat. Worksh. on Theory and Applic. of Graph Transformation.* LNCS 1764, Springer, Berlin 2000

7. M. Große-Rhode et al., *Modeling Distr. Syst. by Modular Graph Transf. based on Refinement via Rule Expressions.* In [   ]

8. S. Gruner et al., *A Visual Modeling Technique for Controlling Graph Transf.* Accepted for the Visualization-Workshop of the ICALP 2000 Conference. To be published by Carleton Scientific, Waterloo 2000

9. S. Gruner, *Diagrammatic Control of Diagrammatic Structure Generation: Additional Information to our* DIAGRAMS2000 *Publication.* Unpublished document, available from the author on demand.

10. M. Kurt, *Entwurf und Implement. einer Kontroll-Komponente für attribut. Graph-Ersetzung.* M.Sc.Thesis, Fac. of Comp. Sc., Techn. Univ. Berlin, 2000   ,   ,
,

11. T. de Marco, *Structured Analysis and Syst. Specification.* Yourdon Pr., N.Y. 1978

12. G. de Micheli et al. (Eds.), *IEEE Transact. on Comp.Aided Design of integrated Circuits and Syst. 16/10.* IEEE Circuits & Syst. Soc., 1997

13. M. Nagl (Ed.), *Building tightly integr. Softw. Developm. Environments.* LNCS 1170, Springer, Berlin 1996

14. M. Nagl et al. (Eds.), *AGTIVE'99 Worksh. on Applic. of Graph Transf. with Industrial Relevance.* LNCS 1779, Springer, Berlin 2000   ,

15. C. Petri, *Kommunikation mit Automaten.* Doct.-Diss., Schriften des Instit. für instrumentelle Mathematik, Bonn 1962

16. G. Rozenberg (Ed.), *Handbook of Graph Grammars and Comp. by Graph Transf. 1 (Foundations).* World Scientific, Singapore 1997   ,

17. G. Rozenberg et al. (Eds.), *Handbook of Graph Grammars and Comp. by Graph Transf. 2 (Specifications and Programming).* World Scientific, Singapore 1999   ,

18. G. Rozenberg et al. (Eds.), *Handbook of Graph Grammars and Comp. by Graph Transf. 3 (Concurrency).* World Scientific, Singapore 1999   ,

19. P. Winston, *Artificial Intelligence.* 3rd ed., Addison Wesley, Reading 1993

20. AGG, http://tfs.cs.tu-berlin.de/agg/   ,   ,

21. GRRR, http://www.cs.ukc.ac.uk/people/staff/pjr6/gdgr/main.html

22. PROGRES, http://www-i3.informatik.rwth-aachen.de/research/progres/

Two-Dimensional Positioning as Visual Thinking

Shingo Takada[1], Yasuhiro Yamamoto[2], and Kumiyo Nakakoji[2,3,4]

[1] Keio University
Yokohama, 223-8522, Japan
michigan@ics.keio.ac.jp
[2] Nara Institute of Science and Technology
Nara, 630-0101, Japan
{yasuhi-y,kumiyo}@is.aist-nara.ac.jp
[3] SRA, Inc
[4] PRESTO, JST

Abstract. People depend on various external representations in various design situations. These external representations are necessary at the time of creation in early stages of a design task, as they help the designer visualize what they are thinking and continue with their task in the process of reflection-in-action. Designers in domains such as architecture have drawn diagrams, or sketches, as the external representations. We take writing and programming as two example domains, and argue that *two-dimensional positioning* serve the same purpose for these domains as *diagrams* do for architectural design. We describe two tools, ART for writing and RemBoard for component-based programming, which help writers or programmers visualize what they are thinking through positioning parts of writing or software components on a two-dimensional space. We examine the issues that are necessary for this, and explore how they were handled in the two tools.

1 Introduction

Design occurs in the first stages of any construction of artifacts. The design task has been characterized as ill-structured [ ], and the problem and solution co-evolve, going through a process of "reflection-in-action"[ ]. Rather than pre-plan what to do during a design task, the designer is engaged in a cycle of producing a representation (e.g. a sketch) and reflecting on them. The representation serves as a "situation" that talks back to the designer, through which the designer asks:

- what "parts" are missing;
- how much the designer is "sure" about a newly created part;
- what the role of this newly created part is in terms of the whole design;
- what the role of this newly created part is in terms of the other parts; and
- which direction the whole design is moving toward and whether the direction is in accordance with the intention behind the design.

M. Anderson, P. Cheng, and V. Haarslev (Eds.): Diagrams 2000, LNAI 1889, pp. 437– , 2000.
© Springer-Verlag Berlin Heidelberg 2000

The representation is useful only for the designer him/herself at the time of its creation [ ]. The designer is free to make conventions that only has meaning to him/herself. Of course, in some cases, the representation may be intended for other people such as the designer's client. But while the actual work (design) is going on, the representation is only for the designer and not for other people, and as a result if a design-in-process is seen by other people, the representation may be interpreted differently from what the designer had intended [ ].

This representation has been called an externalization of what the designer is thinking [ ], or put another way the current situation of the design task. According to Bruner, the externalization of what the designer is thinking is important because it *"relieves us in some measure from the always difficult tasks of 'thinking about our own thoughts' while often accomplishing the same end. It embodies our thoughts and intentions in a form more accessible to reflective efforts"* [ ]. This suggests that the form that the externalization takes is important. If the externalization does not take a suitable form, then it does not help with *'thinking about our own thoughts'*.

In many domains, such as architecture, externalization has often taken the form of sketches, which can be thought of as a type of diagram that is done roughly and will be elaborated later. Designers will make sketches of what they are thinking, reflect on it, and then make changes as necessary. The sketch becomes *"a sort of hypothesis or 'what if' tool"* [ ].

In other domains or situations, it may be more or less difficult to conduct a conventional sketch. For example, consider the situation where someone is writing a document. The writer already has parts of the document typed into a computer and wants to use it. How can the writer "sketch" the document using those existing text fragments? Consider another situation where someone is writing a program using a vast library of software components to choose from. How can the programmer make a "sketch" of the program while keeping in mind the various alternatives that may arise while searching for software components to use?

We consider the two domains of writing and component-based programming, and argue that two-dimensional positioning can serve the same purpose for these domains as "sketches" do for architectural domains. We describe two tools – ART for writing and RemBoard for component-based programming – that help designers visualize what they are thinking by positioning parts of their tasks on a two-dimensional space.

The rest of this paper first begins with an overview of why writing and programming need support for a form of "sketching". We compare these with solving a jigsaw puzzle, and contend that two-dimensional positioning can serve as a "sketching" tool. Section 3 then describes the two tools (ART and RemBoard), and how they helped users in their tasks. Section 4 discusses issues concerning how visualization is supported in the two tools. Section 5 makes concluding remarks.

2 Writing, Programming and Jigsaw Puzzle

We consider writing and component-based programming as two domains that need some form of "sketching". We compare it to solving jigsaw puzzles, and contend that two-dimensional positioning can serve as a "sketch".

2.1 Writing

Designing in writing has conventionally taken place as making an outline of the document to be written. This takes the position that one first identifies a stable structure, then gradually elaborates it from chapter to section to paragraph to sentence.

With the proliferation of computer tools such as word processors, problems have arisen mainly resulting from the ease with which text can be changed [ ]. For example, writers will start typing (writing) a paragraph, but because a word processor is used, they can easily cut-and-paste parts of that paragraph, or copy various notes from previous work. They may do this even though they may not understand how the parts can be connected, or how the parts can be put together.

The writers engage in a trial-and-error process of trying various combinations and organizations of written pieces, and see if they can make sense out of them. They will explore what position the current piece can have within the whole document, what pieces need to be modified, and what has and has not yet been written, etc. The document content itself is not the only information that is important, but information that pertain to the state of the document and its structure, which Kintsch and van Dijk called meta-comments, also play a role in writing [ ].

Thus writers will need to be able to visualize various factors concerned with the current situation, i.e. they will need some method of "sketching".

2.2 Component-Based Programming

Component-based programming is a paradigm where programmers will develop software by "gluing together" various components (or program parts). The task basically consists of (1) finding the necessary components, and (2) making the "glues" necessary to put together the components. We focus on the former.

The search for necessary components in component-based programming is critical as they are the building blocks for the program. Programmers however may not necessarily know exactly what type of components are needed. Although they will at least have conceptually top-level requirements defined for the program, they may not know the details.

The programmers will engage in a trial-and-error process of searching for components that can be used in their program. This process is complicated because the programmers cannot decide which components to use unless they fully understand their detailed behavior. Furthermore, unlike conventional programming paradigms, the component package has an affect on what the programmer will make at the detailed level. Components in a certain package may make certain requirements possible (or impossible) to implement, further complicating the decision of choosing a component. McKinney went as far as to say that the component package should "drive your requirements allocation" [ ].

Thus, the programmers will need to remember intermediate search results, i.e. in order to continue searching, the programmer will need to be able to keep track of what he/she needs, what he/she currently has, and the relationships between what he/she has. In other words, the programmer needs some method of "sketching", allowing them to visualize the current situation.

2.3 Jigsaw Puzzle

We can view both writing and component-based programming as solving a jigsaw puzzle. A person solving a jigsaw puzzle needs to try various combinations of pieces, but needs to see at each instance the state of the task, e.g. what has been finished, what pieces need to be considered, etc. By being able to see the current state, the person can consider the next step to take. Through this process, he/she will obtain a coherent whole, i.e. a completed puzzle.

As we described above, a writer may have several written pieces in hand and need to explore how they can be combined, what needs to be modified, etc. The writer can only do this exploration if he/she can see the current state. Through seeing the current state, the writer can decide on for example what changes to make, ultimately resulting in a coherent whole, i.e. a document.

Similarly, a programmer may need to search through a library of components to develop a program. He/she will need to keep track of what type of component is needed, what he/she already has, and the relationships between what he/she already has. This will let the programmer continue searching for other components, until the goal is reached of obtaining a coherent whole, i.e. a program.

Of course there are differences between writing, programming, and solving jigsaw puzzles, and we will discuss the differences in Section 4.

2.4 Two-Dimensional Positioning

We characterized writing and programming to be similar to solving a jigsaw puzzle. Writers and programmers need to fit the various pieces together so that the document or program will become a coherent whole.

People solving a jigsaw puzzle will often use some space such as a table, and put pieces on it according to some meaning. For example, if a person thinks that a piece corresponds to a certain part of the picture, that person will likely place it at a corresponding place on the table. Other pieces may be grouped together

---

[1] A component package is a library that contains the software components from which the programmer chooses to build software.

according to its color, or whether it is a side piece. The positioning of the pieces has some meaning that will help the person solve the puzzle.

Much like positioning puzzle pieces on a table will help solve a jigsaw puzzle, we take the approach of positioning elements in a two-dimensional space as a means of visualizing the current situation. Elements correspond to the pieces in the jigsaw puzzle. For writing, elements are "parts of writing": phrase, sentence, paragraph, section, etc. In component-based programming, the elements would be the software components themselves.

According to Shipman et al [ ], people use the visual and spatial characteristics of graphical layouts to express relationships between icons and other visual symbols. Thus, people can position elements in a two-dimensional space to help visualize and understand their current situation, from which they can reflect and continue with their task, much like sketches in the architectural domain.

We next describe the two tools we have developed for writing and programming.

3 Two Tools – ART and RemBoard

We describe the two tools that we have developed for the writing and programming domains. We also describe how users were able to use them to visualize and understand the current situation.

3.1 ART

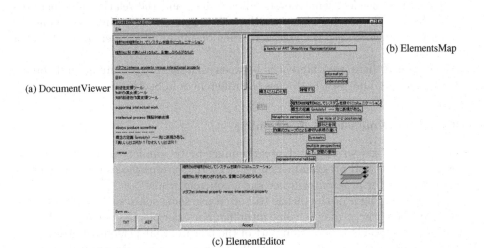

(a) DocumentViewer

(b) ElementsMap

(c) ElementEditor

Fig. 1. ART

The ART system [ ] [ ] (Fig.) supports document construction as a design task allowing users to position segmented text as "elements" in a two-dimensional space. An element is any unit that writers choose to think of as one, such as a phrase, a sentence, a paragraph, or a longer piece of text.

System Overview The ART system consists of the following three main parts: ElementsMap, ElementEditor, and DocumentViewer.

ElementsMap (Fig. (b)) is a two-dimensional space that graphically displays elements that comprise the document. Each element is represented as an icon. An icon does not show the entire content of the element, but only the first ten percent of the element's text; therefore, the size of the element box corresponds to the size of the actual element (unless the user resizes the box). A user can freely change the position of elements by pointing and dragging icons on ElementsMap.

Elements can be created and edited with ElementsMap and ElementEditor (Fig. (c)). The ElementEditor component is a text editor for the contents of an element providing editing functions such as cut, copy, paste, and "spin off," which divides one element into two. Selection of an element in ElementsMap allows a user to modify the content of the element in the ElementEditor. When nothing is selected in ElementsMap, a user can edit text in ElementEditor and create a new element by positioning the newly created icon in ElementsMap. Two or more elements can be merged by selecting multiple elements on ElementsMap.

One point of note is that an element's vertical position in the ElementsMap is interpreted as corresponding to its position in the document sequence, and the DocumentViewer (Fig. (a)) component displays the actual content of the document by sequentially scanning the elements displayed in the ElementsMap from top to bottom. Thus, a user can freely change the order of elements in the whole document by changing the vertical relationship of elements in ElementsMap. Positioning changes and content changes made in ElementsMap and ElementEditor are automatically reflected in all three parts.

Visualizing the Current Situation The ART system provides three types of "views" to understand the current situation. The ElementsMap provides an overview of the entire situation, while ElementEditor provides details of each element. The DocumentViewer displays the context of the element with details of neighborhood elements. The three views are integrated and changes made in one view are dynamically reflected on the other views.

User studies [ ] have shown that positioning on the ElementsMap is the core of using the system. For example, one subject placed elements in the bottom right corner of the ElementsMap. This let the subject know that those elements needed further work. The subject took one of those unfinished elements, made modifications, and then placed them with other elements that already had some work done on them. The subject would then repeat this process. Another subject made a set of elements be the same size and carefully aligned them; this was to indicate that these elements were "finished". Still another subject made some elements particularly larger than others so that it would "call for attention" later in the task.

In one situation, a user felt that two elements "should be related to each other but I don't know how they are related". Instead of aligning them, she overlapped them to let her understand what she was thinking.

In this way, each user visualized various situations in their own way to help them understand their current situation, and let him/her proceed with the task.

3.2 RemBoard

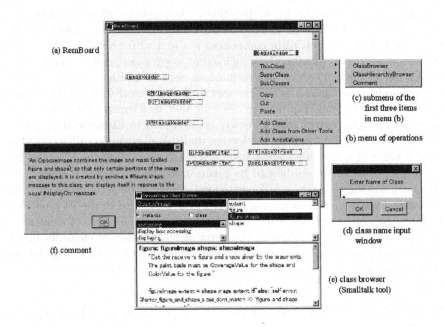

Fig. 2. RemBoard

RemBoard [ ] (Fig. (a)) is a tool for remembering software components that may become necessary while programming in the Smalltalk programming language. More precisely, the components that RemBoard handles are *classes*, although RemBoard can also handle annotations. Smalltalk, which is an object-oriented programming language, allows programmers to reuse classes from a large class library. For example, in the VisualWorks (Smalltalk) environment, programmers can exploit a library of more than 900 reusable classes, which helps increase program quality and productivity.

Finding "necessary" reusable classes from the large library, however, is challenging, making Smalltalk programming hard to learn for novice programmers. It is not easy to understand the whole class hierarchy, and having a hierarchy browser and keyword matching retrieval mechanisms are not enough for novices because it is difficult for them to understand what retrieved classes and methods "really mean" [ ]. There is a need for a mechanism that allows Smalltalk programmers to remember intermediate search results as one cannot decide which classes and objects to use unless one fully understands their detailed behavior. RemBoard provides a representational medium for such programmers to remember what has been retrieved.

System Overview RemBoard is a tool that is added on top of the VisualWorks (Smalltalk) environment. RemBoard is a free two-dimensional space where programmers can place elements, which are classes. Various operations can be performed directly on elements displayed on RemBoard.

Users can put classes on RemBoard by using a window to directly input the class name (Fig. (d)) or by copying from a Smalltalk tool, such as an editor (Fig. (e)) or a tool showing search results. Recorded classes are shown as icons on the two-dimensional space with the class name as its label. Users can also place their own annotations on the two-dimensional space. Users can move, delete, or duplicate elements on RemBoard.

RemBoard also allows users to directly perform operations on iconized classes that are necessary for a programming task: for example, to open an editor (Fig. (e)) on a class, or to show comments that were attached by the original programmer (Fig. (f)).

Visualizing the Current Situation RemBoard uses a two-dimensional space to express relationships among the recorded classes. In the user studies [ ], we have found that subjects used positioning of elements on RemBoard to visualize classes that are under consideration for the task, classes that were thought to be "related" to those classes, and the relationships among these classes.

In one situation, a user positioned a class below another class to show inheritance (the relationship between ImageReader and BMPImageReader in Fig. (a)). This let the user visualize that BMPImageReader inherits various states and behavior from ImageReader, and could keep this in mind while continuing his task. In another situation, a class was placed far from other classes to show that it was not directly related to the other classes but was deemed important to remember (OpaqueImage in Fig. (a)).

Interestingly, the above two situations were taking place simultaneously within a single two-dimensional space without causing any confusion for the subject. When the programmer is understanding the situation, the "meaning" of the positioning depends on what part of the space the programmer is looking at.

4 Discussion

We have compared writing and component-based programming with solving a jigsaw puzzle, and much like using a table to solve a puzzle, we used a two-dimensional space to visualize and understand the current situation, i.e. the two-dimensional space served as a means for "sketching".

There are a few important differences between an actual jigsaw puzzle and writing (or programming):

1. An actual jigsaw puzzle has all the pieces necessary to make the jigsaw puzzle. In writing and programming, not all pieces may exist. The writer (or programmer) will need to make new pieces as necessary.
2. An actual jigsaw puzzle will not have any unnecessary pieces. This is also not true with writing and programming. The writer (or programmer) may have pieces at hand, but they may actually not need it.
3. No changes are necessary for pieces in an actual jigsaw puzzle. Writers (or programmers) may need to make changes to parts of text (or software component) so that they will fit.
4. An actual jigsaw puzzle will have a picture of what the completed puzzle will look like. This also does not hold for writing and programming. There may be a conceptually top-level goal to work for, but there is rarely no detailed map from which to work with.
5. There will only be one solution for a jigsaw puzzle, but a document may be written in many ways, and there may be several components that can satisfy the requirements for a program.

In order to account for these, the writer or programmer will need to be able to:

- make new elements,
- delete elements,
- edit elements, and
- see what they currently have.

We further discuss these issues below. The discussion is augmented by results from user studies of the two tools.

4.1 Overview of the User Studies

In one ART user study [ ], three subjects modified a collection of notes with no coherent structure that concerned the usage of instruments in a laboratory. The subjects' task was to make it into a coherent document. The initial "jigsaw puzzle piece" was one element holding all of the notes. In another ART user study [ ], a subject wrote an actual grant proposal based on three email discussions. In this case, the initial jigsaw puzzle pieces were three elements, each corresponding to one email discussion.

In the RemBoard user study [ ], three subjects modified a small address book application, so that the items displayed in the address book (such as names and addresses) would be aligned. The initial jigsaw puzzle pieces in the Rem-Board user study was the four classes in the application. There was also the basic class library provided by the VisualWorks (Smalltalk) environment.

Overall, the two-dimensional space provided by ART was used more frequently than in RemBoard. This could be attributed to the fact that ART is a stand-alone tool while RemBoard is an add-on tool. The two-dimensional space in ART is the most integral part and the user has to use it to conduct his/her task . On the other hand, RemBoard is an add-on tool and the programmer can write a program without using it.

4.2 Adding, Deleting and Editing Elements

The first three issues of adding, deleting, and editing elements were implemented in both ART and RemBoard. Although elements in both ART and RemBoard could be deleted by choosing the element, and then choosing "delete" or "cut" from a pop-up menu, the implementations of adding and editing elements were done differently.

ART elements are added and edited by using the ElementEditor and ElementsMap, i.e. these two functions are completely integrated within ART.

On the other hand, RemBoard elements are added by using a window to directly input the class name or by copying from a Smalltalk tool. Elements are edited by opening a Smalltalk editor. Since RemBoard is added on top of the VisualWorks (Smalltalk) environment, these two functions are done using the tools provided. This is also a reflection of RemBoard being a tool that supports the search for necessary components, and not a standalone programming environment.

There is another significant difference between ART and RemBoard. An ART element can be split into multiple elements, and multiple elements can be merged into one element. A corresponding operation does not exist in RemBoard. This is the result of the difference between the granularity of elements in ART and RemBoard. An ART element is a piece of text that the writer thinks of as "one", while a RemBoard element is a class. In the ART user studies, users were frequently found to split and merge elements into a granularity which they felt were more useful for their current situation.

4.3 Seeing what They Have

The fourth issue of seeing what the writer or programmer has is most critical. The writer or programmer does not have a complete image of what the document or program will be like when finished. Furthermore, there is no one correct solution for a document or a program; a document may be written in different ways, and several different components may be available as options, each of which satisfies the same requirements. These points mean that understanding the current situation will have much more significance in writing and programming

---

[2] It is of course possible to make one large element.

compared to solving a jigsaw puzzle; the writer or programmer will have to consider various options, each of which may be correct, but "correct" in different ways.

If the writer or programmer is not able to understand the current situation, he/she will not be able to reflect on it and the design process will not proceed well. In other words, the externalization or visualization of the current situation did not go well, and the tool would not function as a "sketching" tool.

Expanding this further, the writer or programmer will need to be able to know the following three types of information:

- what elements they are currently considering,
- the characteristics of the elements they are considering, and
- the relationship between elements they are considering.

They will further need to have some freedom in being able to express the characteristics and relationships.

They should also be able to see the entirety. Because the designer needs to consider many disparate factors, they prefer to be able to view sketches without much head or eye movement [ ]. This led to both ART and RemBoard having a "limited" two-dimensional space. It would have been possible to have the two-dimensional space be larger and have scroll bars, but this would have meant that the designer would need to search within the two-dimensional space.

We now consider each of the three types of information that the writer or programmer will need to know, i.e. those that need to be visualized.

Elements under Consideration We first consider the need to be able to know what elements the writer or programmer is currently considering. We consider this from three viewpoints: labels, element size, and positioning.

Labelling is one way to differentiate elements, and thus be able to know which elements are under consideration. For example, labels of jigsaw puzzle pieces are fixed; it is the picture of the part it is supposed to show. Labels were handled differently for ART and RemBoard. Since each element in RemBoard is a class, the class name was used as a label. It is a definite precise name. In contrast, no such labelling is used in ART. What the writer sees is the first ten percent of the text in the element (unless the writer resizes the element). No explicit labelling mechanism was provided in ART because each element can be thought of differently according to the context. We have, however, observed some users making a pseudo-label by fitting the "label" in the first ten percent of the element's text.

Another way to know what elements are under consideration is by resizing the elements. Both ART elements and RemBoard elements are resizable. One ART user resized several elements so that they would "call for attention" later in the task. Another ART user had resized several elements to the same size to let him know that "those elements are finished". Interestingly, none of the RemBoard users in the user study resized an element for a similar purpose.

Finally, a third way to differentiate elements under consideration is through positioning. Users can position elements in a way so that elements placed in a certain area need work, and others not. One ART user put unelaborated elements at the bottom right corner of the ElementsMap. In RemBoard, a user put unnecessary elements in the top right corner .

Characteristics of Elements Another type of information necessary to know the current situation is to know characteristics of the elements. Size and shape are two example characteristics of a jigsaw puzzle piece. These two characteristics can be used to figure out if one piece fits another. Unlike jigsaw puzzle pieces, ART and RemBoard elements may change, and it can be considered to be useful if such characteristics could be shown. We consider four types of characteristics: size, "last changed", complexity, and origin of element. Note that in a jigsaw puzzle, all four characteristics are fixed, i.e. puzzle pieces are of nearly the same size, they do not change, the complexity of pieces in the same puzzle are the same, and all pieces come from the same jigsaw puzzle package.

Both ART elements and RemBoard elements have *size* as element-specific characteristics. The type of size in ART is limited basically to number of characters and number of words. Although number of sentences is also possible, since an ART element does not necessarily need to be a sentence, it may have no meaning. Currently size in ART is visualized by having the element size be large enough to show the first ten percent of the element's text. In contrast, RemBoard elements have many types of size properties. They include number of methods, number of variables, and number of lines. These also could be visualized through element size.

The second characteristics *"last changed"* denote when that element was last modified. This can be implemented similarly for both ART and RemBoard. Color could be used to show which element was changed last, next to last, etc.

The third characteristics *complexity* can be used for RemBoard elements. Each RemBoard element, or Smalltalk class, belongs in an organized hierarchy, which lets classes inherit behaviors and states from other classes. Each class can also have a use (or used) relationship with other classes. These inheritance and use/used information can be used to compute the complexity of a class [ ], and then visualized through means such as color or size. Although it may be possible to define a similar measure for ART elements, it would basically be a variation of the size characteristics.

The fourth characteristic *origin of element* concerns where the programmer (or writer) "found" the element. RemBoard is integrated within the Smalltalk programming environment which has an extensive class library, i.e. there is a large pool of components from which to choose. Although programmers will develop new classes, keeping track of what classes have been found and how it was found is important. Without such information, the programmer may not be able to, for example, redo part of a keyword search. Thus, RemBoard elements

---

[3] The user could have deleted the element but chose not to because he was "unsure if it really was unnecessary".

have a feature where the programmer can see how it was put on the RemBoard. For example, if keyword search was used, the list of keywords that were used to retrieve that class is shown as part of the label. Because programmers have this information, they can go back and try other combinations of keywords. On the other hand, there are no document database connected to ART, and no information is stored (or visualized) concerning how an element was produced. Although writers can use documents from previous documents that they have used, much of the text is written anew, especially when compared to RemBoard elements. In fact, no ART users have ever expressed an interest in a feature that would let them know how a certain element was produced.

The above characteristics are of course not exhaustive and other characteristics could be beneficial. For example, Yacoub et. al have proposed a set of features characterizing software components [ ].

Relationship between Elements Users will want to view many types of relationships between elements. In jigsaw puzzles, the key relationship is its location compared to the final product. Thus, people will often put puzzle pieces where they think that piece will go, especially if it is a side or corner piece.

Relationships between elements in ART and RemBoard are shown through the positioning in the two-dimensional space. We have already given two examples of global relationships that were observed for ART and RemBoard users: an ART user put unelaborated elements at the bottom right corner of the ElementsMap, and a RemBoard user put unnecessary elements in the top right corner. We now elaborate on other types of relationships.

Between ART and RemBoard, there is only one fixed relationship: the vertical (or top-bottom) orientation in ART. The top-to-bottom ordering corresponds with the order that the elements are concatenated to obtain the document (which is shown in the DocumentViewer). This can be considered to be a natural constraint, as a written document is read from top to bottom. None of the ART users have ever expressed a dissatisfaction with this fixed orientation.

Because of this top-bottom constraint, ART users are basically left with the horizontal (or left-right) orientation to show various relationships. Some of the relationships that were observed were:

- enumeration/itemization: A user positioned elements at the same vertical position, but used horizontal positioning to show that they were items of a group. Interestingly, the same user used the vertical relation to show the same thing in a different situation. She used a different way of visualizing the same relationship.
- need for other elements: A user put some space between two elements to show that he "needed to write something".
- "closeness": All users tended to close the space between elements if he/she thought that those elements were closely related in some way.

RemBoard users expressed relationships such as the following:

- inheritance: A user placed one element below and to the right of another class; for example, BMPImageReader and ImageReader in Fig   (a).
- use/used: A user placed elements next to each other to show that one class used the other class. Interestingly, this was not fixed to just vertical or horizontal positioning. A user would use vertical positioning for one task but horizontal for another. Also, the user was able to remember right away which class was using the other class, without their being an explicit annotation describing it.
- alternative: A user aligned alternatives; for example, GIFReadWriter and JPEGReadWriter is one alternative and GifImageStream and JpegImageStream is another alternative in Fig   (a).
- important but not related: When a user found a class that he thought to be important but could not define its relation with other classes already on RemBoard, he placed it far away from those already on RemBoard; for example, OpaqueImage in Fig.   (a).

Although this was not observed, other relationships could also be possible, such as aggregation which in object-oriented programming means part-whole relationship.

4.4   Sketching and Two-Dimensional Positioning

As we discussed above, the two tools served as a means for the writer or programmer to visualize their current situation. Two-dimensional positioning was a critical component as it allowed the users to express relationships and characteristics of the elements. As with sketching, the users had freedom in how they could express or visualize various situations – except for the vertical relationship in ART, they were free to position elements in any manner they want.

Through visualizing the current situation, the users were able to reflect and consider what the next step should be. For example, the following situations were seen in the user studies:

- An ART user continually moved an element until it reached a state that the user thought "correctly" showed the relationship between it and other elements, i.e. she used the visualization to reflect on the design [ ].
- A RemBoard user put on RemBoard any classes which he thought might be of use later. This helped him visualize what possible combinations might be possible in his programming task [   ].

These observations from the user studies suggest that the two-dimensional positioning served the same purpose as sketches in the architectural domain.

5   Conclusion

We took writing and programming as two example domains, and described two tools – ART for writing and RemBoard for component-based programming – that supports the writing or programming task. The key idea is that two-dimensional positioning can serve the same purpose in these two domains, much like sketches do in the architectural domain, i.e. they serve to visualize what the designer is thinking. We made references to their similarity with solving a jigsaw puzzle, and discussed how visualization occurred.

There are various ways that a user can visualize a certain train of thought. We need to keep this in mind: certain visualization may actually hinder the user's task. A mapping that is "natural" from the user's viewpoint is critical in any type of sketching tool. From this viewpoint, the two tools let users freely position and resize elements; the users are the ones that map a certain situation or thought to a certain representation. The power of attaching meaning to a representation is a powerful one and should be exploited to its fullest, just like sketching in architecture, ART for writing and RemBoard for component-based programming.

References

1. Bruner, J.: The Culture of Education. Harvard University Press, Cambridge, MA. (1996).
2. Chidamber, S. and Kemerer, C.: A Metrics Suite for Object Oriented Design. IEEE Trans. on Software Engineering. 20(6) (1994) 476–493.
3. Kintsch, W. and van Dijk, T.: Towards a model of text comprehension and production. Psychological Review. 85(5) (1978) 363–394.
4. Lawson, B.: Designing with drawings. How Designers Think: The Design Process Demystified. Architectural Press, Oxford, MA. Chapter 14 (1997) 241–259.   ,
5. McKinney, D.: Impact of Commercial Off-The-Shelf (COTS) Software on the Interface Between Systems and Software Engineering. Proc. of 1999 International Conference on Software Engineering. (1999) 627-628.
6. Nakakoji, K., Yamamoto, Y., Takada, S., Reeves, B.: Two-Dimensional Positioning as a Means for Reflection in Design. Conference on Designing Interactive Systems (accepted).   ,   ,
7. Schoen, D. A.: The Reflective Practitioner: How Professionals Think in Action. Basic Books, New York (1983).
8. Shipman, F. M., Marshall, C. C., and Moran, T. P.: Finding and Using Implicit Structure in Human-Organized Spatial Layouts of Information. Human Factors in Computing System (CHI'95). (1995) 346–353.
9. Simon, H. A.: The Sciences of the Artificial (Third ed.). MIT Press, Cambridge, MA (1996).
10. Takada, S., Nakakoji, K., and Torii, K.: Using 2D Space for Understanding What "Search Options" We Have Taken in Exploring a Class Library. Technical Report ccc-98-9, Cognitive Science Lab, NAIST (1998).   ,   ,   ,
11. Takada, S., Otsuka, Y., Nakakoji, K. and Torii, K.: Strategies for Seeking Reusable Components in Smalltalk. Proc. of 5th International Conf. on Software Reuse. (1998) 66–74.
12. Yamamoto, Y., Takada, S., and Nakakoji, K.: Representational Talkback: An Approach to Support Writing as Design. Proc. of 3rd Asia-Pacific Computer Human Interaction Conf. (1998) 125–131.   ,

13. Williams, N.: New Technology, New Writing, New Problems? Intellect Books, England (1992).
14. Yacoub, S., Ammar, H. and Mili, A.: Characterizing a Software Component. Proc of International Workshop on Component-Based Software Engineering. (1999) 151–158.

Reordering the Reorderable Matrix as an Algorithmic Problem

Erkki Mäkinen and Harri Siirtola*

Department of Computer and Information Sciences
P.O. Box 607, FIN-33014 University of Tampere, Finland
{em,hs}@cs.uta.fi

Abstract. The Reorderable Matrix is a visualization method for tabular data. This paper deals with the algorithmic problems related to ordering the rows and columns in a Reorderable Matrix. We establish links between ordering the matrix and the well-known and much studied problem of drawing graphs. First, we show that, as in graph drawing, our problem allows different aesthetic criterions which reduce to known NP-complete problems. Second, we apply and compare two simple heuristics to the problem of reordering the Reorderable Matrix: a two-dimensional sort and a graph drawing algorithm.

1 Introduction

Bertin's Reorderable Matrix [,] is a simple visualization method to explore multivariate or multidimensional data. The basic principle is to transform a multidimensional data set into a 2D interactive graphic. The graphical presentation of a data set closely resembles the underlying data table in that it contains rows and columns. These rows and columns can be permuted, allowing different views of the data set. The actual data values are replaced with circles that have a size relative to the actual data value. The smallest value is represented as a 0-sized circle and the largest value as a circle filling the available area. While interacting with the visual presentation, the user has a chance to detect patterns in the presentation and to gain insight into the data. Human vision is well equipped for this kind of pattern recognition.

Figure contains a small Reorderable Matrix [, p. 32] having binary values only. In this simple example we have sixteen townships $A, B, C, \ldots, P$ for which we know the presence or absence of nine characteristics, $1, 2, 3, \ldots, 9$. The question is whether the same planning decisions should be applied to all these townships?

Processing the Reorderable Matrix in Figure involves bringing together similar rows and columns. This in turn involves dragging rows and columns, one by one, and can take a while. We leave the intermediate steps out and represent the result in Figure .

From the arrangement in Figure it is quite easy to see that there are obvious groups in townships. Townships H, K have similar characteristics and could be

---

* HCI Group TAUCHI, http://www.cs.uta.fi/hci/

M. Anderson, P. Cheng, and V. Haarslev (Eds.): Diagrams 2000, LNAI 1889, pp. 453–   , 2000.
© Springer-Verlag Berlin Heidelberg 2000

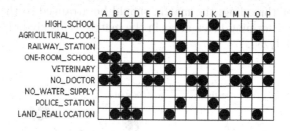

Fig. 1. The townships planning decisions example

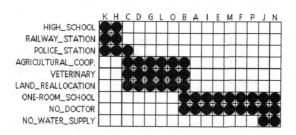

Fig. 2. An arrangement that reveals the correct decisions

labeled column-wise as CITIES and row-wise as URBAN. Similarly, townships N, J, P, M, I, F, E and A could be classified column-wise as VILLAGES and row-wise as RURAL. The remaining set could be called TOWNS, as an intermediate level between CITIES and countryside.

Even for this simple example there is a large number of *arrangements* or possible row and column permutations to be explored. For the townships example the number of possible arrangements is

$$(\# \; of \; rows)! \times (\# \; of \; columns)! > 7.5 \times 10^{18}.$$

Many of the arrangements are isomorphic as patterns – e.g., mirroring the matrix vertically or horizontally does not change the patterns – although the arrangements are not the same. Still, even the number of different patterns is high. In this example the size of the matrix was small, 16×9, when Bertin suggests that the Reorderable Matrix should be usable with sizes up to $x \times y = 10,000$. The largest experiment he writes about was $415 \times 76 = 31,850$. It is obvious that finding the interesting patterns calls for automation.

1.1 Exploring the Reorderable Matrix

The Reorderable Matrix can assist in a knowledge acquisition task in a number of ways. The general principle is to thread the matrix either horizontally or vertically according to, a column or a row, respectively. The chosen column or row should be one that seems to portray a phenomenon that has general influence. After threading, the matrix should be arranged so that similar rows appear together, forming black areas in the matrix. This kind of arrangement is ready to be analyzed.

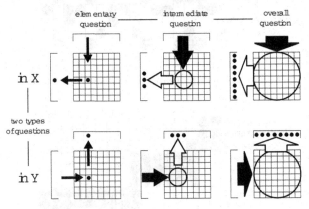

Fig. 3. Three levels of information: elementary, intermediate and overall [, p. 13]

There are three kind of questions that can be answered with an appropriately arranged matrix: questions about certain objects or characteristics, questions about subsets of objects or characteristics and questions about overall object or characteristic sets. Figure   illustrates these three levels of information.

1.2 Motivation for This Work

As far as we know, ordering the Reorderable Matrix has not been regarded before as an algorithmic problem in the literature. Only [ ] contains a remark on the possible computational difficulty of the problem.

There exists a great variety of algorithmic problems concerning the Reorderable Matrix. The situation resembles that of graph drawing algorithms: depending on the aesthetic criterions applied, the graph drawing problem may reduce to various known combinatorial and other problems. We can minimize the number of edge crossings, display as much symmetries as possible, or draw the vertices evenly over the drawing area, just to mention a few possibilities (for details and other possibilities, see [ ]). One of our main goals here is to show that the same applies to the Reorderable Matrix, i.e., depending on the view chosen, we can model the task of reordering the matrix by using various different combinatorial problems. Notice that both in graph drawing and in ordering the Reorderable Matrix, the different aesthetic criterions are often conflicting, i.e., it is not usually possible to simultaneously fulfill more than one criterion.

Note that we do not give an exact mathematical definition for the problem of reordering the Reorderable Matrix. As an implication, we cannot (mathematically) argue that the exactly formulated problems match to the original reordering problem. However, it is obvious that the exact problems to be discussed later in this paper are closely related to relevant subproblems of reordering the Reorderable Matrix. The situation is analogous to the one of drawing graphs "nicely": we have to always define what we mean by "nicely", and the exact mathematical definitions given do not cover all aspects of nice drawings.

2 Preliminaries

We assume a familiarity with the rudiments of the theory of computational complexity and NP-complete problems as given in [ ].

The Reorderable Matrix can be regarded as an $m \times n$ matrix with entries from the set $\{0, 1, \ldots, e\}$. The entry in row i and column j in matrix M is denoted by M_{ij}. The submatrix containing the entries M_{ij}, where $1 \leq i \leq p$ and $1 \leq j \leq q$, is said to be the *upper* $(p, q) - submatrix$ of M. Similarly, the entries M_{ij}, where $p \leq i \leq m$ and $q \leq j \leq n$, form the *lower* $(p, q) - submatrix$ of M (see Figure).

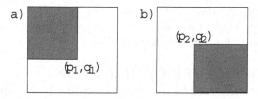

Fig. 4. (a) Upper (p_1, q_1)-submatrix and (b) lower (p_2, q_2)-submatrix

Graph bandwidth minimization [, ,] is perhaps the most well-known problem type involving matrices and row and column permutations. In bandwidth minimization our goal is to permute the rows and columns of a given matrix such that the non-zero entries of the matrix form as thin a "band" as possible along the main diagonal (see Figure).

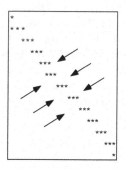

Fig. 5. The purpose of bandwidth minimization (* stands for a non-zero entry)

However, the link between graphs and matrices implies that in these problems row and columns permutations are performed simultaneously, while in reordering the Reorderable Matrix row and column permutations are independent of each other. Despite the differences in formulation, the NP-completeness of various bandwidth problems gives us a hint that most reasonable problems concerning the Reorderable Matrix must also be NP-complete.

Another well-known matrix operation closely resembling the Reorderable Matrix permutation is Gaussian elimination, where rows and columns are multiplied by and subtracted from each other. Contrary to the bandwidth problems, we now handle rows and columns independently. On the other hand, in Gaussian elimination we do not *permute* rows and columns, but perform arithmetic operations between them. For a NP-complete problem formulation related to Gaussian elimination (DIRECTED ELIMINATION ORDERING), see [,].

At least the following known NP-complete problems have a close connection with the Reorderable Matrix.

MATRIX DOMINATION [ ]

Instance: An $n \times n$ matrix M with entries from $\{0,1\}$, and a positive integer K.
Question: Is there a subset $C \subseteq \{1,2,\ldots,n\} \times \{1,2,\ldots,n\}$ with at most K elements such that $M_{ij} = 1$ for all $(i,j) \in C$, and whenever $M_{ij} = 1$, then there exists an (i',j') in C for which either $i = i'$ or $j = j'$?

RECTILINEAR PICTURE COMPRESSION [ ]

Instance: An $n \times n$ matrix M with entries from $\{0,1\}$, and a positive integer K.
Question: Is there a collection of K or fewer quadruples (a_i, b_i, c_i, d_i), $1 \leq i \leq K$, where $a_i \leq b_i$, $c_i \leq d_i$, such that for every pair (i,j), $1 \leq i,j \leq n$, $M_{ij} = 1$ if and only if there exists k, $1 \leq k \leq K$, such that $a_k \leq i \leq b_k$ and $c_k \leq j \leq d_k$?

3 Reordering Aesthetics

The problem formulations presented in the previous section use square matrices. The Reorderable Matrix is not usually square, but for notational convenience and with no loss of generality, we will only deal with square matrices.

A typical approach in ordering the Reorderable Matrix is to arrange the rows and columns such that there are "black areas" in the top left and bottom right corners implying "white areas" in the top right and bottom left corners. As we assume that the matrix entries are from the set $\{0, 1, \ldots, e\}$, we can define 'white' to be any of the values $0, 1, \ldots, c$, and 'black' to be any of the values $c+1, c+2, \ldots, e$, with an appropriate constant c (the value of which is not fixed here). This aesthetic can be formalized as follows.

PROBLEM1

Instance: An $n \times n$ matrix M with non-negative entries, and an integer K.
Question: Is it possible to perform K or less row permutations and K or less

column permutations such that all the non-white entries appear in the upper (K, n)-submatrix or in the lower $(K + 1, n + 1)$-submatrix?

Since PROBLEM1 clearly is in NP, its NP-completeness can be proved by defining a polynomial transformation from MATRIX DOMINATION.

Theorem 1. *PROBLEM1 is NP-complete.*

Proof. Consider an instance of MATRIX DOMINATION with an $n \times n$ matrix P and an integer k. The corresponding instance of PROBLEM1 consists of an $(n + k) \times (n + k)$ matrix and an integer $K = k$. The new matrix M has the form shown in Figure . The $n \times n$ submatrix is the original matrix P, while the new upper (k, n)-submatrix, the new lower $(k+1, n+1)$-submatrix and the new $k \times k$ matrix in the upper right corner contain zeros only.

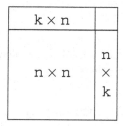

Fig. 6. The matrix in the instance of PROBLEM1

Suppose the instance of MATRIX DOMINATION is related with "yes" answer, i.e. there are K or less non-zero entries dominating P. Arbitrarily order the dominating entries: $P_{i(1)j(1)}, P_{i(2)j(2)}, \dots, P_{i(\kappa)j(\kappa)}$, where $\kappa \leq k$. For each dominating entry $P_{i(t)j(t)}$, $t = 1, \dots, \kappa$, permute rows t and $k + i_t$ and columns j_t and $n + t$ in M. Since each non-zero entry of P is dominated by some of the entries $P_{i(t)j(t)}$, $t = 1, \dots, \kappa$, the permutations done move all the non-white elements of M to the upper (k, n)-submatrix or to the lower $(k + 1, n + 1)$-submatrix. On the other hand, if such permutations are possible in M, then P must be dominated by $k = K$ or less entries. This completes the proof.

Wilf [, Section 2.4.], has posed open the completeness status of a problem somewhat similar to PROBLEM1. In Wilf's problem one is asked to find row and column permutations such that the resulting matrix is triangular.

PROBLEM1 is related to a case where we expect that there are two "clusters" of positively correlating factors. In general, there can be any number of such clusters, i.e., any number of "black areas" in the matrix. This, in turn, can be modeled by RECTILINEAR PICTURE COMPRESSION. Again, we consider binary matrices and leave open the definitions of "white" and "black" entries.

PROBLEM2

Instance: An $n \times n$ matrix M with entries from $\{0, 1\}$, and a positive integer K.
Question: Is it possible to perform row and column permutations in M such that

there eventually is a collection of K or fewer quadruples (a_i, b_i, c_i, d_i), $1 \leq i \leq K$, where $a_i \leq b_i$, $c_i \leq d_i$, $1 \leq i \leq K$, such that for every pair (i,j), $1 \leq i, j \leq n$, $M_{ij} = 1$ if and only if there exist a k, $1 \leq k \leq K$, such that $a_k \leq i \leq b_k$ and $c_k \leq j \leq d_k$?

As an example, consider the matrix in Figure . Its black areas can be represented by five rectangles in several different ways. One of the solutions with five rectangles is $\{(1,2,1,3), (3,3,3,3), (3,8,4,6), (8,16,7,8), (15,16,9,9)\}$.

Theorem 2. *PROBLEM2 is NP-complete.*

Proof. PROBLEM2 has RECTILINEAR PICTURE COMPRESSION as a subproblem. Since PROBLEM2 clearly is in NP, the theorem follows immediately by restriction [ ](pp. 63–66) from the NP-completeness of RECTILINEAR PICTURE COMPRESSION.

Also the following known NP-complete problems could be used as the basis for formulating a reordering aesthetic:

MATRIX COVER [ ]
Instance: An $n \times n$ matrix M with non-negative entries, and an integer K.
Question: Is there a function $f : \{1, 2, \ldots, n\} \to \{+1, -1\}$ such that

$$\sum_{1 \leq i,j \leq n} M_{ij} \cdot f(i) \cdot f(j) \leq K?$$

TRIE COMPACTION [,]
Instance: A multiset $\{X(1), X(2), \ldots, X(n)\}$ of bit strings of length m and an integer K.
Question: Is it possible to place the strings such that overlapping is possible if in overlapping positions at most one of the strings has a non-null content and such that the total length of the resulting bit string is at most K?

However, we omit the details of these aesthetics here.

4 Methods for Reordering

We have shown that certain subproblems of reordering the Reorderable Matrix are closely related to well-known NP-complete problems. Hence, it is reasonable to consider heuristic approaches to the problem.

We present two methods for reordering the Reorderable Matrix. The former is a simple two dimensional (2D) sort and the latter is a variant of Sugiyama's graph layout algorithm. Both algorithms were implemented and tested with a number of matrices (see web page http://www.cs.uta.fi/~hs/iv99/). We present two of these test cases in the following discussion.

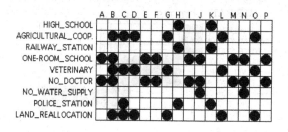

Fig. 7. The townships matrix

4.1 The Townships Example

The first test case is the townships matrix we presented earlier. It is a simple binary matrix with obvious 'interesting' sets of arrangements. The initial arrangement is displayed again in Figure .

In Figure   we have one of the 'interesting' arrangements for the townships example. Subsets in the data with similar characteristics can be easily seen.

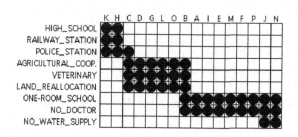

Fig. 8. One of the 'optimal' arrangements for the townships example

The set of 'interesting' arrangements in this example is quite large, as Figure illustrates. Most of the rows and columns inside subsets CITIES, TOWNS and VILLAGES could be in any order and the information would still be the same.

4.2 The Hotel Example

The second test case contains monthly data compiled from an imaginary hotel [, pp. 1–11]. The object set is twelve months and the characteristics are properties that hotel management could use for business planning.

The data contains various numbers describing the clientele, the average price of room and the average length of stay. Most of the numbers are percent figures, except for the average length of stay, the average price of rooms and the information whether or not there was a convention during that month. The average length of stay is given in days and the average price of room is given in the local

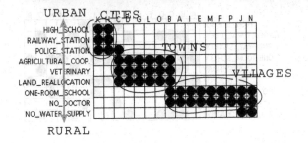

Fig. 9. The interesting subsets for the townships example

J	F	M	A	M	J	J	A	S	O	N	D		
26	21	26	28	20	20	20	20	20	40	15	40	1	% CLIENTELE FEMALE
69	70	77	71	37	36	39	39	55	60	68	72	2	% —//— LOCAL
7	6	3	6	23	14	19	14	9	6	8	8	3	% —//— U.S.A.
0	0	0	0	8	6	6	4	2	12	0	0	4	% —//— SOUTH AMERICA
20	15	14	15	23	27	22	30	27	19	19	17	5	% —//— EUROPE
1	0	0	8	6	4	6	4	2	1	0	1	6	% —//— M EAST, AFRICA
3	10	6	0	3	13	8	9	5	2	5	2	7	% —//— ASIA
78	80	85	86	85	87	70	76	87	85	87	80	8	% BUSINESSMEN
22	20	15	14	15	13	30	24	13	15	13	20	9	% TOURISTS
70	70	75	74	69	68	74	75	68	68	64	75	10	% DIRECT RESERVATIONS
20	18	19	17	27	27	19	19	26	27	21	15	11	% AGENCY —//—
10	12	6	9	4	5	7	6	6	5	15	10	12	% AIR CREWS
2	2	4	2	2	1	1	2	2	4	2	5	13	% CLIENTS UNDER 20 YEARS
25	27	37	35	25	25	27	28	24	30	24	30	14	% —//— 20-35 —//—
48	49	42	48	54	55	53	51	55	46	55	43	15	% —//— 35-55 —//—
25	22	17	15	19	19	19	19	19	20	19	22	16	% —//—MORE THAN 55—//—
163	167	166	174	152	155	145	170	157	174	165	158	17	PRICE OF ROOMS
1.65	1.71	1.65	1.91	1.90	2.00	1.54	1.60	1.73	1.82	1.66	1.44	18	LENGTH OF STAY
67	82	70	83	74	77	56	62	90	92	78	55	19	% OCCUPANCY
		X	X	X				X	X	X	X	20	CONVENTIONS

Fig. 10. The hotel example data table [, pp. 1–11]

currency. Hotel management would use this kind of data to design marketing, to define price structure, and to plan services offered to the customers.

The data table in Figure is displayed as a Reorderable Matrix in Figure . This matrix is not a binary one, but contains numbers from very different domains.

4.3 2D Sort Method

The 2D sort method is a simple heuristics that tries to build black areas to the top left and the bottom right corners of the matrix. This is done by comparing weighted row and column sums and sorting the matrix repeatedly according to these values.

The 2D sort method first arranges the matrix into ascending order according to weighted row sums. The weights are the column positions of each cell. The next phase arranges the matrix again in a similar manner, but now in column-wise direction. These two phases are repeated until no row or column exchanges occur. This termination condition requires that the sort algorithm must be stable.

Fig. 11. The hotel data table as a Reorderable Matrix

Generally, when we want to find interesting arrangements, we can have some kind of hypothesis to start with. Usually we want to specify a row or a column that we believe to have overall influence in the data set. In 2D sort this can be accomplished by sorting the matrix according to a single row or a column and then running the 2D sort. This reference row or column will not necessarily stay where we place it – or even remain unchanged during the 2D sort – but it will certainly be one of the driving factors in the sort. This feature can be used to generate different arrangements from the matrix.

4.4 2D Sort: The Townships Example

In Figure   we have the townships matrix sorted with the 2D sort. As can be seen, the subsets CITIES, TOWNS and VILLAGES were formed, but townships C, B, J and N caused problems. These townships do not fall completely inside classifications, but have some characteristics that the other similar townships lack. However, for a user it would be a simple task to 'correct' the arrangement to the one preferred by human readers (as in Figure   ).

The 2D sort could be further tuned by experimenting with the weight distribution. Possibly a non-linear distribution of weights could 'draw' the black areas together even better.

4.5 2D Sort: The Hotel Example

Figure   shows the result of arranging the hotel example with the 2D sort. This particular operation was initiated by threading the matrix according to the

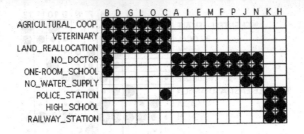

Fig. 12. The townships matrix sorted with the 2D sort

row %BUSINESSMEN and then issuing the sort command. As can be seen, the reference row is not the top row in the resulting arrangement, but it is still part of the black area appearing in the top left corner.

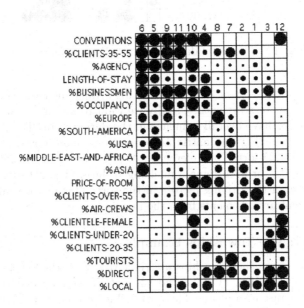

Fig. 13. The hotel matrix sorted with the 2D sort

What can be seen from or how should we interpret the arrangement in Figure   ? At least the following facts are easily found:

- Businessmen are mainly from the age group 35 to 55 years old.
- Businessmen make their room reservations through an agency.
- Businessmen stay at the hotel when there is a convention in town.
- Businessmen usually stay longer than other guests.

Choosing another reference row not appearing in this arrangement's black areas will probably produce new observations from the data set.

4.6 Sugiyama's Algorithm

The problem of drawing bipartite graphs with as few edge crossings as possible is a much studied subproblem of graph drawing. It is assumed that the vertices are drawn in horizontal lines such that separate vertex sets are placed in different horizontal lines and edges are drawn as straight lines between the sets. The drawing problem reduces to the problem of ordering the vertices in the lines such that the number of edge crossings is minimized.

There are actually two separate problems: we can fix the order in one of the horizontal lines and ask the optimal order of vertices in the other line, or we can order the vertices in both vertex sets of the bipartite graph in question, i.e., order the vertices in both of the horizontal lines. Even the former problem is known to be NP-complete [ ]. Next, we will discuss the latter, somewhat harder problem. Since there is no possibility of confusion, we call it simply the *drawing problem*.

A well-known heuristic approach to the drawing problem is Sugiyama's algorithm [ ] which is based on the *average heuristic* (sometimes called the barycenter heuristic). In the average heuristic we order the vertices according to the averages of their adjacent vertices in the opposite vertex set. By repeating this ordering process in turns in the two vertex sets, we (hopefully) reach orderings of vertices which minimize the number of edge crossings. Figure   illustrates a sample bipartite graph whose vertices are ordered according to this heuristic. For further information concerning the drawing problem, consult [, ,].

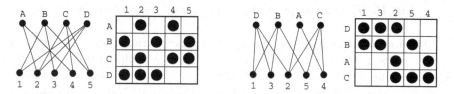

Fig. 14. Applying the average heuristic to a simple bipartite graph

Figure   shows how the adjacency matrices of the bipartite graphs are changed when applying the averaging heuristic. It is evident that when applying the heuristic, the corresponding adjacency matrices have the tendency to be reordered so that there are "black areas" in the top left and bottom right corners. This is just what we wanted to establish in connection with PROBLEM1. In what follows we try to make use of the averaging heuristic when ordering the Reorderable Matrix.

A clear difference between the drawing problem and the ordering of the Reorderable Matrix is that while in the former problem we have binary values only, the latter one deals with the set $\{0, 1, \ldots, e\}$. Our first solution to this problem is very simple: we suppose that the possible values of the matrix are divided into two categories: "blacks" and "whites" or 1's and 0's. Although in some sense self-evident, this approach has also a "deeper" rationale: the user of the Reorderable Matrix can define different "threshold values" for different objects and their attributes. For some object in an application, a certain value is "black" if the user considers it significant enough.

4.7 Sugiyama's Algorithm: The Townships Example

Sugiyama's algorithm will produce different arrangements depending on the number of up and down iterations. With some data sets the arrangement will converge and become stationary, but with some other data sets the arrangement may converge for a moment and then pulsate between a number of states. So, the problem in producing 'interesting' arrangements with Sugiyama's algorithm is with these stationary layouts – you get exactly one arrangement, no matter what the initial setting is.

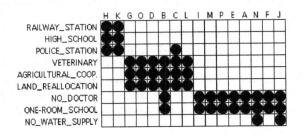

Fig. 15. The townships example drawn with Sugiyama's algorithm

The townships example is almost a stationary arrangement. The only elements in the matrix that do not converge are the townships that do not fit into classifications. The arrangement in Figure   is remarkably close to the optimal human-arranged one: moving three columns will produce the same arrangement as in Figure   .

4.8 Sugiyama's Algorithm: The Hotel Example

Instead of dividing the matrix entries into "blacks" and "whites" we can also deal with the original value set $\{0, 1, \ldots, e\}$. In that case we have to replace the normal averages by weighted ones. As above, we deal with the rows and columns of the matrix in turns. Since the entries of an $m \times n$ matrix M can have values $0, 1, \ldots, k$, we count the sums $\sum_{j=1}^{n} j \cdot M_{ij}$ (for rows) and $\sum_{i=1}^{m} i \cdot M_{ij}$ (for columns), and order the rows and columns according to the sums obtained.

In Figure   we have chosen on purpose an arrangement that is close to the arrangement presented in Figure   , the 2D sort version of the hotel example.

The difference in the layout is clear: in this version the black area will occur on the diagonal and close to the middle of the matrix. However, the same kind of observations from the data as made with the 2D sort version can also be made from this arrangement.

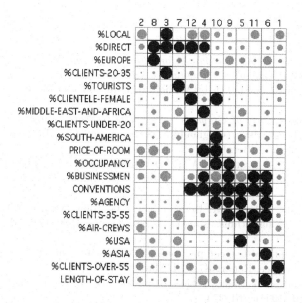

Fig. 16. The hotel example drawn with Sugiyama's algorithm

In the implementation that produced the arrangement in Figure , we used normalized values in the range $[0, 1]$. The threshold value for considering cells 'black' was 0.89 and was chosen by experimenting with various settings. In Figure the cells that have a value above the threshold value are black and the cells below the threshold value are grey.

5 Concluding Remarks

We have proved that various problems related to the reordering aesthetics of the Reorderable Matrix are NP-complete. This suggests that it is reasonable to try heuristic reordering approaches.

We have shown that the averaging heuristic for minimizing the number of edge crossings when drawing bipartite graphs is well suited reordering the Reorderable Matrix. We have implemented both a binary version and its generalization which works with all entry values.

We believe that both the 2D sort method and Sugiyama's algorithm are useful in assisting the user while exploring the Reorderable Matrix. However, the usability of these ideas must be verified with user testing. We have already done usability experiments with the basic Reorderable Matrix [ ].

Acknowledgements

This work was supported by the Academy of Finland (project 35025) and by the National Technology Agency TEKES (project 20287).

References

1. J. Bertin. *Graphics and Graphic Information Processing.* Walter de Gruyter & Co., Berlin, 1981. (Originally *La graphique et le traitemente graphique de l'information,* 1967, translated in English by William J. Berg and Paul Scott). , , ,

2. J. Bertin. *Semiology of Graphics – Diagrams Networks Maps.* The University of Wisconsin Press, 1983. (Originally *Sémiologue graphique,* 1967, translated in English by William J. Berg).

3. P. Z. Chinn, J. Chvátalová, A. K. Dewdney, and N. E. Gibbs, The bandwidth problem for graphs and matrices – a survey. *J. Graph Theory* **6** (1992), 223–254.

4. G. Di Battista, P. Eades, R. Tamassia and I. G. Tollis, Annotated bibliography on graph drawing algorithms. *Comput. Geom. Theory Appl.* **4** (1994) 235–282. ,

5. J. M. Dill, Optimal trie compaction is NP-complete. Cornell University, Dept. of Computer Science, Report **87-814**, March 1987.

6. P. Eades and N. Wormald, Edge crossings in drawings of bipartite graphs. *Algorithmica* **10** (1994), 361–374.

7. M. R. Garey, R. L. Graham, D. S. Johnson, and D. E. Knuth, Complexity results for bandwidth minimization. *SIAM J. Appl. Math.* **34** (1978), 477–495.

8. M. R. Garey and D. S. Johnson, *Computers and Intractability: A Guide to the Theory of NP-completeness.* W. H. Freeman, 1979. , ,

9. H. Hinterberger and C. Schmid, Reducing the influence of biased graphical perception with automatic permutation matrices. In *Proceedings of the Seventh Conference on Scientific Use of Statistic-Software, SoftStat93,* Heidelberg. Gustav Fischer Verlag, Stuttgart, 1993, pages 285–291.

10. M. Jünger and P. Mutzel, 2-layer straightline crossing minimization: performance of exact and heuristic algorithms. *J. Graph Algorithms and Applications* **1** (1997), 1–25.

11. J. Katajainen and E. Mäkinen, A note on the complexity of trie compaction. *EATCS Bull.* **41** (1990), 212–216.

12. D. J. Rose and R. E. Tarjan, Algorithmic aspects of vertex elimination of directed graphs. *SIAM J. Appl. Math.* 34 (1978), 176–197.

13. H. Siirtola. Interaction with the Reorderable Matrix. In E. Banissi, F. Khosrowshahi, M. Sarfraz, E. Tatham, and A. Ursyn, editors, *Information Visualization IV'99,* pages 272–277. Proceedings International Conference on Information Visualization, IEEE Computer Society, July 1999. http://www.cs.uta.fi/~hs/iv99/siirtola.pdf.

14. K. Sugiyama, S. Tagawa, and M. Toda, Methods for visual understanding of hierarchical system structures. *IEEE Trans. Syst. Man Cybern.* **SMC-11** (1981), 109–125.

15. H. S. Wilf, On crossing numbers, and some unsolved problems. B. Bollobás and A. Thomason (eds), *Combinatorics, Geometry, and Probability: A Tribute to Paul Erdős. Papers from the Conference in Honor of Erdős' 80th Birthday Held at Trinity College*, Cambridge University Press, 1997, pages 557-562.

Clouds: A Module for Automatic Learning of Concept Maps

Francisco Câmara Pereira and Amílcar Cardoso

Centro de Informática e Sistemas da Universidade de Coimbra (CISUC)
Polo II – Pinhal de Marrocos, 3030 Coimbra
{camara,amilcar}@dei.uc.pt

Abstract. There are currently several interesting works on interactive concept map construction. This simple representation of knowledge - the concept maps - is widely accepted as a promising device for helping in complex tasks such as planning and learning. Moreover, several psychologists (mainly from the constructivist stream) argue that the use of concept maps in teaching can bring relevant improvements in students. Nevertheless, as far as we know, these tools for interactive construction of concept map diagrams have a passive role in the sense that their main concerns rely upon interface and generality. If a Machine Learning based module was added to such frameworks, the computer could have an active role in participating in the concept map construction.

This paper presents Clouds, a module that uses Inductive Learning methods to help a user build her own concept maps. It uses each new entry on the map as an input for the learning algorithms, which can be used later for suggesting new concepts and relations.

1 Introduction

A concept map is a very simple diagrammatic representation of the way concepts relate to each other in a domain. It consists on a directed graph in which arcs are relations and nodes are concepts. Although elementary, this representation has already been studied and applied in education with success [1]. Given its value and applicability, several commercial concept map construction tools are already available (like Inspiration [2] or Decision Explorer [3]), which are very interesting and diverse programs that help a user organise and represent graphically his/her own concept maps. Their behaviour is nevertheless essentially passive in the sense that there is no initiative in suggesting concepts and relations to the user. If we get a mechanism for the apprehension of the way one relates concepts, then we will be able to predict and have an active role in the interaction.

Integrated in a MSc project [4], we developed a system, named Clouds, that uses Machine Learning techniques to help a user to build concept maps. It uses each new

M. Anderson, P. Cheng, and V. Haarslev (Eds.): Diagrams 2000, LNAI 1889, pp. 468-470, 2000.

entry (in the form of new relations between concepts) as an input to two Inductive Learning algorithms. Both aim to understand what characterises each relation. Then, as the interaction goes on, it starts applying the learnt knowledge to ask and suggest for new concepts and relations. This is an Artificial Intelligence based module that we believe can be used for support in other applications that depend in structures similar to concept maps.

2   Using Inductive Learning Algorithms in Concept Map Construction

Since the user enters information in the form of relations between concepts (e.g. *eat(monkey, banana), property(sun, yellow)*), it makes sense to focus learning on getting what characterises each relation, both in terms of its arguments and of the context that surrounds it. We use two Inductive Learning algorithms to extract each of these features. The first algorithm program aims at finding, for each relation, for pairs of categories that it typically links (e.g. "trees *typically* produce fruit"). In order to do it, Clouds relies on a taxonomic isa-list to find generalizations and specializations. The algorithm is very simple: each time it receives an observation, it calculates the isa-lists involved and joins the current hypothesis with the new observation. The leftmost intersections of both lists yields a generalization.

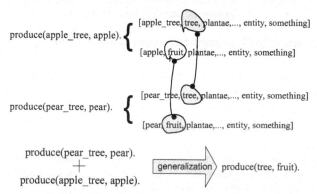

Fig. 1. Intersecting the lists of "produce(pear_tree, pear)" and "produce(apple_tree, apple)", and selecting the leftmost elements (tree and fruit), yields a generalization "produce(tree, fruit)"

The specialization occurs when a negative example is given. In this case, Clouds searches down in the tree for the most general specializations that "avoid" this new observation. The result of this, conversely to generalization, is to split the space into new hypothesis. This divide and conquer results in a number of binary predicates that represents the pairs of categories of the arguments that cover the positive examples and avoid the negative ones.

The second algorithm is based on Inductive Logic Programming [5]. We implemented a relation explanation generator that concentrates on the context of each argument.

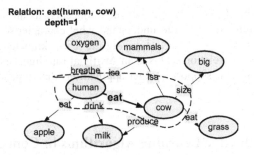

Fig. 2. The relation "eat(human, cow)" and its context

This method enables Clouds to understand each relation in terms of what usually characterizes its context, opening the way for applying deduction and abduction.

The result of this algorithm has the form of Prolog expressions, as in the example shown in fig. 3

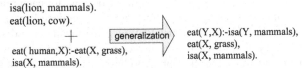

Fig. 3 – Generalization of the first argument of eat/2 based on observations "isa(lion, mammals" and "eat(lion, cow)"

We refer the reader to [5] to know more about this algorithm.

References

1. Novak, J. D. and Gowin, D. B.; Learning How To Learn. New York, Camb. University Press. 1984.
2. Inspiraton. Software Developed by *Inspiration Software Inc.* http://www.inspiration.com
3. Decision Explorer. Developed by *Banxia Software*. http://www.banxia.com/demain.html
4. Pereira, F. C. *Construção Interactiva de Mapas Conceptuais*, M.Sc. Dissertation, University of Coimbra, Portugal, 2000
5. S. Muggleton and C. Feng (1990) Efficient Induction of Logic Programs. In Proceedings of the First Conference on Algorithmic Learning Theory, Tokyo, Ohmsa Publishers. Reprinted by Ohmsa Springer Verlag.

A Diagrammatic Notation for Interval Algebra*

Zenon Kulpa

Institute of Fundamental Technological Research of the PAS,
PL-00-049 Warsaw, Poland
zkulpa@ippt.gov.pl

Abstract. In this paper, a two-dimensional, diagrammatic representation of the space of intervals, called an *MR*-diagram, is presented, together with another diagrammatic notations based on it, like the so-called *W*-diagram and some other auxiliary notations. Examples of the use of the notation in the algebra of interval relations, in interval arithmetic, and for solving a simple common-sense problem involving time intervals, are given.

1 Introduction

Years after the pioneering works on interval arithmetic [13, 10] and time-interval reasoning in AI [1], interval analysis becomes now a well-established field providing mathematical and computational tools for various applications. Until now, diagrams were practically not used for the development of this field, except for a few isolated cases [2, 3, 12], without further continuation. However, it seems that the field may benefit from the use of diagrammatic methods in a similar manner as the complex-plane diagrams helped to develop complex analysis in the past [11]. With a preliminary report [4], a systematic investigation of interval space diagrams and their applications has been started, resulting recently with the diagrammatic system for basic interval arithmetic [7, 8]. In this paper, the basic elements of the developed diagrammatic language for interval analysis are introduced. Due to space limitations, the presentation is restricted mostly to a few drawings, with only minimal explanation. For more details, see [5-9].

2 Intervals and Interval Diagrams

A *real interval* is a continuous bounded subset of the set of reals $\boldsymbol{R}$. It can be characterised either by its *endpoints*, or by *midpoint* and *radius*: $u = [u^-, u^+] = \breve{u} \pm \hat{u}$, where $u^- \leq u^+$ (i.e., $\hat{u} \geq 0$). Most of the occurrences of interval diagrams in the literature [13, 10, 12] used the endpoints coordinate space (*E-diagram*). However, for practical and theoretical reasons, the *midpoint-radius*

---

* The research reported here has been partially supported by the grant No. *8T11F00615* (for years 19982001) from State Committee for Scientific Research (KBN).

M. Anderson, P. Cheng, and V. Haarslev (Eds.): Diagrams 2000, LNAI 1889, pp. 471– , 2000.

diagram (*MRdiagram*) is preferable, see Fig.1. Until now, similar diagrams appeared only in [13, 3], and their systematic investigation started only with the works of this author [5–9]. An extension of the space of real intervals into the space of *directed intervals* (without restriction on û, i.e., occupying also the lower half-plane in the *MR*-diagram) is also in use [3].

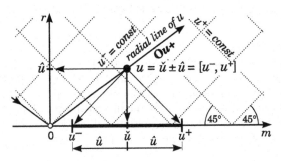

Fig. 1. The representation of intervals in a *MR*-diagram

For research on *interval relations* [1, 2], a *W-diagram* [5, 6, 9], which is a version of the *MR*-diagram able to represent images of (*basic*) *interval relations*, is useful (Fig.2a). A *lattice diagram* (of the lattice of basic interval relations) and a *conjunction diagram* are also of use here – see Fig. 2b and [5, 6, 9].

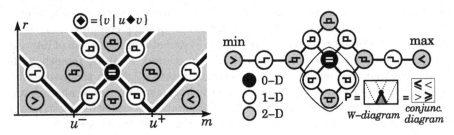

Fig. 2. The *W*-diagram (*left*) and lattice and conjunction diagrams (*right*) for interval relations

Constructions in the *MR*-diagram can also represent arithmetical operations on intervals. Constructions for *multiplication* and *division* are quite nontrivial. The basic cases of them are shown in Fig.3 (see [7, 8] for details). They play a crucial role in diagrammatic analysis of systems of linear interval equations [8].

Some common-sense interval problems studied in AI [1, 12] can be also easily represented and solved in the *MR*-diagram [6]. The problem in Fig.4 is formulated as: "I want to meet someone *during my lunch break* (1). I need *at least 20 minutes* (2) for a talk with the person, and he/she *must depart before 12:45pm*

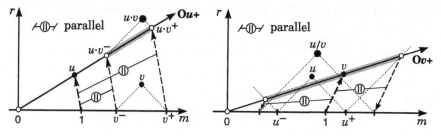

Fig. 3. Constructions for interval multiplication (*left*) and division (*right*)

(3). My lunch break may start *not earlier than 11:30am* (4), must *end at 1pm at the latest* (5) and may last *from half an hour* (6) *to an hour* (7). When can I arrange the meeting?"

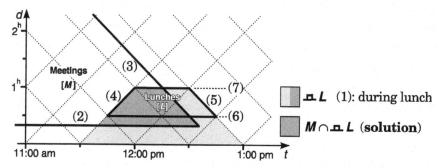

Fig. 4. Solving a simple problem involving time intervals in a *MR*-diagram; the solution (*marked in dark gray*) is: "The meeting may start between 11:30am and 12:25pm, and will end between 11:50am and 12:45pm, lasting from 20 minutes to an hour."

References

1. Allen, J. F.: Maintaining Knowledge about Temporal Relations. Comm. ACM 26 (1983) 832843
2. Freksa, C.: Temporal Reasoning Based on Semi-intervals. Artif. Intell. 54 (1992) 199-227
3. Kaucher, E.: Über metrische und algebraische Eigenschaften einiger beim numerischen Rechnen auftretender Räume, PhD Thesis, Universität Karlsruhe, Karlsruhe (1973)
4. Kulpa, Z.: Two-Dimensional Representation of Interval Relations: Preliminaries. IFTR PAS Internal Report, Warsaw (1995) 8 pp.
5. Kulpa, Z.: Diagrammatic Representation of Interval Space in Proving Theorems about Interval Relations. Reliable Computing 3 (1997) 209-217

6. Kulpa, Z.: Diagrammatic Representation for a Space of Intervals. Machine Graphics & Vision 6 (1997) 524

7. Kulpa, Z.: Diagrammatic Representation of Interval Space; Part I: Basics; Part II: Arithmetic. IFTR PAS Internal Report, Warsaw (1999) pp. 27+39

8. Kulpa, Z.: Diagrammatic Representation for Interval Arithmetic. Accepted for Linear Algebra Appl. (2000)

9. Kulpa, Z., Le, T. L.: Characterization of Convex and Pointisable Interval Relations by Diagrammatic Methods. Machine Graphics & Vision 9 (2000) 221-231

10. Moore, R. E.: Interval Analysis. Prentice Hall, Englewood Cliffs, NJ (1966)

11. Needham, T.: Visual Complex Analysis. Clarendon Press, Oxford (1997)

12. Rit, J.-F.: Propagating Temporal Constraints for Scheduling. In: Proc. Fifth National Conf. on AI (AAAI-86). Morgan Kaufmann, Los Altos, CA (1986) 383388

13. Warmus, M.: Calculus of Approximations. Bull. Acad. Polon. Sci., Cl. III, IV (1956) 253259

Animation of Diagrams: An Aid to Learning?

Richard Lowe

Faculty of Education
Curtin University of Technology
Australia
r.lowe@educ.curtin.edu

Static graphics are often used to present key aspects of dynamic instructional content (e.g. chemical reaction diagrams in Chemistry and weather maps in Meteorology). However, because static graphics represent the dynamics of a situation by implication only, students sometimes find them difficult to interpret properly. In contrast, animated graphics can represent dynamics quite explicitly. Studying relevant animations may help students become more adept at learning from related static representations (see Lowe, 1995a). In the case of beginning students of meteorology, being able to predict a subsequent pattern of meteorological markings from a given weather map is a very important skill (but one they learn slowly and imperfectly by studying static weather maps).

This poster reports a study of how non-meteorologists interrogated a weather map animation designed to develop their skills in using static weather maps. The animation shows the complexities of weather map dynamics and for meteorological novices, its study probably imposes such a high cognitive load that their attention would have to be applied very selectively. The present investigation focussed upon the nature of the dynamic information extracted from the animation and subsequently retained. In previous research, a major deficiency in beginning meteorology students' drawing of weather map predictions was their lack of differentiation among various types of meteorological features in terms their dynamic character (Lowe, 1999a). This deficiency was ascribed to their scanty and simplistic knowledge of meteorological dynamics. Mental model theory (Johnson-Laird, 1983) suggests that predictions result from running forwards a mental model of the referent situation. Accordingly, the quality of the mental model that is run will influence the accuracy of the prediction generated. The animation used in the present investigation was designed as a resource for increasing learners' capacities to construct higher quality mental models by helping them develop an improved knowledge base about meteorological dynamics.

A previous study using the weather map animation (Lowe, 1995b) indicated that its effect on non-meteorologists' predictive capacities was limited. An initial investigation of possible reasons for this lack of effectiveness indicated that subjects' extraction of information from the animation was dominated by perceptual salience rather than meteorological relevance (Lowe, 1999b). Information about graphic entities that exhibited pronounced dynamic changes (particularly in position) during the course of the animation tended to be selectively extracted at the expense of less dynamic aspects of the presentation. In that study, there was little evidence for either fine differentiation between features in terms of their dynamics or the extraction of information about meteorologically-significant relationships amongst features. The

M. Anderson, P. Cheng, and V. Haarslev (Eds.): Diagrams 2000, LNAI 1889, pp. 475-478, 2000.
© Springer-Verlag Berlin Heidelberg 2000

present study is a more detailed exploration of how non-meteorologists deal with the information available in the animation.

1   Method and Data Analysis

Subjects were 12 undergraduate students from the Faculty of Education at Curtin University with no special knowledge of meteorology. They undertook two tasks in close succession; a learning task then an application task. In both tasks, subjects were given an A-4 sized Australian summer weather map (an 'original') and asked to draw on a separate blank map the pattern of meteorological markings they would predict to occur 24 hours later than the original. Although a different original was used for each of the tasks (for example, only one contained a cyclone), they both depicted highly typical summer patterns. For the learning task, subjects used an animated weather map sequence to help them develop predictions. This animation was user-controllable via video-like controls and depicted meteorological changes occurring over a typical seven-day summer period. Neither of the originals formed part of this animation. In the application task, subjects drew predictions for the second original but without the animation. Subjects' drawing actions during both tasks were recorded on video as were their interrogations of the animation. After these prediction tasks, subjects rated on a five-point scale the relative extent to which they expected changes to occur in the (a) position and (b) form of each of the meteorological markings over a 24 hour period. A further group of subjects drawn from the same population who had not otherwise participated in the study also completed the rating task.

It was assumed that the order in which markings were built up on the blank maps during the learning and application tasks respectively were indications of how readily different components of the information set were extracted from the animation then retrieved from memory. The extent to which relevant dynamic information was extracted or retained was gauged from measurements of changes in the position and form of markings between originals and predictions. Subjects' ratings of expected changes in position and form provided an indication of the effectiveness of the animation in improving the dynamic component of their weather map knowledge base. Analysis of subjects' behaviour and output was based upon a set of meteorologically-significant macro features (determined from a consultant meteorologist's 'model answers'), each of which was typically comprised of a number of individual graphic elements. The first appearance of any component element of a particular feature was used to assign a production order to that feature. Measurements of changes in position and form were made for a subset of the total group of map markings selected as indices for key meteorological features. Measurements were based on a bounding box constructed for each marking.

2   Results and Discussion

When drawing predictions for both the learning and application tasks, subjects produced the component meteorological markings as chains of subgroups with the individual subgroups separated from each other by substantial pauses. The subgroups typically consisted of a set of several individual graphic elements that were closely associated with each other in terms of their position, form, or dynamic behaviour. The subgroups that emerged from the subjects' prediction drawing process were identified as Primary meteorological features such as fronts and pressure cells; the markings drawn by subjects were not sufficiently coordinated in terms of their fine structure to identify clearly defined troughs and ridges (a weather map's Secondary features; see Lowe, 1999b).

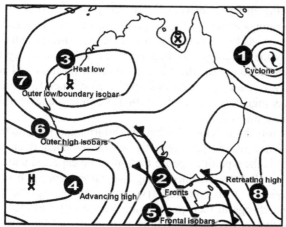

Fig. 1. Original map for learning task showing
prediction subgroups and mean drawing sequence

The mean order in which meteorological features were produced suggests that the perceptual characteristics of those features played a major role in their sequencing (Figure 1). For purposes of discussion, features are divided into those that were drawn early in the sequence and those that were drawn late. For the learning task sequence, features that were relatively persistent, perceptually more coherent (in gestalt terms), and that had characteristics differing substantially from those of other aspects of the overall meteorological pattern tended to occur earlier than those that were more fleeting, less well defined or less distinctive. The early-drawn features were distinctive in terms of both their form and their dynamic character (relative to the 'background texture' of other markings). Their dynamics were distinctive either because they moved rapidly or because they were relatively stationary compared with their highly mobile surroundings. Later-drawn markings tended to be isobars occupying space between or around the early-drawn features. For the application task, the relationship between feature characteristics was less clear-cut. However, features that were well defined, formed distinctive graphic 'objects' and underwent the greatest position change still tended to be produced earlier.

Comparisons of the position and form changes produced during the learning and application tasks were made via six index markings common to both original maps. For both tasks, mean changes in the locations of features were more pronounced along the east-west axis than along the north-south axis. This appropriately reflects the tendency for zonal movement of weather systems to the south of Australia due to global-scale meteorological influences. There was considerable diversity in the extent to which individual features changed their positions (most pronounced for the learning predictions). At a more inclusive scale, there was some limited evidence for localised coordinated treatment of multiple markings, both within a particular meteorological feature and among a number of adjacent features. The diversity in position changes from the learning task was paralleled by diversity in changes in the form of markings. Again, these changes in form occurred at the level of whole features and for the different markings that comprised a particular feature. The position and form ratings indicated that the group who had worked with the animation made greater distinctions between the dynamic characteristics of individual markings than were made by the control group. In general, the animation group tended to give higher change expectations than the controls for both position and form (with the exception of relatively stationary features such as the cyclone). However, results from the prediction tasks suggest that these distinctions occur at the level of Primary rather than Secondary features.

This study supports the previous finding that information is extracted on the basis of perceptual salience rather than thematic relevance. Such reliance on domain-general processing is the type of response that might be expected under conditions of high cognitive load. Results obtained using the current approach extend previous findings by revealing that 'objects' (visually-coherent groups of markings) undergoing a rapid change in their position are not the only features that tend to be noticed. Features that move relatively little also attract attention if they are on a visually-active background. In addition, changes of form also appear to be influential in determining which aspects are noticed and which escape notice. Taken together, these findings suggest that various forms of dynamic contrast with the visual context strongly influenced what information subjects extracted from the animation and (to a lesser extent) retained in memory.

References

1. Lowe, R. K. (1995a, July). *Developing basic chart reading skills: using interactive animation for building effective mental models.* Paper presented at the 2éme Conférence Internationale sur l'Enseignement Assisté par Ordinateur et l'Enseignement á Distance en Météorologie, Toulouse, France.
2. Lowe, R. K. (1995b, August). *Supporting conceptual change in the interpretation of meteorological diagrams.* Paper presented at the 6th European Conference for Research on Learning and Instruction, Nijmegen, Netherlands.
3. Lowe, R. K. (1999a). Domain-specific constraints on conceptual change in knowledge acquisition from diagrams. In W. Schnotz, S. Vosniadou & M. Carretero (Eds.), *New perspectives on conceptual change.* (pp. 223-245) Amsterdam: Elsevier.
4. Lowe, R. K. (1999b). Extracting information from an animation during complex visual learning. *European Journal of Psychology of Education, 14,* 225-244.

Diagrams as Components of Multimedia Discourse: A Semiotic Approach

John H. Connolly

Department of Computer Science,
Loughborough University, LE11 3TU, UK

1   Introduction

There are many different ways of looking at diagrams. Here, however, we consider them in relation to human communication. In particular, we draw upon some concepts taken from semiotics and linguistics, with a view to showing how a semiotically based approach can help to make more explicit the way in which diagrams contribute to the interpersonal communication process which is known as discourse.

2   The Semiotic Approach

Semiotics is the study of *signs*, which are entities (such as words) that are used in communication in order to stand for something other than themselves. There exist different kinds of sign, and in particular, it is accepted that a distinction should be drawn

In Diagram 3 you will find the fingerings for three easy notes on the D whistle.

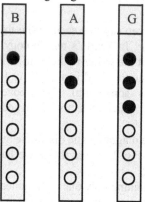

Diagram 3: The fingerings for the notes B, A and G

Fig 1: Illustrative excerpt from a multimedia discourse

M. Anderson, P. Cheng, and V. Haarslev (Eds.): Diagrams 2000, LNAI 1889, pp. 479-482, 2000.
© Springer-Verlag Berlin Heidelberg 2000

between *symbols* (signs that do not resemble what they denote) and *icons* (signs that do perceptibly resemble what they designate). Let us apply these concepts to an example diagram. Imagine that Fig 1 is an excerpt from a book aimed at teaching people to play the musical instrument known as the D Whistle.

In the diagram within Fig 1 there are various signs: the title and the labels representing the notes, and the drawing of the whistle, which can be divided into smaller components, such as the rectangle representing the mouthpiece, the circles representing the holes, and so on. The title and the labels are symbolic in nature, whereas the drawing and its component elements are iconic. The diagram as a whole. is a *complex sign*, constructed out of the various symbols and icons of which it consists.

The central topic of the present paper is *diagrammatic discourse* — that is, communication in which diagrams play a significant role in the conveyance of meaning. Diagrammatic discourse typically contains both graphical and natural-language (NL) parts. Let us call the NL-based elements of a diagram the *annotatory text*. This encompasses the title, labels and any other alphanumeric elements that belong to the diagram itself, as opposed to the *accompanying text* — the surrounding material in which the diagram is embedded.

Whenever diagrammatic discourse exhibits a combination of the media of NL text and graphics, it is necessarily *multimedia* in nature. The application of semiotics to multimedia has previously been suggested by authors such as Purchase [2]. However, we can carry this approach further by incorporating some ideas from linguistic discourse analysis (DA).

To begin with, let us express the content of Fig 1 purely as a NL text:

(1) Now let's learn how to produce some notes on the D whistle. By covering just the hole nearest the mouthpiece and blowing reasonably gently you can play a B. Then cover this hole and also the one immediately below it, and play an A. Next, if you cover these two holes and the next one below as well, you can produce a G.

Using terms from DA, we can say that each of the four sentences in (1) consists of a communicative act (CA). In Searle's terminology [3], the first two CAs are *representatives* (which commit the writer to the truth of the propositions which they express), as is the fourth, while the third is a *directive* (an attempt by the writer to cause the addressee to do something). Moreover, a coherent group of communicative acts like these, which collectively advance the discourse, is termed a *move*.

However, in the case of the diagrammatic discourse in Fig 1, the situation is more complicated, as is shown in Fig 2, where the move has been split into two *sub-moves*:

a *linguistic sub-move*, containing only one communicative act ('In Diagram 3 you will find the fingerings for three easy notes on the D whistle'), as compared with the four in (1), and constituting the accompanying text to a *diagrammatic sub-move*, basically graphical in nature, but also containing some annotatory text.

What we have in Fig 2, then, is a representation of a piece of multimedia discourse. Not only does it include both symbolic and iconic elements, but also, and very importantly, every significant element within the entirety of Fig 2 is a sign of some kind, and is therefore accommodated within the semiotic description of the discourse.

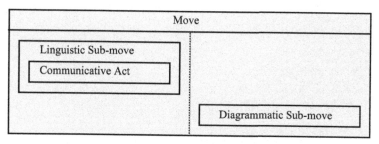

Fig 2: Description of the multimedia discourse in the body of Fig 1

It may be asked whether the diagrammatic sub-move can be analysed into individual acts, as the linguistic sub-move can. This is, indeed, generally possible to some extent. First of all, the title of the diagram is usually a representative CA. Furthermore, it may well be possible to identify CAs within the body of the diagram. For instance, the leftmost drawing and associated label 'B' in Fig 1 can be regarded as a representative CA in much the same way as the second sentence in (1).

However, speaking more generally, there is a problem, namely that iconic representations often embody so much information that they generate an unmanageably large number of potential acts. Consequently, although we can allow for the possibility of identifying individual CAs within diagrams, an exhaustive analysis of diagrams into their constituent CAs is not in general to be recommended, as it tends to be subject to diminishing returns. Other problems arise too — for example, how to formalise the semantics of diagrammatic discourse, perhaps using a framework such as Situation Theory; see Devlin [1] — but these will require further research.

3   Conclusion

Although further work remains, it seems that a semiotically based analysis of diagrams as components of tmultimedia discourse can help to reveal the details of how those diagrams fulfil their function.   Hopefully, also, whatever enriched understanding may emerge from such a theoretical appreciation will find some practical application in areas such as the design of multimedia documents, the teaching of the principles of multimedia to those who are new to the field, and the design of intelligent computing systems with the capacity for processing multimedia input and output.

482   John H. Connolly

References

1. Devlin, K.: Logic and Information. Cambridge University Press, Cambridge (1991).
2. Purchase, H. C.: Defining Multimedia. IEEE Multimedia 5, no. 1 (1998), 58-65.
3. Searle, J.: The classification of illocutionary acts. Language in Society 5 (1976), 1-24.

Formalising the Essence of Diagrammatic Syntax

Corin Gurr and Konstantinos Tourlas

Division of Informatics, The University of Edinburgh
2 Buccleuch Place, Edinburgh EH8 9LW, UK
corin@cogsci.ed.ac.uk
kxt@dcs.ed.ac.uk

Diagrams have a long history as visual aids which assist in structuring and simplifying potentially complex reasoning tasks. Recent years have witnessed a rapid, ongoing popularisation of diagrammatic notations in the specification, modelling and programming of computing systems, leading to diagrammatic languages requiring increasingly complex semantic interpretations. A general theory of diagrammatic languages, following those of more typical text-based languages, requires as a minimum an account of diagram syntax, of semantics, and of interpretations: the relationships between syntax and semantics. A satisfactory theory of diagrams must also account for more cognitive aspects, notably the ways in which pragmatic features of diagrams contribute to their effectiveness for human users, and how individual cognitive differences affect human interpretation of diagrams.

We present preliminary work and ideas in the direction of providing a mathematical understanding of the essential features and idiosyncrasies of diagrammatic syntax. In particular, we are investigating the applicability of methods and devices drawn from the theory of *categories* [ ], from which we adapt the notions of *sketch* [ ] and *morphism* in order to capture the constituent elements of diagrams as well as relations among diagrams in the same class.

1 Elements of Diagrammatic Syntax

Previous accounts of diagram syntax have commonly modelled diagrams in terms of sets, whose elements stand for visual objects, and (mathematical) relations capturing primitive visual relations among objects. Consider, for instance, a typical Petri net diagram consisting of circles, called *places*, and boxes, called *transitions*, linked by arrows and subject to the condition that no objects of the same kind may be linked directly. Such a diagram may be modelled in terms of two distinct sets of visual objects, or *sorts*: places and transitions. Two relations model how places are linked to transitions and vise versa. So, each such diagram is modelled as a tuple $\langle P, T, F_1, F_2 \rangle$ where P is the set of places, T the set of transitions, whereas $F_1 \subseteq P \times T$ and $F_2 \subseteq T \times P$ are relations.

This view of diagrams underpins Barwise and Etchemendy's account of *homomorphic representations* [ ], Harel's higraph foundation of Statecharts [ ], Gurr's abstract worlds [ ] and the graphical signatures of Wang et al. [ ]. However, the various formal structures—typically tuples—arising from this view often seem

M. Anderson, P. Cheng, and V. Haarslev (Eds.): Diagrams 2000, LNAI 1889, pp. 483–   , 2000.
© Springer-Verlag Berlin Heidelberg 2000

disparate, owing partly to the lack of any formal notion of "signature" accounting for the common form underlying all tuples arising from diagrams in any given class. Additionally, this lack of signature information presents a serious obstacle in providing sufficiently general and precise definitions of *homomorphism* between diagrams in the same class and, for that matter, between diagrams and the semantic structures they represent.

Our suggestion, overviewed in the following section, is that the concept of *sketch*, being itself a powerful generalisation of the signatures in universal algebra, may provide the right level of generality required to conceptualise the constitution of structures arising in the modelling of diagrams.

2 Sketches for Diagrams

Like signatures, sketches are formal specifications of mathematical structures, however they are based on *graphs*, i. e. collections of *objects* (nodes) and *arrows* (directed edges), rather than being linguistic in nature.

In terms of our running example, a sketch for the class of Petri net diagrams contains two objects Pl and Tr, one for each sort of visual objects (places and transitions). The graph also contains arrows $f_1 : \text{Pl} \to \text{Tr}$ (the notation meaning that f_1 is an arrow from Pl to Tr) and $f_2 : \text{Tr} \to \text{Pl}$, capturing the arities of the "directly-linked-to" relations:

$$\text{Pl} \quad \underset{f_2}{\overset{f_1}{\rightleftarrows}} \quad \text{Tr}$$

Let us refer to this sketch as *PNsketch*.

So far, we have only demonstrated how a sketch may capture the visual elements in a class of diagrams "of the same type". In order to obtain the (mathematical) structures describing actual diagrams in the class, one must map the objects of the sketch to sets and the arrows to (binary) relations; that is, one must *interpret* the sketch.

In general, interpretations (commonly known as *models*) of sketches are *graph homomorphisms* from the sketch, say S, to another graph G. (A graph homomorphism h from S to G maps respectively the objects and arrows of S to those of G, preserving the sources and targets of arrows; i.e. whenever $f : A \to B$ is an arrow of S then $h(f) : h(A) \to h(B)$ is an arrow in G.) A suitable graph into which to interpret our sketch *PNsketch* is **Rel**$_\mathsf{f}$, the graph having as objects all finite sets and as arrows between any two objects, say A and B, all (binary) relations from the set A to the set B. Then, a graph homomorphism M from *PNsketch* to **Rel**$_\mathsf{f}$ provides exactly what one regards as a *model*, or description, of an Petri net diagram: finite sets $M(\text{Pl})$ and $M(\text{Tr})$, whose elements stand for visual objects of the sorts "place" and "transition", together with relations $M(f_1) \subseteq M(\text{Pl}) \times M(\text{Tr})$, $M(f_2) \subseteq M(\text{Tr}) \times M(\text{Pl})$. Alternatively, M may be seen essentially as a tuple $\langle M(\text{Pl}), M(\text{Tr}), M(f_1), M(f_2) \rangle$ of exactly the kind identified in the previous section.

3   Gains, Prospects and Problems

We have presented only the simplest possible kind of sketch here, and illustrated how it may play the role of signature for structures capturing diagrammatic syntax, the structures themselves formalised as graph homomorphisms.

In more advanced sketches one may specify a variety of conditions constraining the interpretation of objects and arrows, in a way exceeding the expressiveness of traditional notions of signature, thus allowing for the "tight" specification of structures. On the downside, a sketch for a syntactically complex class of diagrams may grow prohibitively large. We regard sketches mostly as a *conceptual* tool, allowing for the modelling of diagrammatic notations in the abstract, thus providing guidance on how to structure more informal definitions.

In particular, our approach suggests an appealing way of conceptualising other important aspects such as "emergent relations" in diagrams (i.e. visual relations arising from the synthesis of primitive ones). A sketch S accounting only for primitive relations may be suitably "completed" to a sketch $T(S)$ with emergent features added, the completion given by a mapping (or rule) T. On any interpretation M of S, this rule induces a completed interpretation M_T accounting for the emergent relations in the diagram described by M.

Another concept under investigation is that of *morphism* between structures capturing diagrams in the same class. Morphisms formalise relations between diagrams, such as sub-diagram relations or those arising by regarding a diagram as an abstraction of another. We have succeeded in deriving notions of morphism as refinements of *natural transformations* [ ], the standard way of relating interpretations of the same sketch.

On the whole, our preliminary results in applying category theoretic notions to the formalisation of diagrammatic syntax appear encouraging. We believe such notions may prove instrumental in underpinning and conceptualising the essence of diagrammatic representations.

References

1. M. Barr and C. Wells. Category Theory for Computing Science. Prentice-Hall, 1990.   ,
2. J. Barwise and J Etchemendy. Heterogeneous logic. In J. Glasgow, N. H. Narayan, and B. Chandrasekaran, editors, Diagrammatic Reasoning: Cognitive and Computational Perspectives, pages 211-234. MIT Press, 1995.
3. Corin A. Gurr. On the isomorphism, or lack of it, of representations. In Kim Marriot and Bemd Meyer, editors, Visual Language Theory, chapter 10, pages 293-305. Springer, 1998.
4. David Harel. On visual formalisms. In Diagrammatic Reasoning: Cognitive and Computational Perspectives, pages 235-272. MIT Press, 1995.
5. Saunders MacLane. Categories for the Working Mathematician, volume 5 of Graduate Texts in Mathematics. Springer-Verlag, 1971.
6. D. Wang, J. Lee, and H. Zeevat. Reasoning with diagrammatic representations. In Diagrammatic Reasoning: Cognitive and Computational Perspectives, pages 211-234. MIT Press, 1995.

Using Grids in Maps[*]

Alexander Klippel and Lars Kulik

University of Hamburg, Department for Informatics
Vogt-Kölln-Str. 30, D-22527 Hamburg
{klippel,kulik}@informatik.uni-hamburg.de

Abstract. Our approach aims at a general description that is common to all types of grids used in diagrammatic representations despite their individual differences. Based on our analysis, we specify different types of spatial knowledge and single out in which way a particular type of grid represents a particular type of spatial knowledge. This specification identifies the various contributions of grids to diagrammatic representations. It turns out that grids in maps and especially in schematic maps have two complementary functions. First, they enable inferences that are not possible using only the spatial map features. Second, they provide additional design freedom, as important information that is not represented in the schematic map itself, can be encoded in the grid structure.

1 Introduction

Grid structures are widely used in diagrammatic representations. As search grids they partition a representational medium independent of the spatial properties of the representation, as images of geographic coordinates they reveal the structure of the underlying projection of a map, and as spatial layers they add additional information to a map, for instance, on distances or travel time.

Although there is already a growing body of research considering grids as discrete global grids in geographical information systems (cf. Kimerling et al., 1999), these investigations are primarily concerned with technical applications like automated map generation. There are considerably fewer investigations focusing on the issue which type of qualitative inferences can be drawn by employing grids in maps or, more generally, diagrammatic representations.

Since we are interested in the use of diagrammatic representations by humans our work concentrates on qualitative spatial reasoning. The indicated formalization specifies a geometric structure that can be embedded in a geometry as specified by Hilbert (1956). Under this perspective we pursue two objectives. On the one hand, we

---

[*] The research reported in this paper has been supported by the Deutsche Forschungsgemeinschaft (DFG) in the project 'Räumliche Strukturen in Aspektkarten' (Fr 806/8) and 'Axiomatik räumlicher Konzepte' (Ha 1237/7). We thank Carola Eschenbach, Christian Freksa, Christopher Habel, and Heike Tappe.

M. Anderson, P. Cheng, and V. Haarslev (Eds.): Diagrams 2000, LNAI 1889, pp. 486-489, 2000.
© Springer-Verlag Berlin Heidelberg

analyze the possibilities to represent qualitative knowledge in grid structures. On the other hand, we examine in which way grid structures enrich spatial representations like schematic maps, for instance subway maps.

2   Applying Grids to Maps

Different kinds of spatial knowledge induce different representational functions of grids and determine the properties of visualized grid lines that can be formally described. Due to space constraints we focus on three case studies to reveal the representational functions and to give an idea of the underlying concepts required for a formalization (cf. Kulik & Klippel, 1999). These case studies illustrate the partitioning properties of grid structures, the ordering information on grid structures, and the shape information of grid cells.

Partitioning Properties and Ordering Information of Grid Structures.   The betweenness property and the connectedness of grid lines (cf. Eschenbach et al., 1998) are necessary basic concepts for characterizing the ordering structure of a visualized grid. Generally, cartographic features like the north arrow or coordinates distinguish the ordering of grid lines and grid cells, respectively. On the other hand, grids partition the representational medium into discrete cells and thereby restrict the search space. The properties of grid cells can be employed to represent several kinds of spatial knowledge:

1. The cells can be ordered in two directions. This function of the grid is independent of the represented aspect of the world and applies only to the medium, as for example for search grids in city maps. A special case is illustrated in Fig. 1a using the inherent order information of natural numbers and the alphabet.
2. Neighborhood relations of grid cells can be used to represent qualitative spatial distances. It is possible to represent this aspect in a diagrammatic representation without changing the basic representation (Fig. 1b). To employ neighboring concepts the notion of 'same-side' and 'different side' are essential. Depending on the context 4- or 8-neighborhoods are appropriate.

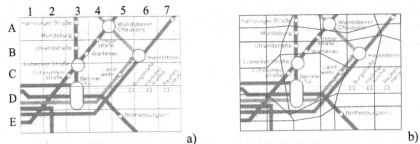

Fig. 1. a) The left figure depicts a part of the underground map of Hamburg with an additional representationally independent search grid that partitions the medium and, therefore, restricts the search space. To locate 'Rothenburgsort' the grid restricts the search space to cell 'E5'. b) The right figure depicts a part of the underground map of Hamburg. The grid indicates spatial distortions of subway routes by using polygonal curves. Stations in 'reasonable walking distance' are in the same or in a neighboring grid cell

Shape of Grid Cells. If a grid is assumed in the represented world, the grid is usually understood as a partitioning of the represented area in an evenly manner. Hence, the prototype of a grid consists of grid cells of the same shape and size. Therefore, the corresponding grid of the representing world reveals by the variation of its cells the underlying spatial structure of the representation (cf. Hassen & Beard, 1998). The shape of a grid on a map resulting from an image of a projection of parallels and meridians determines whether maps, like in Fig. 2, preserves distance, angle, or area information (cf. Maling, 1992). The characterization of shape features require the concept of congruence that applies to segments of visualized grid lines and their enclosed angles. Congruence information enables the comparison of two quantities without measuring them.

Fig. 2. Different projections maintain different kinds of spatial knowledge, which leads to different forms of grid cells

3   Conclusion

There are several ways to use grids to convey knowledge in a diagrammatic repre-sentations. Corresponding to our analysis we differentiate three representational functions that grids accomplish: Grids can partition the representational medium, they can be used to provide additional knowledge on already represented knowledge; and they can represent a further aspect of the represented world. The last case provides the opportunity of design freedom: If important knowledge is represented in the visualized grid, we obtain the possibility to bypass localization constraints. Grid structures enhance the possibilities of qualitative spatial reasoning on diagrammatic representations. The opportunities for applications in cartographic design, especially for schematic maps, await further investigation and empirical testing.

Reference

1. Eschenbach, C., Habel, C., Kulik, L., Leßmöllmann, A. (1998). Shape nouns and shape concepts: A geometry for 'corner'. In C. Freksa, C. Habel & K.F. Wender (Eds.), *Spatial Cognition* (pp. 177–201). Berlin: Springer.
2. Hassen, K. & Beard, K. (1998). Visual evaluation of GIS algorithms using a reference grid. *Cartography and Geographic Information Systems*, 25(1), 42–50.
3. Hilbert, D. (1956). *Grundlagen der Geometrie*. Stuttgart: Teubner.
4. Kimerling, A. J., Sahr, K., White, D., & Song, L. (1999). Comparing geometrical properties of global grids. *Cartography and Geographic Information Science*, 26(4), 271–288.
5. Kulik, L. & Klippel, A. (1999). Reasoning about cardinal directions using grids as qualitative geographic coordinates. In C. Freksa & D.M. Mark (Eds.), *Spatial Information Theory* (pp. 205–220). Berlin: Springer.
6. Maling, D. H. (1992): *Coordinate systems and map projections*. Oxford: Pergamon Press.

Case Analysis in Euclidean Geometry: An Overview

Nathaniel Miller

Cornell University Department of Mathematics
nat@math.cornell.edu

Abstract. This paper gives a brief overview of **FG**, a formal system for doing Euclidean geometry whose basic syntactic elements are geometric diagrams, and which has been implimentented as the computer system **CDEG**. The computational complexity of determining whether or not a given diagram is satisfiable is also briefly discussed.

1   A Diagrammatic Formal System

To begin with, consider Euclid's first proposition, which says that an equilateral triangle can be constructed on any given base. While Euclid wrote his proof in

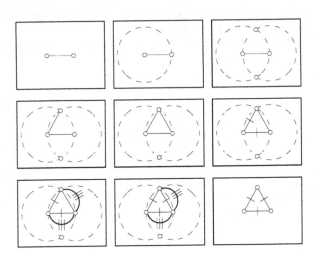

Fig. 1. Euclid's first proposition

Greek with a single diagram, the proof that he gave is essentially diagrammatic, and is shown in Figure . Diagrammatic proofs like this are common in informal treatments of geometry, and the diagrams in Fig.   follow standard conventions: points, lines, and circles in a Euclidean plane are represented by drawings of dots and different kinds of line segments, which do not have to really be straight, and

M. Anderson, P. Cheng, and V. Haarslev (Eds.): Diagrams 2000, LNAI 1889, pp. 490–   , 2000.
© Springer-Verlag Berlin Heidelberg 2000

line segments and angles can be marked with different numbers of slash marks to indicate that they are congruent to one another. In this case, the dotted segments in these diagrams are supposed to represent circles, while the solid segments represent pieces of straight lines.

It has often been asserted that proofs like this, which make crucial use of diagrams, are inherently informal. The comments made by Henry Forder in *The Foundations of Euclidean Geometry* are typical: "Theoretically, figures are unnecessary; actually they are needed as a prop to human infirmity. Their sole function is to help the reader to follow the reasoning; in the reasoning itself they must play no part." [, p.42] Traditional formal proof systems are sentential—that is, they are made up of a sequence of sentences. Usually, however, these formal sentential proofs are very different from the informal diagrammatic proofs. A natural question, then, is whether or not diagrammatic proofs like the one in Fig. can be formalized in a way that preserves their inherently diagrammatic nature.

The answer to this question is that they can. In fact, the derivation contained in Fig. is itself a formal derivation in a formal system called **FG**, which has also been implemented in the computer system **CDEG**. These systems are based a precisely defined syntax and semantics of Euclidean diagrams. We can define a diagram to be a particular type of geometric object satisfying certain conditions; this is the syntax of our system. We can give a formal definition of which arrangements of lines, points, and circles in the plane are represented by a given diagram; this is the semantics. Finally, we can give precise rules for manipulating the diagrams—rules of construction and inference. All of these rules can be made entirely precise and implemented on a computer. The details are too long and technical to include here, but the interested reader can find them in [ ]. A crucial idea, however, is that all of the meaningful information given by a diagram is contained in its topology, in the general arrangement of its points and lines in the plane. Another way of saying this is that if one diagram can be transformed into another by stretching, then the two diagrams are essentially the same. This is typical of diagrammatic reasoning in general.

2 Case Analysis

One of the most interesting aspects of these systems is how they treat case analysis. Any system for doing Euclidean geometry must deal somehow with geometric constructions. For example, Euclid's first postulate says that a line segment can be constructed joining any two given points. These formal systems have a corresponding rule that was used in the third and fourth steps of the derivation in Fig. . How does the computer know what diagrams could result from connecting the two dots in the third diagram in Fig. ? The answer is that it returns all diagrams in which the two dots are connected by a new segment. In this particular case, there is only one such possible diagram, because of the restrictions placed on what a diagram can look like. Often, however, there will be more than one possibility, and in that case the computer will return them all.

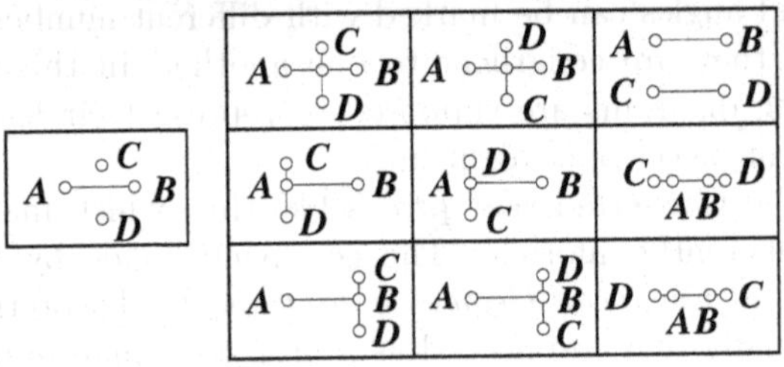

Fig. 2. (a and b) What can happen when points C and D are connected?

For example, consider the situation shown in Fig. a. How many different ways can points C and D be connected? As it turns out, there are nine different topologically distinct possibilities, as **CDEG** confirms, which are shown in Fig. b. Thus, in both the formal system **FG** and in its computer implementation **CDEG**, when you apply the rule that says you can connect any two given points by a line segment to points C and D in the diagram in Fig. a, you get a disjunctive array of all nine of the diagrams in Fig. b.

3   Satisfiable and Unsatisfiable Diagrams

An important requirement for the construction rules in a formal system for doing geometry is that they be *sound*—that is, that they really do return every single physically realizable possibility. As one would expect, the construction rules used in **FG** and **CDEG** are indeed sound. We might also hope, however, that they would also be *frugal*—that they would never return any extra, unsatisfiable cases. This, unfortunately, is not the case. Our definition of what a diagram is allows us to draw diagrams that do not represent any physically realizable situation. For example, one could draw a diagram in which two different lines intersected in more than one place, since the lines in the diagram are not required to be straight. This particular example is eliminated by our definition of what constitutes a well-formed diagram, but there are other more complicated examples which are not, and sometimes some of these unsatisfiable diagrams may be returned by one of the construction rules.

Is it possible to eliminate these extra cases? Yes, but not efficiently. It is possible to determine whether or not a given diagram is satisfiable, by writing down a formula of first order logic that is satisfiable over the real numbers if and only if the given diagram is satisfiable; and we can determine if this formula is satisfiable by Tarski's famous decision procedure. It turns out that in this particular case we can determine if the formula is satisfiable in polynomial space. However, it also turns out that the problem of determining whether or not a given diagram is satisfiable is at least NP-hard, which roughly means that any possible procedure for deciding if a given diagram is satisfiable is too inefficient for practical use. For further details, see [ ].

References

1. Euclid, *Elements*, T. L. Heath, ed., second edition, New York: Dover, 1956.
2. Forder, Henry George, *The Foundations of Euclidean Geometry*, Cambridge: Cambridge University Press, 1927.
3. Miller, Nathaniel, "Case Analysis in Euclidean Geometry," available at http://www.math.cornell.edu/~nat/diagrams. ,

Experimenting with Aesthetics-Based Graph Layout

Helen C. Purchase, David Carrington, and Jo-Anne Allder

School of Computer Science and Electrical Engineering, The University of Queensland
Brisbane 4072, Australia
{hcp,davec,joanne}@csee.uq.edu.au

1 Introduction

Many automatic graph layout algorithms have been implemented to display relational data in a graphical (usually node-arc) manner. The success of these algorithms is typically measured by their computational efficiency and the extent to which they conform to aesthetic criteria (for example, minimising the number of crossings, maximising symmetry). Little research has been performed on the usability aspects of such algorithms: do they produce graph drawings that make the embodied information easy to use and understand? Is the computational effort expended on conforming to the assumed aesthetic criteria justifiable with respect to better usability? This paper reports on usability studies to investigate automatic graph layout algorithms with respect to human use.

2 Syntactic Performance

The first two experiments investigated human performance with respect to subjects' ability to answer abstract questions about the structure of a graph (e.g. How long is the shortest path between two given nodes?)

2.1 Individual Aesthetics

The first study considered the effects of five individual aesthetics: *minimize bends, minimize edge crossings, maximise the minimum angles between adjacent edges from a node orthogonality* and *symmetry.*

For each aesthetic being considered, two drawings of the same graph were produced (representing a strong or weak presence). The experiment was conducted online: subjects had the relevant terminology and the three questions explained, and a worked example was presented. They then answered three graph theoretic questions chosen as a means of measuring syntactic performance on the sixteen graph drawings.

Two dependent variables were recorded: the response time to answer each question, and the correctness of each answer. A within-subjects analysis method was used. Fifty-five second-year computer science students took part.

M. Anderson, P. Cheng, and V. Haarslev (Eds.): Diagrams 2000, LNAI 1889, pp. 498–   , 2000.
© Springer-Verlag Berlin Heidelberg 2000

ANOVA analysis and Tukey's WSD pairwise comparison procedure were performed for both errors and response time. The results indicate support for reducing the number of crossings and bends, and increasing the display of symmetry. However, no support was found for maximising the minimum angle or increasing the orthogonality.

2.2 Algorithms

The second study considered the effects of eight different graph drawing algorithms, all implemented in the GRAPHED system (Himsolt, 1990)). The algorithms included three force-directed algorithms, three planar orthogonal grid algorithms with bends minimisation, a planar grid drawing with straight line edges, and a compressed straight-line planar grid algorithm.

Each algorithm was applied to a given graph, resulting in eight experimental graph drawings. The experimental method was identical to that of the syntactic aesthetics experiment. Fifty-five third-year computer science students took part.

ANOVA analysis and Tukey's WSD pairwise comparison procedure were performed for both errors and response time. The investigation gave inconclusive results, making it difficult to say that one algorithm is 'better' than another in the context of syntactic understanding of the abstract graph structure. The compressed straight-line planar grid algorithm gave slightly poorer results with respect to errors.

3   Semantic Preference

This experiment explored semantic preferences for graph layouts. Two Unified Modeling Language (UML) (Booch, Rumbaugh and Jacobson, 1998) diagram types were chosen as the semantic domain. Class diagrams provide a static view of a system by describing the types of objects in the system and the relationships between them. Collaboration diagrams provide a dynamic view of a system by showing the interactions between objects that send and receive messages.

Six aesthetics were evaluated in both the class and the collaboration diagrams: *minimize bends, minimize edge crossings, orthogonality, width of layout, text direction* and *font type*. Two domain-specific aesthetics were also evaluated.

This experiment used preference as the method of usability measurement. By asking whether they prefer one UML drawing to another, subjects are likely to be anticipating the use of these drawings for a software engineering task: their responses are therefore likely to be related to their perceived usefulness of the diagrams.

A UML class diagram and a UML collaboration diagram were created. The class diagram had 14 classes and 18 relationships, and the collaboration diagram had 12 objects and 17 messages. Each diagram was drawn 16 times, twice for

---

[1] Full statistical details can be found in Purchase (1997).

[2] Full statistical details can be found in Purchase (1998).

each aesthetic or secondary notational feature under consideration. Each pair provided a contrast for one particular layout aesthetic.

Each subject was presented with an individual evaluation booklet containing instructions, a brief tutorial on UML class diagrams, a questionnaire requesting information about the subject, eight preference questions (one for each aesthetic) and a ranking question.

Each preference question presented the participant with a contrasting pair of UML class diagrams. Subjects were asked to mark their preferred diagram and to provide an explanation for their choice. The ranking question at the end of the booklet requested that subjects rank the top three (most preferred) and the bottom three (least preferred) diagrams from the sixteen different diagrams.

Seventy student volunteers from the University of Queensland participated in the class diagram evaluation. Ninety students participated in the corresponding collaboration diagram evaluation with the same style of booklet and experimental methodology.

We concluded that the priority order of the aesthetics common to both diagram types is: arc crossing, orthogonality, information flow, arc bends, text direction, width of layout and font type.

4 Discussion

These three experiments have revealed much about the potential usability of graph drawings produced by automatic layout algorithms. The evidence is overwhelmingly in favour of reducing the number of arc crossings as the most important aesthetic to consider. While the results of the syntactic experiments did not highlight orthogonality as being important, in the domain of UML diagrams, this aesthetic moves up the priority list to second place: this is a clear signal that algorithms that are designed for abstract graph structures, with no consideration of their ultimate use, will not necessarily produce useful visualisations of semantic information.

Acknowledgements

Thanks are due to the students who participated in the experiments, P. Eades, R. Cohen, J. McCreddon, M. James, D. Leonard and the Australian Research Council.

References

1. G. Booch, J. Rumbaugh, and I. Jacobson. *The Unified Modeling Language User Guide*. Addison-Wesley, 1998.
2. M. Himsolt. GRAPHED user manual. Universität Passau, 1990.
3. M. Petre. Why looking isn't always seeing: Readership skills and graphical programming. *CACM*, 38(6):33–44, June 1995.

4. H. C. Purchase. Which aesthetic has the greatest effect on human understanding? In G. Di Battista, editor, *Proceedings of Graph Drawing Symposium 1997*, pages 248–259. Springer-Verlag, Rome, Italy, 1997. LNCS, 1353.

5. H. C. Purchase. Performance of layout algorithms: Comprehension, not computation. *Journal of Visual Languages and Computing*, 9:647–657, 1998.

Author Index